STALLCUP'S®

Illustrated Code Changes, 2011 Edition

Based on the NEC® and Related Standards

James G. Stallcup

James W. Stallcup

JONES & BARTLETT
LEARNING

World Headquarters
Jones & Bartlett Learning
40 Tall Pine Drive
Sudbury, MA 01776
978-443-5000
info@jblearning.com
www.jblearning.com

Jones & Bartlett Learning Canada
6339 Ormindale Way
Mississauga, Ontario L5V 1J2
Canada

Jones & Bartlett Learning International
Barb House, Barb Mews
London W6 7PA
United Kingdom

National Fire Protection Association
1 Batterymarch Park
Quincy, MA 02169
www.NFPA.org

Jones & Bartlett Learning books and products are available through most bookstores and online booksellers. To contact Jones & Bartlett Learning directly, call 800-832-0034, fax 978-443-8000, or visit our website, www.jblearning.com.

Substantial discounts on bulk quantities of Jones & Bartlett Learning publications are available to corporations, professional associations, and other qualified organizations. For details and specific discount information, contact the special sales department at Jones & Bartlett Learning via the above contact information or send an email to specialsales@jblearning.com.

Production Credits
Chairman, Board of Directors: Clayton Jones
Chief Executive Officer: Ty Field
President: James Homer
Sr. V.P., Chief Operating Officer: Don W. Jones, Jr.
V.P., Design and Production: Anne Spencer
V.P., Manufacturing and Inventory Control: Therese Connell
Executive Publisher: Kimberly Brophy

Acquisitions Editor–Electrical: Martin Schumacher
Associate Production Editor: Lisa Cerrone
Senior Marketing Manager: Stephen Hickey
Design, Graphics, and Layout: Billy G. Stallcup
Cover Design: Scott Moden
Printing and Binding: Courier Kendallville
Cover Printing: Courier Kendallville

Copyright © 2012 by Grayboy, Inc.

Publication of this work is for the purpose of circulating information and opinion among those concerned for fire and electrical safety and related subjects. While every effort has been made to achieve a work of high quality, neither the publisher, the NFPA, the authors, or the contributors to this work guarantee the accuracy or completeness of or assume any liability in connection with the information and opinions contained in this work. The publisher, NFPA, authors, and contributors shall in no event be liable for any personal injury, property, or other damages of any nature whatsoever, whether special, indirect, or consequential, or compensatory, directly or indirectly resulting from the publication, use of or reliance upon this work.

This work is published with the understanding that the publisher, NFPA, authors, and contributors to this work are supplying information and opinion but are not attempting to render engineering or other professional services. If such services are required, the assistance of an appropriate professional should be sought.

National Electrical Code® and NEC® are registered trademarks of the National Fire Protection Association, Inc.

Library of Congress Cataloging-in-Publication Data
Stallcup, James G.
  Stallcup's illustrated code changes : based on the NEC and related standards / James G. Stallcup with James W. Stallcup. — 2011 ed.
    p. cm.
  ISBN 978-0-7637-9094-3 (pbk.)
  1. National Fire Protection Association. National Electrical Code (2011) 2. Electrical engineering–Insurance requirements–United States. I. Stallcup, James W. II. Title. III. Title: Illustrated code changes.
  TK260.S765 2011
  621.319'24021873–dc22
                                                                                                    2010037763
6048

Printed in the United States of America
14 13 12 11 10   10 9 8 7 6 5 4 3 2 1

Foreword

As in the past, the changes that have been made to the newest edition of the *National Electrical Code* have been numerous and, as always, prompted a need for a publication that not only designates these changes in a well-designed format, but provides a detailed and visually accepted presentation for the easy comprehension of the reader.

Now, once again, the 2011 *National Electrical Code* contains numerous changes. Learning what these changes are and the reasoning behind them should not be the only reason for a user to learn them.

As in all Stallcup publications, the correlation and interactivity of various codes and standards is considered mandatory in the electrical industry. Therefore, it is imperative that there be a *Code* by which other codes and standards can reference as the *Code* of standard practice. The *National Electrical Code* is recognized as this *Code* and therefore, for installation and maintenance of electrical systems, as well as for safety issues, keeping oneself aware of changes not only in the *National Electrical Code,* but in the electrical industry as a whole, is a must.

The Stallcups

Introduction

The 2011 edition of the *National Electrical Code*® (NEC®) contains many comprehensive revisions pertaining to specific *National Electrical Code* rules and regulations. Electrical personnel have an immediate and awesome task in not only learning, but implementing these revisions in their everyday design, installation, and inspection of electrical systems.

The material in this book, if read and studied carefully in a continuous and enthusiastic manner, will provide a proper update on the revisions in the 2011 *National Electrical Code*. However, even though it is true that only time and discussion among electrical personnel will provide the answers on how to interpret and apply some of these rules, one can use this book to get a head start.

Illustrated Code Changes explains the major changes in the 2011 *National Electrical Code* and can be used as a guide for fast and easy reference. These changes are presented in numerical order to correlate with the Articles and Sections as they appear in the 2011 *National Electrical Code* and are also illustrated to give a more detailed description. Where appropriate, reasons for revisions and new articles are given, along with what kind of impact such changes will have on manufacturers, designers, installers, and inspectors.

For every proposal made, there was a reason, and hopefully, this book will provide some of the reasons why a change resulted.

CORRELATING THE NEC WITH OTHER STANDARDS

Type of Change	New Definition			Committee Change	Accept in Principle			2008 NEC	-
ROP	pg. 61	# 4-3	log: 3502	ROC	pg. -	# -	log: -	UL	-
Submitter: James J. Rogers				Submitter: -				OSHA	-
NFPA 70B	-			NFPA 70E	-			NFPA 79	-

With each change, references are made to well-known standards that are used in the industry, However, the ROP and ROC that are referenced to in the table are documents that are directly involved in the code-making process and are explained below, along with the different types of changes and Code-Making Panel actions.

NEC PROCESS

ROP (Report on Proposals)

The information found in this document not only specifies the proposed changes, but also includes who submitted the change and substantiation for such proposal.

ROC (Report on Comments)

All of the actions that were taken by the Code-Making Panels and the Technical Correlating Committee at the meetings prior to the NFPA annual meeting and the adoption to the 2011 *National Electrical Code* are found in this document.

Type of Change:

New Article
New Section
New Subsection
New Subdivision
New Exception
New Informational Note
Revision
Deletion

Panel Action:

Accept
Accept in Principle
Accept in Part
Accept in Principle in Part
Reject

This table was developed by the authors with the intention of enhancing the perception of the change while correlating the *National Electrical Code* with other standards. Hopefully, the instructor and/or student of the *National Electrical Code* will find this information useful, if not interesting.

Chapter 7
Special Conditions

Chapter 8
Communications Systems

Article 90
Introduction

Chapter 1
General

Chapter 2
Wiring and Protection

Chapter 3
Wiring Methods and Materials

Chapter 4
Equipment for General Use

Chapter 5
Special Occupancies

Chapter 6
Special Equipment

Chapter 7
Special Conditions

Chapter 8
Communications Systems

Table of Contents

Introduction and General

Article 90 and **Chapter 1** of the *National Electrical Code* have always been referred to as the "get acquainted" material that every designer, installer, electrician, apprentice, inspector, and maintenance person must review and understand before the other chapters, articles, and sections of the *National Electrical Code* can really be understood and applied.

The first article in the *National Electrical Code* is **Article 90**, which contains the Introduction. **Article 90** covers the purpose of the *National Electrical Code,* along with other pertinent information that is applicable throughout each chapter of the *National Electrical Code*.

Chapter 1 acquaints the user of the *National Electrical Code* with definitions and clearance rules that are mandatory to ensure the safety of the general public and personnel working in, near, or on wiring methods and equipment.

Users as well as students of the *National Electrical Code* must review and become acquainted with **Article 90** and **Chapter 1** before attempting to study, learn, and apply the other articles and chapters to a particular design or installation.

It is this concept of study that will make interpretations and applications of the many requirements in the *National Electrical Code* much easier to understand.

Type of Change	Revision			Committee Change	Accept in Principle			2008 NEC	90.2(B)(5)
ROP	pg. 24	# 1-29	log: 3136	ROC	pg. 16	# 1-23	log: 1753	UL	-
Submitter: Neil F. LaBrake, Jr.				Submitter: James T. Dollard, Jr.				OSHA	1910.302(a)(2)
NFPA 70B	-			NFPA 70E	90.2(B)(5)			NFPA 79	-

90.2(B) Not Covered. This *Code* does not cover the following:

(5) Installations under the exclusive control of an electric utility where such installations

d. Are located by other written agreements either designated by or recognized by public service commissions, utility commissions, or other regulatory agencies having jurisdiction for such installations. These written agreements shall be limited to installations for the purpose of communications, metering, generation, control, transformation, transmission, or distribution of electric energy where legally established easements or rights-of-way cannot be obtained. These installations shall be limited to federal lands, native American reservations through the U.S. Department of the Interior Bureau of Indian Affairs, military bases, lands controlled by port authorities and state agencies and departments, and lands owned by railroads.

Stallcup's Comment: A revision has been made to clarify that there are areas in which an easement or rights-of-way cannot be legally obtained. This revision provides text to permit "other written agreements" for those installations as well as a prescriptive list of the locations.

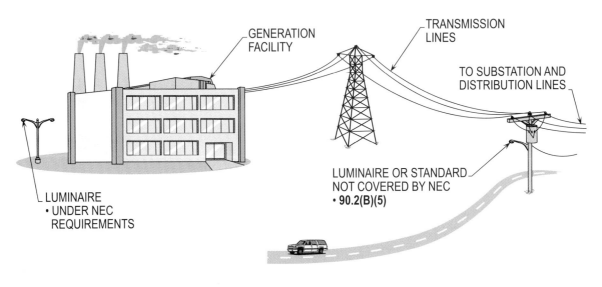

Stallcup's Note: Installations under control of utilities are limited to:
• federal lands • native American reservations • military bases • railroad land, etc. per **90.2(B)(5)d**.

GENERATION FACILITY

TRANSMISSION LINES

TO SUBSTATION AND DISTRIBUTION LINES

LUMINAIRE OR STANDARD NOT COVERED BY NEC
• **90.2(B)(5)**

LUMINAIRE
• UNDER NEC REQUIREMENTS

**NOT COVERED
NEC 90.2(B)(5)d**

Purpose of Change: To clarify that, where an easement or rights-of-way cannot be obtained, other written agreements shall be permitted for installations under the control of utilities.

Type of Change	Revision			Committee Change	Accept			2008 NEC	90.5(C)
ROP pg. 30	# 1-37a	log: CP 100		ROC pg. -	# -	log: -		UL	-
Submitter: Code Making Panel 1				Submitter: -				OSHA	-
NFPA 70B -				NFPA 70E	90.5(C)			NFPA 79	-

90.5 Mandatory Rules, Permissive Rules, and Explanatory Material.

(D) Informative Annexes. Nonmandatory information relative to the use of the *NEC* is provided in informative annexes. Informative annexes are not part of the enforceable requirements of the *NEC*, but are included for information purposes only.

Stallcup's Comment: A new subsection has been made to clarify and more clearly delineate the adoptable and enforceable requirements of the *National Electrical Code*. The *National Electrical Code* contains notes that are enforceable requirements, such as table notes. Many standards now contain normative and informative annexes. Normative annexes are requirements and informative annexes are not.

EXPLANATORY MATERIAL
BELOW MAY PERTAIN TO
ITEMS (a) THROUGH (j)

• INFORMATIONAL NOTES ARE
  NOT ENFORCEABLE

• INFORMATIVE ANNEXES
  ARE NOT ENFORCEABLE

• BRACKETED REFERENCES
  REFER USERS OF CODE
  TO OTHER NFPA
  DOCUMENTS.

Stallcup's Note: All FPNs are now called informational notes and annexes are called informative annexes.

SWITCHBOARD
FRAMES AND
STRUCTURES
(a)

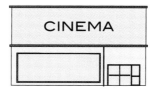
GARAGES, THEATERS, AND
MOTION PICTURE STUDIOS
(b)

LUMINAIRES
(c)

ENCLOSURES FOR
MOTOR
CONTROLLERS
(d)

MOTION PICTURE
PROTECTION EQUIPMENT
(h)

ELECTRIC
SIGNS
(i)

POWER-LIMITED REMOTE-
CONTROL, SIGNALING, AND
FIRE ALARM CIRCUITS
(k)

MOTOR
FRAMES
(j)

INFORMATIVE ANNEXES
NEC 90.5(D)

Purpose of Change: To clarify the re-identification between mandatory rules and explanatory material, such as informational notes, brackets, and informative annexes.

Type of Change	Revision			Committee Change	Accept			2008 NEC	Article 100
ROP	pg. 36	# 1-54	log: 206	ROC	pg. -	# -	log: -	UL	-
Submitter: Glossary of Terms TAC				Submitter: -				OSHA	1910.399
NFPA 70B	-			NFPA 70E	Article 100			NFPA 79	-

Article 100 Definitions.

Automatic. Performing a function without the necessity of human intervention.

Stallcup's Comment: A revision has been made to generate consistent definitions and minimize the number of duplicate definitions in the NFPA Glossary of Terms in accordance with the scope of the NFPA Glossary of Terms Technical Advisory Committee.

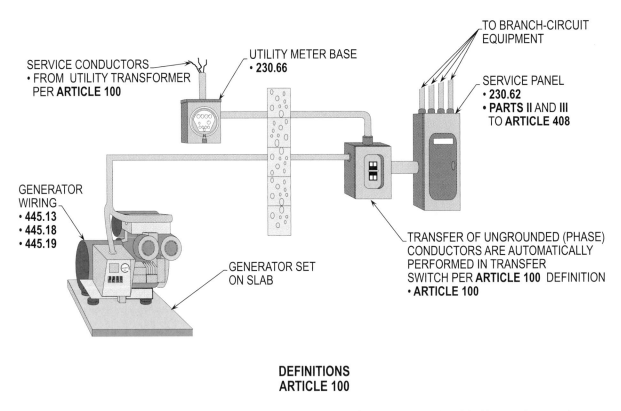

DEFINITIONS
ARTICLE 100

Purpose of Change: To define that "automatic" function is performed without the aid of human intervention.

Type of Change	Revision			Committee Change	Accept in Part		2008 NEC	Article 100
ROP pg. 41	# 2-5		log: 2246	**ROC** pg. 22	# 2-1	log: 1949	**UL**	-
Submitter: Lorenzo Adam				Submitter: D. Jerry Flaherty			**OSHA**	1910.399
NFPA 70B -				**NFPA 70E** -			**NFPA 79**	-

Article 100 Definitions.

Bathroom. An area including a basin with one or more of the following: a toilet, a urinal, a tub, a shower, a bidet, or similar plumbing fixtures.

Stallcup's Comment: A revision has been made to clarify what constitutes a bathroom.

For example, an area including a basin and a urinal will require a receptacle outlet to be installed within 3 ft (900 mm) of the basin.

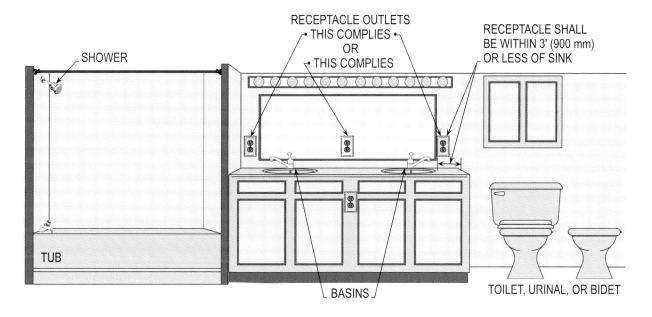

DEFINITIONS
ARTICLE 100

Purpose of Change: To clarify what is considered a bathroom.

Type of Change	Definition Relocated			Committee Change	Accept			2008 NEC	250.2
ROP	pg. 53	# 5-10	log: 3040	ROC	pg. -	# -	log: -	UL	-
Submitter: Mike Holt				Submitter: -				OSHA	-
NFPA 70B	-			NFPA 70E	Article 100			NFPA 79	-

Article 100 Definitions.

Ground Fault. An unintentional, electrically conducting connection between an ungrounded conductor of an electrical circuit and the normally non–current-carrying conductors, metallic enclosures, metallic raceways, metallic equipment, or earth.

Stallcup's Comment: This definition has been relocated to **Article 100** since it is used in two or more articles of the *National Electrical Code*.

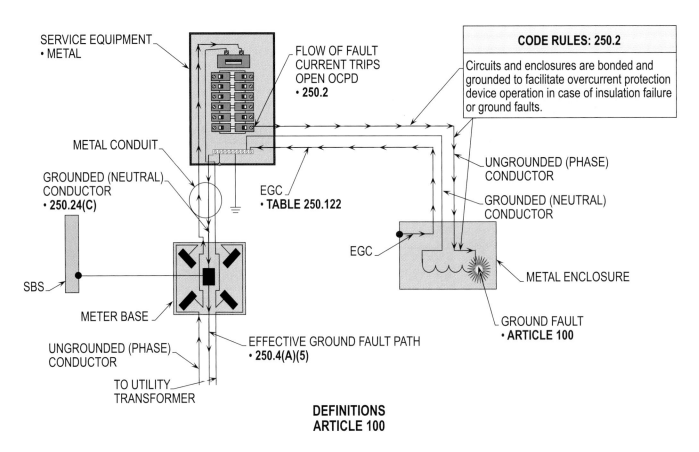

**DEFINITIONS
ARTICLE 100**

Purpose of Change: To relocate the definition of "ground fault" from **Article 250** to **Article 100** due to its use in the *National Electrical Code*.

Type of Change	Revision			Committee Change	Accept			2008 NEC	Article 100
ROP	pg. 61	# 1-110a	log: CP 101	ROC	pg. -	# -	log: -	UL	-
Submitter: Code-Making Panel 1				Submitter: -				OSHA	-
NFPA 70B	-			NFPA 70E	-			NFPA 79	-

Article 100 Definitions.

Nonautomatic. Requiring human intervention to perform a function.

Stallcup's Comment: A revision has been made to generate consistent definitions and minimize the number of duplicate definitions in the NFPA Glossary of Terms in accordance with the scope of the NFPA Glossary of Terms Technical Advisory Committee.

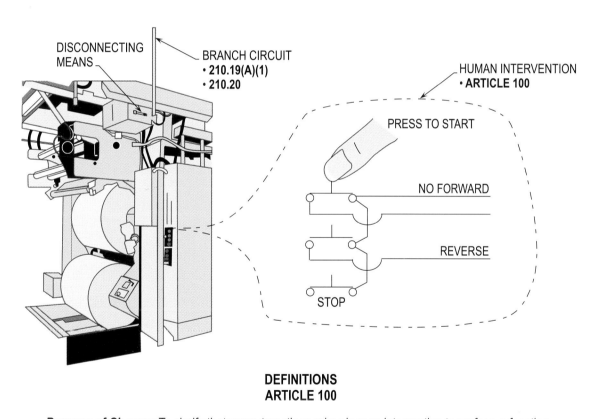

**DEFINITIONS
ARTICLE 100**

Purpose of Change: To clarify that nonautomatic requires human intervention to perform a function.

Type of Change	New Definition			Committee Change	Accept in Principle		2008 NEC	-
ROP	pg. 61	# 4-3	log: 3502	ROC	pg. -	# -	log: -	UL	-
Submitter: James J. Rogers				Submitter: -			OSHA	-
NFPA 70B	-			NFPA 70E	-		NFPA 79	-

Article 100 Definitions.

Service Conductors, Overhead. The overhead conductors between the service point and the first point of connection to the service-entrance conductors at the building or other structure.

Stallcup's Comment: A new definition has been added to clarify that overhead service conductors are between the service point and the service-entrance conductors.

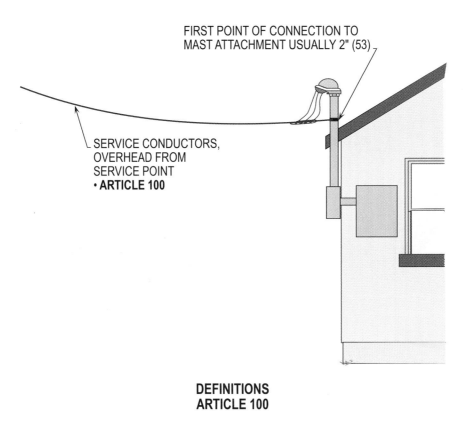

FIRST POINT OF CONNECTION TO
MAST ATTACHMENT USUALLY 2" (53)

SERVICE CONDUCTORS,
OVERHEAD FROM
SERVICE POINT
• ARTICLE 100

DEFINITIONS
ARTICLE 100

Purpose of Change: To clarify what are considered overhead service conductors.

Type of Change	New Definition		Committee Change	Accept		2008 NEC	-
ROP	pg. 6/	# 1 15	log: 3504	ROC	pg. 37	# 4-2	log: 1327	UL	-
Submitter: James J. Rogers			Submitter: Glossary of Terms TAC			OSHA	-
NFPA 70B	-			NFPA 70E	-		NFPA 79	-

Article 100 Definitions.

Service Conductors, Underground. The underground conductors between the service point and the first point of connection to the service-entrance conductors in a terminal box, meter, or other enclosure, inside or outside the building wall.

Informational Note: Where there is no terminal box, meter, or other enclosure, the point of connection is considered to be the point of entrance of the service conductors into the building.

Stallcup's Comment: A new definition has been added to clarify that undergound service conductors are between the service point and the service-entrance conductors.

Stallcup's Note: If there is no terminal box, meter, or other enclosure, point of connection is the point of entrance per **Article 100** and **IN**.

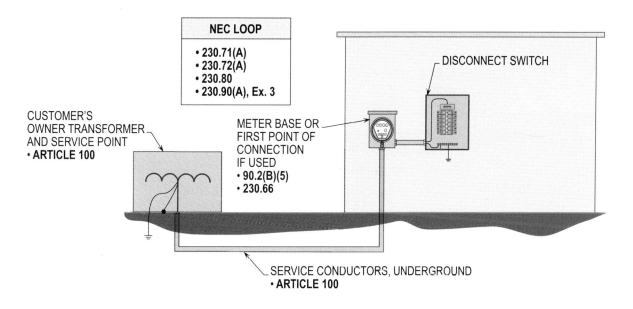

NEC LOOP
- **230.71(A)**
- **230.72(A)**
- **230.80**
- **230.90(A), Ex. 3**

DISCONNECT SWITCH

CUSTOMER'S OWNER TRANSFORMER AND SERVICE POINT
- **ARTICLE 100**

METER BASE OR FIRST POINT OF CONNECTION IF USED
- **90.2(B)(5)**
- **230.66**

SERVICE CONDUCTORS, UNDERGROUND
- **ARTICLE 100**

**DEFINITIONS
ARTICLE 100**

Purpose of Change: To clarify what are considered underground service conductors.

Type of Change	Revision			Committee Change	Accept			**2008 NEC**	Article 100
ROP	pg. 67	# 4-16	log: 3505	**ROC**	pg. 37	# 4-5	log: 2257	**UL**	-
Submitter: James J. Rogers				Submitter: Roger D. McDaniel				**OSHA**	-
NFPA 70B	-			**NFPA 70E**	Article 100			**NFPA 79**	-

Article 100 Definitions.

Service Lateral. The underground conductors between the utility electric supply system and the service point.

Stallcup's Comment: A revision has been made to clarify where service lateral conductors begin and end in reference to the *National Electrical Code*. If service lateral conductors are installed by electric service utility companies by established agreements or easements, then the *National Electrical Code* does not apply and the electric service utility can define these conductors.

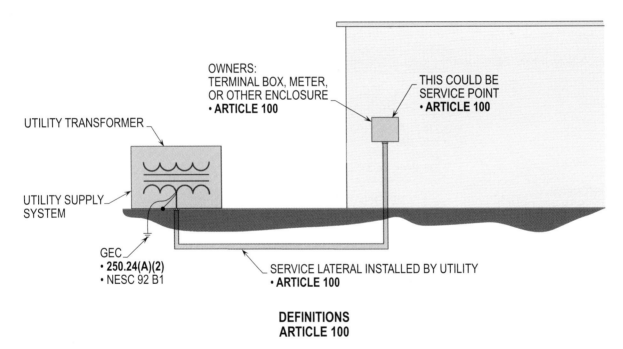

OWNERS:
TERMINAL BOX, METER,
OR OTHER ENCLOSURE
• **ARTICLE 100**

THIS COULD BE
SERVICE POINT
• **ARTICLE 100**

UTILITY TRANSFORMER

UTILITY SUPPLY
SYSTEM

GEC
• **250.24(A)(2)**
• NESC 92 B1

SERVICE LATERAL INSTALLED BY UTILITY
• **ARTICLE 100**

DEFINITIONS
ARTICLE 100

Purpose of Change: To clarify what is considered a service lateral.

Type of Change	New Definition			Committee Change	Accept		2008 NEC	-
ROP	pg. 1234	# 12-3	log: 4102	ROC	pg. 1234	# 12-2	log: 1326	UL	-
Submitter: Ray Stanko				Submitter: Glossary of Terms TAC			OSHA	-
NFPA 70B	-			NFPA 70E	-		NFPA 79	-

Article 100 Definitions.

Uninterruptible Power Supply. A power supply used to provide alternating current power to a load for some period of time in the event of a power failure.

Informational Note: In addition, it may provide a more constant voltage and frequency supply to the load, reducing the effects of voltage and frequency variations.

Stallcup's Comment: A new definition has been added to generate consistent definitions and minimize the number of duplicate definitions in the NFPA Glossary of Terms in accordance with the scope of the NFPA Glossary of Terms Technical Advisory Committee.

The new definition is consistent with National Standards for uninterruptible power supply equipment.

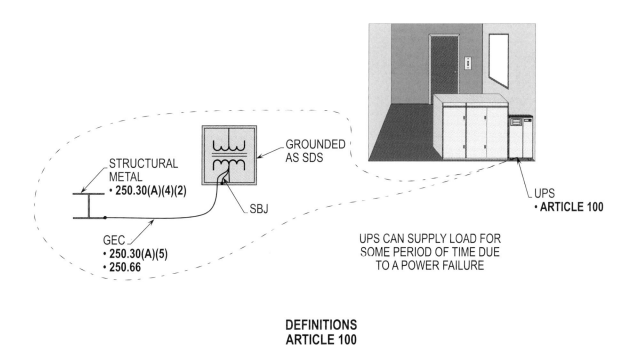

STRUCTURAL METAL
• 250.30(A)(4)(2)

GROUNDED AS SDS

SBJ

GEC
• 250.30(A)(5)
• 250.66

UPS
• ARTICLE 100

UPS CAN SUPPLY LOAD FOR SOME PERIOD OF TIME DUE TO A POWER FAILURE

**DEFINITIONS
ARTICLE 100**

Purpose of Change: To clarify what is considered an uninterruptible power supply (UPS).

Type of Change	Revision			Committee Change	Accept		**2008 NEC**	110.10
ROP	pg. 74	# 1-130	log: 4403	**ROC**	pg. 41	# 1-90	log: 2812	**UL**	-
Submitter: Jay Tamblingson				Submitter: Jay Tamblingson			**OSHA**	1910.303(b)(5)
NFPA 70B	-			**NFPA 70E**	120.3(D)		**NFPA 79**	-

110.10 Circuit Impedance, Short-Circuit Current Ratings, and Other Characteristics.

The overcurrent protective devices, the total impedance, the equipment short-circuit current ratings, and other characteristics of the circuit to be protected shall be selected and coordinated to permit the circuit protective devices used to clear a fault to do so without extensive damage to the electrical equipment of the circuit. This fault shall be assumed to be either between two or more of the circuit conductors or between any circuit conductor and the equipment grounding conductor(s) permitted in 250.118. Listed equipment applied in accordance with their listing shall be considered to meet the requirements of this section.

Stallcup's Comment: A revision has been made to the heading to clarify that equipment short-circuit current ratings are included in the requirements for electrical installations to provide increased awareness of these ratings.

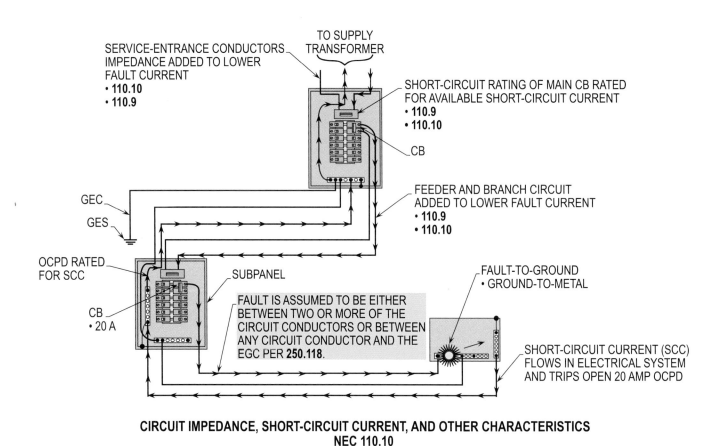

CIRCUIT IMPEDANCE, SHORT-CIRCUIT CURRENT, AND OTHER CHARACTERISTICS
NEC 110.10

Purpose of Change: To clarify that circuit impedance, short-circuit ratings, and other characteristics shall be permitted to be used to lower the flow of short-circuit current, etc.

Type of Change	Revision		Committee Change	Accept in Principle		2008 NEC	110.14
ROP	pg. 79	# 1-149	log: 1739	ROC	pg. 43	# 1-101	log: 1599	UL	486A-B
Submitter: Robert A. McCullough			Submitter: Jim Pauley			OSHA	1910.303(c)
NFPA 70B	-		NFPA 70E	-		NFPA 79	3.3.106

110.14 Electrical Connections

Because of different characteristics of dissimilar metals, devices such as pressure terminal or pressure splicing connectors and soldering lugs shall be identified for the material of the conductor and shall be properly installed and used. Conductors of dissimilar metals shall not be intermixed in a terminal or splicing connector where physical contact occurs between dissimilar conductors (such as copper and aluminum, copper and copper-clad aluminum, or aluminum and copper-clad aluminum), unless the device is identified for the purpose and conditions of use. Materials such as solder, fluxes, inhibitors, and compounds, where employed, shall be suitable for the use and shall be of a type that will not adversely affect the conductors, installation, or equipment.

Connectors and terminals for conductors more finely stranded than Class B and Class C stranding as shown in Chapter 9, Table 10, shall be identified for the specific conductor class or classes.

Stallcup's Comment: A revision has been made to clarify that finely stranded cable shall be connected using proper terminals. A new **Table 10** in **Chapter 9** has been added that is consistent with the product standards, UL 486A-B.

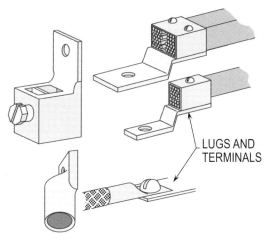

LUGS AND
TERMINALS

CONNECTORS, TERMINALS,
AND CONDUCTORS
• 110.14

CONNECTORS AND TERMINALS FOR CONDUCTORS
MORE FINELY STRANDED THAN CLASS B AND CLASS C
STRANDING AS SHOWN IN **CHAPTER 9, TABLE 10**
SHALL BE SO IDENTIFIED
• **110.14**

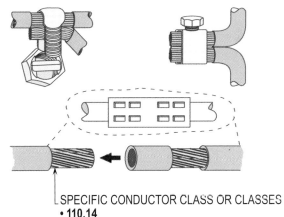

SPECIFIC CONDUCTOR CLASS OR CLASSES
• 110.14

**ELECTRICAL CONNECTIONS
NEC 110.14**

Purpose of Change: To clarify how connectors and terminals more finely stranded than outlined in **Table 10** to **Chapter 9** shall be identified.

Type of Change	Revision			Committee Change	Accept			2008 NEC	110.16
ROP	pg. 1234	# 1-162	log: 1591	ROC	pg. 1234	# 1-105	log: 57	UL	-
Submitter: Russell LeBlanc				Submitter: TCC				OSHA	-
NFPA 70B	-			NFPA 70E	130.3(C)			NFPA 79	-

110.16 Arc-Flash Hazard Warning.

Electrical equipment, such as switchboards, panelboards, industrial control panels, meter socket enclosures, and motor control centers, that are in other than dwelling units, and are likely to require examination, adjustment, servicing, or maintenance while energized shall be field marked to warn qualified persons of potential electric arc flash hazards. The marking shall be located so as to be clearly visible to qualified persons before examination, adjustment, servicing, or maintenance of the equipment.

Stallcup's Comment: A revision has been made to clarify that an arc-flash hazard warning shall be required for a large multifamily dwelling that could have the same or larger size electric service as a commercial office building.

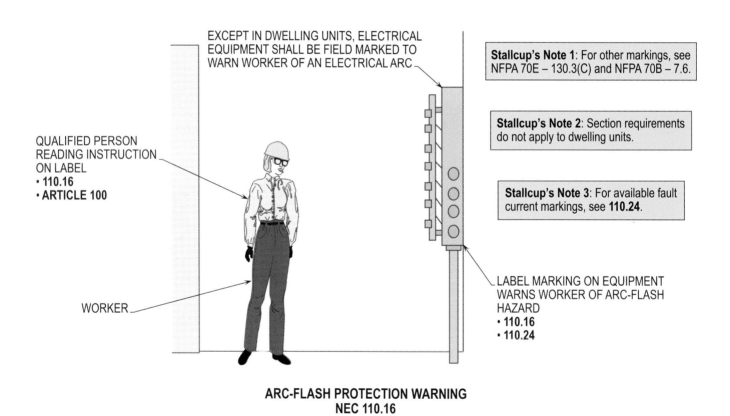

ARC-FLASH PROTECTION WARNING
NEC 110.16

Purpose of Change: To add units to dwelling so reference in section is to "in other than dwelling units."

Type of Change	New Section			Committee Change	Accept			2008 NEC	-
ROP pg. 86	# 1-183	log: 4783		**ROC** pg. 46	# -	log: -		**UL**	-
Submitter: Michael J. Johnson				Submitter: Donald R. Cook				**OSHA**	1910.303(b)(4)
NFPA 70B -				**NFPA 70E** -				**NFPA 79**	4.8

110.24 Available Fault Current.

(A) Field Marking. Service equipment in other than dwelling units shall be legibly marked in the field with the maximum available fault current. The field marking(s) shall include the date the fault current calculation was performed and be of sufficient durability to withstand the environment involved.

(B) Modifications. When modifications to the electrical installation occur that affect the maximum available fault current at the service, the maximum available fault current shall be verified or recalculated as necessary to ensure the service equipment ratings are sufficient for the maximum available fault current at the line terminals of the equipment. The required field marking(s) in 110.24(A) shall be adjusted to reflect the new level of maximum available fault current.

Exception: The field marking requirements in 110.24(A) and 110.24(B) shall not be required in industrial installations where conditions of maintenance and supervision ensure that only qualified persons service the equipment.

Stallcup's Comment: A new section has been added that requires service equipment in other than dwelling units to be field-marked with the amount of available short-circuit current when installed. An interrupting rating or short-circuit current rating for electrical equipment shall be equal to or greater than the available fault current. Any equipment operating with ratings less than the available fault current is a violation of the *National Electrical Code* and creates a potentially unsafe condition.

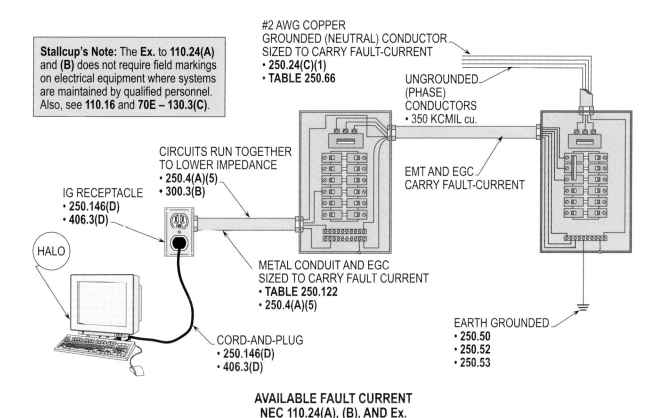

AVAILABLE FAULT CURRENT
NEC 110.24(A), (B), AND Ex.

Purpose of Change: To verify that markings and modifications are frequently checked to clarify that posted information on label is accurate, except where system is maintained by qualified personnel.

Type of Change	Revision			Committee Change	Accept			2008 NEC	110.26(A)(3) and 110.26(E)
ROP	pg. 92	# 1-207	log: 2381	ROC	pg. -	# -	log: -	UL	-
Submitter: David G. Humphrey				Submitter: -				OSHA	1910.303(g)(1)(i)[C]
NFPA 70B	-			NFPA 70E	205.4			NFPA 79	-

110.26(A)(3) Height of Working Space.

The work space shall be clear and extend from the grade, floor, or platform to a height of 2.0 m (6 1/2 ft) or the height of the equipment, whichever is greater. Within the height requirements of this section, other equipment that is associated with the electrical installation and is located above or below the electrical equipment shall be permitted to extend not more than 150 mm (6 in.) beyond the front of the electrical equipment.

Exception No. 1: In existing dwelling units, service equipment or panelboards that do not exceed 200 amperes shall be permitted in spaces where the height of the working space is less than 2.0 m (6 1/2 ft).

Stallcup's Comment: A revision has been made to clarify the height of working space. This revision separates the requirements of **110.26(E)** for headroom space and **110.26(A)(3)** for height of working space.

The **exception** in **110.26(E)** has been moved to **110.26(A)(3)** to clarify the height of working space in existing dwelling units.

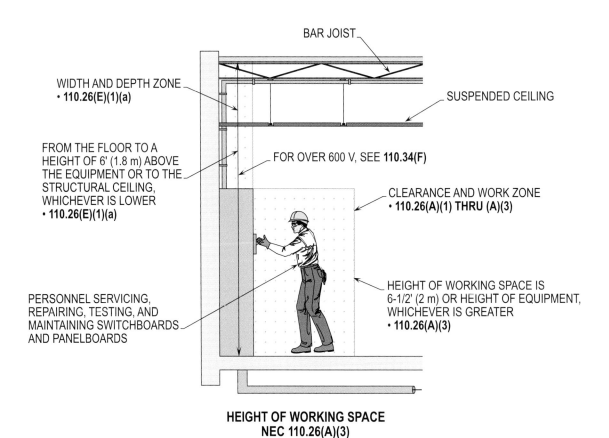

**HEIGHT OF WORKING SPACE
NEC 110.26(A)(3)**

Purpose of Change: To relocate the height of working space from **110.26(E)** to **110.26(A)(3)**.

Type of Change	Revision			Committee Change	Accept			2008 NEC	110.20
ROP	pg. 83	# 1-171	log: 4388	ROC	pg. -	# -	log: -	UL	-
Submitter: John R. Kovacik				Submitter: -				OSHA	1910.305(e)
NFPA 70B	-			NFPA 70E	205.6			NFPA 79	6.2.2

110.28 Enclosure Types.

Enclosures (other than surrounding fences or walls) of switchboards, panelboards, industrial control panels, motor control centers, meter sockets, enclosed switches, transfer switches, power outlets, circuit breakers, adjustable-speed drive systems, pullout switches, portable power distribution equipment, termination boxes, general-purpose transformers, fire pump controllers, fire pump motors, and motor controllers, rated not over 600 volts nominal and intended for such locations, shall be marked with an enclosure-type number as shown in Table 110.28.

Table 110.28 shall be used for selecting these enclosures for use in specific locations other than hazardous (classified) locations. The enclosures are not intended to protect against conditions such as condensation, icing, corrosion, or contamination that may occur within the enclosure or enter via the conduit or unsealed openings.

Stallcup's Comment: A revision has been made to clarify the enclosures that shall be marked with an enclosure-type number.

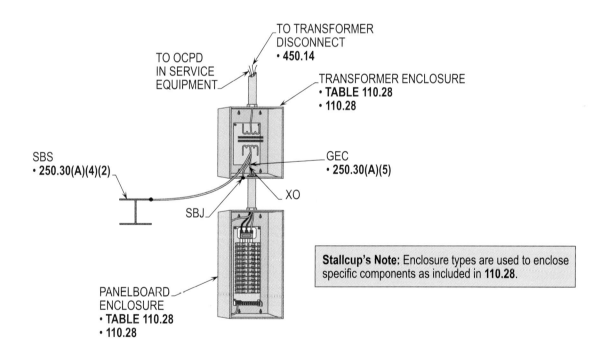

ENCLOSURE TYPES
NEC 110.28 AND TABLE 110.28

Purpose of Change: To include specific components that are required to be enclosed by a particular enclosure.

Type of Change	Revision			Committee Change	Accept			2008 NEC	110.31(A)
ROP	pg. 102	# 1-252	log: 3747	ROC	pg. 55	# 1-142	log: 2005	UL	-
Submitter: Jim Pauley				Submitter: Leo F. Martin, Jr.				OSHA	1910.303(h)2)(i)
NFPA 70B	-			NFPA 70E	-			NFPA 79	-

110.31(A) Electrical Vaults.

Where an electrical vault is required or specified for conductors and equipment operating at over 600 volts, nominal, the following shall apply.

(1) Walls and Roof. The walls and roof shall be constructed of materials that have adequate structural strength for the conditions, with a minimum fire rating of 3 hours. For the purpose of this section, studs and wallboard construction shall not be permitted.

(2) Floors. The floors of vaults in contact with the earth shall be of concrete that is not less than 102 mm (4 in.) thick, but where the vault is constructed with a vacant space or other stories below it, the floor shall have adequate structural strength for the load imposed on it and a minimum fire resistance of 3 hours.

(3) Doors. Each doorway leading into a vault from the building interior shall be provided with a tight-fitting door that has a minimum fire rating of 3 hours. The authority having jurisdiction shall be permitted to require such a door for an exterior wall opening where conditions warrant.

Exception to (1), (2), and (3): Where the vault is protected with automatic sprinkler, water spray, carbon dioxide, or halon, construction with a 1-hour rating shall be permitted.

(4) Locks. Doors shall be equipped with locks, and doors shall be kept locked, with access allowed only to qualified persons. Personnel doors shall swing out and be equipped with panic bars, pressure plates, or other devices that are normally latched but open under simple pressure.

(5) Transformers. Where a transformer is installed in a vault as required by Article 450, the vault shall be constructed in accordance with the requirements of Part III of Article 450.

Informational Note No. 1: For additional information, see ANSI/ASTM E119-1995, *Method for Fire Tests of Building Construction and Materials*, NFPA 251-2006, *Standard Methods of Tests of Fire Resistance of Building Construction and Materials,* and NFPA 80-2010, *Standard for Fire Doors and Other Opening Protectives.*

Informational Note No. 2: A typical 3-hour construction is 150 mm (6 in.) thick reinforced concrete.

Stallcup's Comment: A revision has been made to clarify the requirements for an electrical vault. The title has been changed to cover all electrical vaults, not only fire resistance of electrical vaults. The revision to this section now covers walls and roof, floors, doors, locks, and transformers.

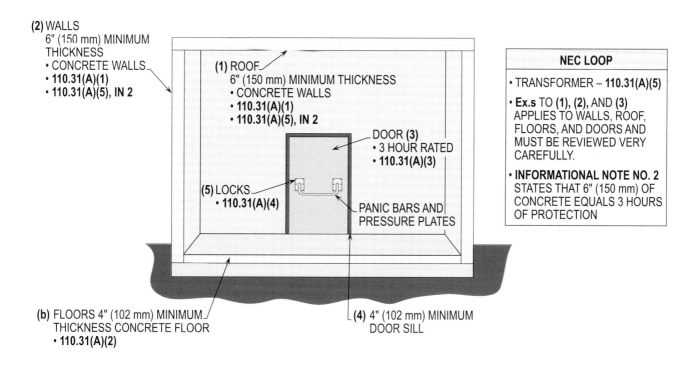

(2) WALLS
6" (150 mm) MINIMUM THICKNESS
• CONCRETE WALLS
• **110.31(A)(1)**
• **110.31(A)(5), IN 2**

(1) ROOF
6" (150 mm) MINIMUM THICKNESS
• CONCRETE WALLS
• **110.31(A)(1)**
• **110.31(A)(5), IN 2**

DOOR **(3)**
• 3 HOUR RATED
• **110.31(A)(3)**

(5) LOCKS
• **110.31(A)(4)**

PANIC BARS AND
PRESSURE PLATES

NEC LOOP

• TRANSFORMER – **110.31(A)(5)**

• **Ex.s** TO **(1)**, **(2)**, AND **(3)** APPLIES TO WALLS, ROOF, FLOORS, AND DOORS AND MUST BE REVIEWED VERY CAREFULLY.

• **INFORMATIONAL NOTE NO. 2** STATES THAT 6" (150 mm) OF CONCRETE EQUALS 3 HOURS OF PROTECTION

(b) FLOORS 4" (102 mm) MINIMUM THICKNESS CONCRETE FLOOR
• **110.31(A)(2)**

(4) 4" (102 mm) MINIMUM
DOOR SILL

ELECTRICAL VAULTS
NEC 110.31(A)(1) THRU (A)(5)

Purpose of Change: To incorporate the requirements for doors, locks, and transformers with exception and informational notes to **110.31(A)**.

Name   Date

Chapter 1   Section   Answer
Introduction and General

1. Nonmandatory information relative to the use of the *National Electrical Code* is provided in: _____   _____

(A) Advisory annexes   **(B)** Informative annexes
(C) Fine print annexes   **(D)** None of the above

2. A bathroom is an area including a basin with one or more of the following: _____   _____

(A) A urinal   **(B)** A shower
C) A bidet   **(D)** All of the above

3. The overhead service conductors are: _____   _____

(A) The overhead conductors between the service point and the first point of connection to the service-entrance conductors at the building or structure.
(B) The conductors from the service point to the service disconnecting means.
(C) The overhead conductors between the utility electric supply system and the service point.
(D) The service conductors between the terminals of the service equipment and a point usually outside the building, clear of building walls, where joined by tap or splice to the service drop or overhead service conductors.

4. A service lateral is: _____   _____

(A) The underground conductors between the service point and the first point of connection to the service-entrance conductors in a terminal box, meter, or other enclosure, inside or outside the building wall.
(B) The service conductors between the terminals of the service equipment an the point of connection to the service lateral or underground service conductors.
(C) The underground conductors between the utility electric supply system and the service point.
(D) The point of connection between the facilities of the serving utility and the premises wiring.

5. Service equipment in other than dwelling units shall be legibly marked in the field with the: _____   _____

(A) Maximum available fault current   **(B)** Date of the fault current calculation
(C) All of the above   **(D)** None of the above

6. The work space shall be clear and extend from the grade, floor, or platform to a height of _____   _____   _____
ft or the height of the equipment, whichever is greater.

(A) 6   **(B)** 6-1/2
(C) 7   **(D)** 7-1/2

7. In existing dwelling units, service equipment or panelboards that do not exceed _____ amperes   _____   _____
shall be permitted in spaces where the height of the working space is less than 6-1/2 ft.

(A) 100   **(B)** 150
(C) 175   **(D)** 200

Section   Answer

_____ _____ **8.** Which of the following enclosures shall be marked with enclosure-type numbers:

(A) Transfer switches   **(B)** Adjustable-speed drive systems
(C) Fire pump controllers   **(D)** All of the above

_____ _____ **9.** The walls and roof of an electrical vault shall be constructed of materials that have adequate structural strength for the conditions, with a minimum fire rating of _____ hours.

(A) 1   **(B)** 2
(C) 3   **(D)** 6

_____ _____ **10.** Each doorway leading into a electrical vault from the building interior shall be provided with a tight-fitting door that has a minimum fire rating of _____ hours.

(A) 1-1/2   **(B)** 3
(C) 3-1/2   **(D)** 4

Wiring and Protection

Chapter 2 of the *National Electrical Code* has always been referred to as the "Designing Chapter" and is used by engineers, electrical contractors, and electricians who have the responsibility of calculating loads and sizing the elements of the electrical system.

Chapter 2 is the starting point to begin calculating ampacities for branch circuits and feeders. Even the ampacity for sizing service-entrance conductors is calculated by applying the rules of **Article 220**, which are found in **Parts I, II, and III,** as well as **Part IV** of **Article 230**.

The key number for finding the requirements necessary for calculating loads in **Chapter 2** is 200. In other words, all articles and sections will be identified by using a 200 series number. When designing electrical systems, **Chapter 2** and the 200 series are utilized with other pertinent articles and sections. It is nearly impossible for a designer to calculate loads and determine the size of various elements of the electrical system, if he or she is not properly acquainted with the calculation requirements of **Chapter 2**.

> **For example,** if the user wanted to calculate the load in amps for a motor feeder, he or she would refer to **220.14(C)**, and this section references **430.24** since **Article 220** does not list the rules for calculating loads for motor circuits. If **Article 220** does not contain the rules for calculating the load, it will refer the user to the required section in other articles of the *National Electrical Code*.

When calculating the load to size the conductors, use **Table 220.3**, and if the overcurrent protection device must be sized larger than the ampacity of the supply conductors and the equipment served, see **Table 240.4(G)** and verify that such equipment is listed.

Type of Change	Revision			Committe Change	Accept			2008 NEC	200.6(A)
ROP	pg. 105	# 5-34	log: 1489	ROC	pg. -	# -	log: -	UL	467
Submitter: Darryl Hill				Submitter: -				OSHA	1910.304(a)(1)
NFPA 70B	14.1.6.25			NFPA 70E	Article 100			NFPA 79	3.3.49

200.6 Means of Identifying Grounded Conductors.

(A) Sizes 6 AWG or Smaller.

An insulated grounded conductor of 6 AWG or smaller shall be identified by one of the following means:

(1) A continuous white outer finish.

(2) A continuous gray outer finish.

(3) Three continuous white stripes along the conductor's entire length on other than green insulation.

(4) Wires that have their outer covering finished to show a white or gray color but have colored tracer threads in the braid identifying the source of manufacture shall be considered as meeting the provisions of this section.

(5) The grounded conductor of a mineral-insulated, metal-sheathed cable shall be identified at the time of installation by distinctive marking at its terminations.

(6) A single-conductor, sunlight-resistant, outdoor-rated cable used as a grounded conductor in photovoltaic power systems, as permitted by 690.31, shall be identified at the time of installation by distinctive white marking at all terminations.

(7) Fixture wire shall comply with the requirements for grounded conductor identification as specified in 402.8.

(8) For aerial cable, the identification shall be as above, or by means of a ridge located on the exterior of the cable so as to identify it.

Stallcup's Comment: A revision has been made to format the means for identifying grounded conductors into a list. This revision provides clarity, uniformity, and usability to **200.6(A)**.

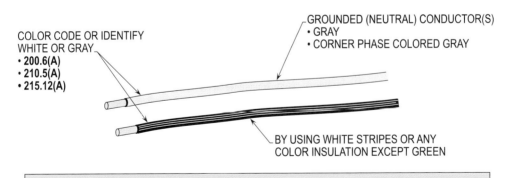

COLOR CODE OR IDENTIFY WHITE OR GRAY
• 200.6(A)
• 210.5(A)
• 215.12(A)

GROUNDED (NEUTRAL) CONDUCTOR(S)
• GRAY
• CORNER PHASE COLORED GRAY

BY USING WHITE STRIPES OR ANY COLOR INSULATION EXCEPT GREEN

Stallcup's Note: The grounded (neutral) conductor 6 AWG or smaller shall have a continuous white or gray outer finish, or continuous white strips along its entire length on other than green insulation.

SIZES 6 AWG OR SMALLER
200.6(A)(1) THRU (A)(8)

Purpose of Change: To clarify the proper method to color code or identify the grounded conductor in a list format.

Type of Change	Revision			Committe Change		Accept in Principle		2008 NEC	200.6(D)
ROP	pg. 106	# 5-51	log: 2258	ROC	pg. -	# -	log: -	UL	467
Submitter: Russell LeBlanc				Submitter:				OSHA	1910.304(a)(1)
NFPA 70B	14.1.6.25			NFPA 70E	Article 100			NFPA 79	3.3.49

200.6 Means of Identifying Grounded Conductors.

(D) Grounded Conductors of Different Systems.

Where grounded conductors of different systems are installed in the same raceway, cable, box, auxiliary gutter, or other type of enclosure, each grounded conductor shall be identified by system. Identification that distinguishes each system grounded conductor shall be permitted by one of the following means:

(1) One system grounded conductor shall have an outer covering conforming to 200.6(A) or (B).

(2) The grounded conductor(s) of other systems shall have a different outer covering conforming to 200.6(A) or 200.6(B) or by an outer covering of white or gray with a readily distinguishable colored stripe other than green running along the insulation.

(3) Other and different means of identification as allowed by 200.6(A) or (B) that will distinguish each system grounded conductor.

The means of identification shall be documented in a manner that is readily available or shall be permanently posted where the conductors of different systems originate.

Stallcup's Comment: A revision has been made to clarify that the identification requirement is not limited to panelboards and switchboards only.

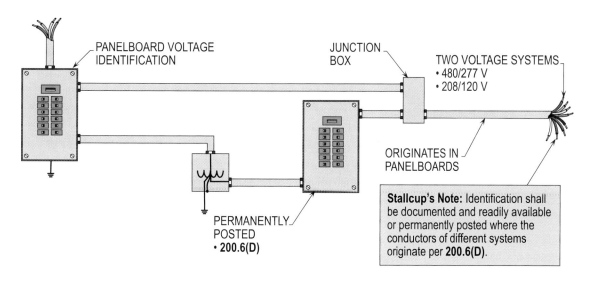

GROUNDED CONDUCTORS OF DIFFERENT SYSTEMS
200.6(D)

Purpose of Change: To require proper documentation and make it readily accessible or permanently posted where conductors originate.

Type of Change		Revision		Committe Change		Accept		2008 NEC	210.8
ROP	pg. 118	# 2-77	log: 1735	ROC	p. 61	# 2-29	log: 1717	UL	943
Submitter: Jared Boone				Submitter: Phil Simmons				OSHA	1910.304(a)(3)
NFPA 70B	13.1.3.1			NFPA 70E	110.9(C)			NFPA 79	-

210.8 Ground-Fault Circuit-Interrupter Protection for Personnel.

Ground-fault circuit-interruption for personnel shall be provided as required in 210.8(A) through (C). The ground-fault circuit-interrupter shall be installed in a readily accessible location.

Informational Note: See 215.9 for ground-fault circuit-interrupter protection for personnel on feeders.

Stallcup's Comment: A revision has been made to clarify that a ground-fault circuit-interrupter shall be installed in a readily acessible location.

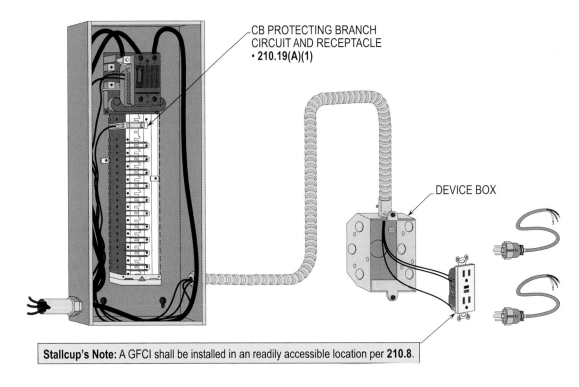

CB PROTECTING BRANCH
CIRCUIT AND RECEPTACLE
• **210.19(A)(1)**

DEVICE BOX

Stallcup's Note: A GFCI shall be installed in an readily accessible location per **210.8**.

GROUND-FAULT CIRCUIT-INTERRUPTER PROTECTION FOR PERSONNEL
210.8

Purpose of Change: To clarify that a ground-fault circit-interrupter (GFCI) receptacle shall be installed in an readily accessible location.

Type of Change	Revision			Committe Change		Accept			2008 NEC	210.8(A)(7)
ROP	pg. 122	# 2-103	log: 1610	ROC	pg. 63	# 2-42	log: 1590		UL	943
Submitter: David Shields				Submitter: Jim Pauley					OSHA	1910.304(a)(3)
NFPA 70B	13.1.3.1			NFPA 70E	-				NFPA 79	-

210.8 Ground-Fault Circuit-Interrupter Protection for Personnel.

(A) Dwelling Units. All 125-volt, single-phase, 15- and 20-ampere receptacles installed in the locations specified in 210.8(A)(1) through (8) shall have ground-fault circuit-interrupter protection for personnel.

(7) Sinks – located in areas other than kitchens where receptacles are installed within 1.8 m (6 ft) of the outside edge of the sink

Stallcup's Comment: A revision has been made to clarify that kitchens are covered in **210.8(A)(6),** and sinks in other areas require receptacles that are installed within 6 ft (1.8 m) of the outside edge of the sink to be GFCI protected.

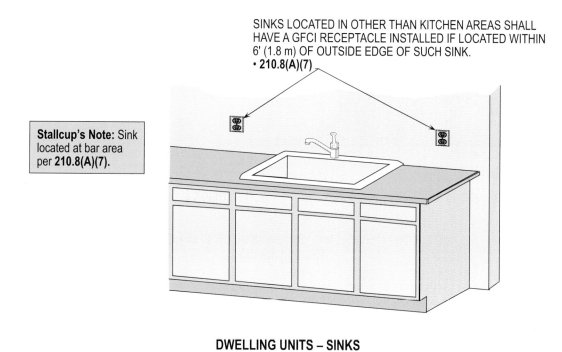

SINKS LOCATED IN OTHER THAN KITCHEN AREAS SHALL HAVE A GFCI RECEPTACLE INSTALLED IF LOCATED WITHIN 6' (1.8 m) OF OUTSIDE EDGE OF SUCH SINK.
• **210.8(A)(7)**

Stallcup's Note: Sink located at bar area per **210.8(A)(7).**

DWELLING UNITS – SINKS
210.8(A)(7)

Purpose of Change: A GFCI receptacle is required if located within 6 ft (1.8 m) of a sink located in other than kitchens.

Type of Change	New Subdivision			Committe Change		Accept in Principle		2008 NEC	-
ROP	pg. 122	# 2-105	log: 299	ROC	pg. -	# -	log: -	UL	943
Submitter: Christine Porter				Submitter: -				OSHA	1910.304(a)(3)
NFPA 70B	13.1.3.1			NFPA 70E	-			NFPA 79	-

210.8 Ground-Fault Circuit-Interrupter Protection for Personnel.

(B) Other Than Dwelling Units. All 125-volt, single-phase, 15- and 20-ampere receptacles installed in the locations specified in 210.8(B)(1) through (8) shall have ground-fault circuit-interrupter protection for personnel.

(6) Indoor wet locations

Stallcup's Comment: A new subdivison has been added to require ground-fault circuit-interrupter protection for indoor wet locations.

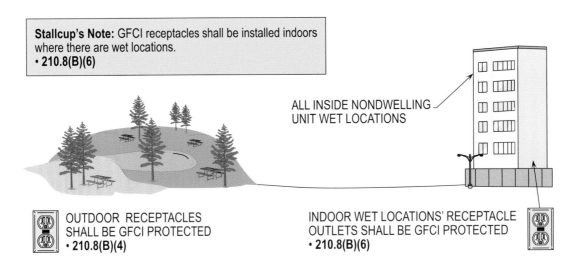

Stallcup's Note: GFCI receptacles shall be installed indoors where there are wet locations.
• **210.8(B)(6)**

ALL INSIDE NONDWELLING UNIT WET LOCATIONS

OUTDOOR RECEPTACLES SHALL BE GFCI PROTECTED
• **210.8(B)(4)**

INDOOR WET LOCATIONS' RECEPTACLE OUTLETS SHALL BE GFCI PROTECTED
• **210.8(B)(6)**

OTHER THAN DWELLING UNITS – INDOOR WET LOCATIONS
NEC 210.8(B)(6)

Purpose of Change: To require inside receptacles of nondwelling units installed in wet locations to be GFCI protected.

Type of Change	New Subdivision			Committe Change		Accept		2008 NEC	-
ROP	pg. 123	# 2-110	log: 4182	ROC	pg. 65	# 2-52	log: 2258	UL	943
Submitter: Richard A. Janoski				Submitter: William Benard				OSHA	1910.304(a)(3)
NFPA 70B	13.1.3.1			NFPA 70E	-			NFPA 79	-

210.8 Ground-Fault Circuit-Interrupter Protection for Personnel.

(B) Other Than Dwelling Units. All 125-volt, single-phase, 15- and 20-ampere receptacles installed in the locations specified in 210.8(B)(1) through (8) shall have ground-fault circuit-interrupter protection for personnel.

(7) Locker rooms with associated showering facilities

Stallcup's Comment: A new subdivison has been added to require ground-fault circuit-interrupter protection for locker rooms with associated showering facilities.

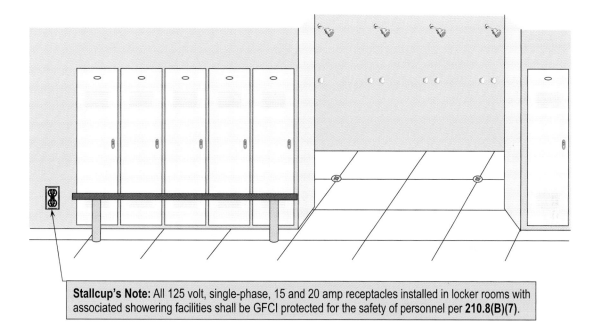

Stallcup's Note: All 125 volt, single-phase, 15 and 20 amp receptacles installed in locker rooms with associated showering facilities shall be GFCI protected for the safety of personnel per **210.8(B)(7)**.

OTHER THAN DWELLING UNITS – LOCKER ROOMS
NEC 210.8(B)(7)

Purpose of Change: To require GFCI protection for locker rooms with associated showering facilities.

Type of Change	New Subdivision			Committe Change	Accept			2008 NEC	-
ROP	pg. 125	# 2-122	log: 4178	ROC	pg. 64	# 2-45	log: 2241	UL	943
Submitter: Timothy D. Curry				Submitter: John Williamson				OSHA	1910.304(a)(3)
NFPA 70B	13.1.3.1			NFPA 70E	-			NFPA 79	-

210.8 Ground-Fault Circuit-Interrupter Protection for Personnel.

(B) Other Than Dwelling Units. All 125-volt, single-phase, 15- and 20-ampere receptacles installed in the locations specified in 210.8(B)(1) through (8) shall have ground-fault circuit-interrupter protection for personnel.

(8) Garages, service bays, and similar areas where electrical diagnostic equipment, electrical hand tools, or portable lighting equipment are to be used

Stallcup's Comment: A new subdivison has been added to require ground-fault circuit-interrupter protection for garages, service bays, and similar areas where electrical diagnostic equipment, electrical hand tools, or portable lighting equipment are to be used.

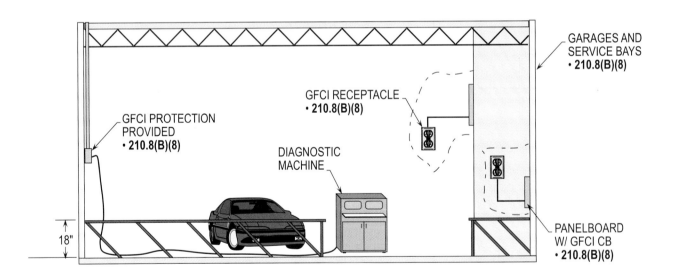

> **Stallcup's Note:** Electric diagnostic equipment, electrical hand tools, or portable lighting shall be provided with GFCI protection per **210.8(B)(8)**.

OTHER THAN DWELLING UNITS – GARAGES, SERVICE BAYS, AND SIMILAR AREAS
210.8(B)(8)

Purpose of Change: To clarify the procedure of providing GFCI protection for receptacles in garages, service bays, and similar areas.

Type of Change	New Subsection		Committe Change	Accept in Principle in Part		2008 NEC	210.12(B)
ROP	pg. 138	# 2-179	log: 4876	ROC	pg. 76	# 2-90	log: 2393	UL	1699
Submitter: David Zinck			Submitter: Donald R. Cook			OSHA	-
NFPA 70B	-		NFPA 70E	-		NFPA 79	-

210.12 Arc-Fault Circuit-Interrupter Protection.

(B) Branch Circuit Extensions or Modifications – Dwelling Units. In any of the areas specified in 210.12(A), where branch-circuit wiring is modified, replaced, or extended, the branch circuit shall be protected by one of the following:

(1) A listed combination-type AFCI located at the origin of the branch circuit

(2) A listed outlet branch-circuit type AFCI located at the first receptacle outlet of the existing branch circuit

Stallcup's Comment: A new subsection has been added to require AFCI protection for a branch circuit that is modified, replaced, or extended. The branch circuit that is modified, replaced, or extended shall be protected by a listed combination AFCI located at the origin of the branch circuit or by a listed branch-circuit AFCI located at the first receptacle outlet of the existing branch circuit.

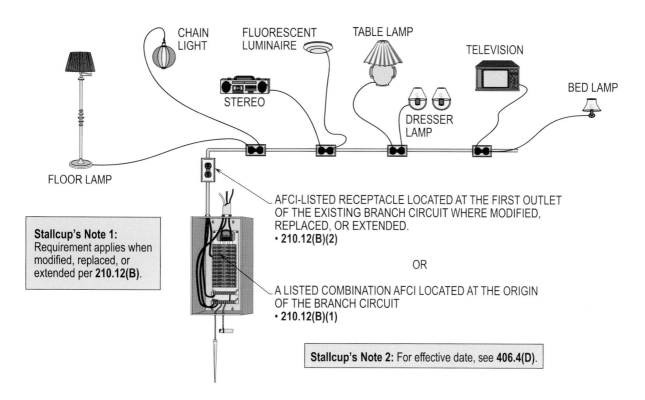

BRANCH CIRCUIT EXTENSIONS OR MODIFICATIONS – DWELLING UNITS
210.12(B)(1) AND (B)(2)

Purpose of Change: To clarify procedures of providing AFCI protection for receptacles in dwelling units.

Type of Change	New Subsection			Committe Change	Accept in Principle in Part			2008 NEC	-
ROP	pg. 147	# 2-223	log: 2962	ROC	pg. 85	# 2-114	log: 1225	UL	498
Submitter: Joseph Whitt				Submitter: Vince Baclawski				OSHA	-
NFPA 70B	-			NFPA 70E	-			NFPA 79	-

210.52 Dwelling Unit Receptacle Outlets.

(I) Foyers. Foyers that are not part of a hallway in accordance with 210.52(H) and that have an area that is greater than 5.6 m² (60 ft²) shall have a receptacle(s) located in each wall space 900 mm (3 ft) or more in width and unbroken by doorways, floor-to-ceiling windows, and similar openings.

Stallcup's Comment: A new subsection has been added to require receptacle(s) to be added in foyers that are not part of a hallway in accordance with **210.52(H)** and that have an area that is greater than 60 ft² (5.6 m²). The receptacle(s) shall be located in each wall space 3 ft (900 mm) or more in width and unbroken by doorways, floor to-ceiling windows, and similar openings.

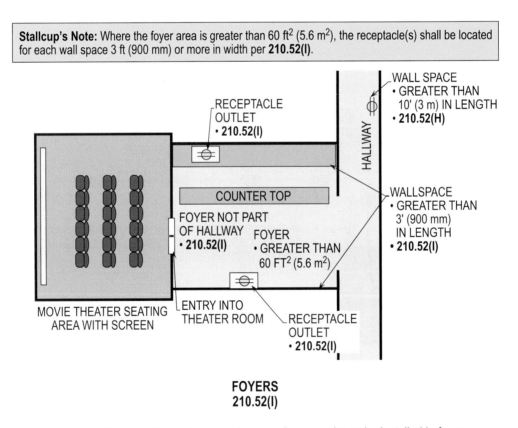

Stallcup's Note: Where the foyer area is greater than 60 ft² (5.6 m²), the receptacle(s) shall be located for each wall space 3 ft (900 mm) or more in width per **210.52(I)**.

FOYERS
210.52(I)

Purpose of Change: To require properly spaced receptacles to be installed in foyers.

Type of Change	New Subdivision			Committe Change	Accept in Part			2008 NEC	-
ROP	pg. 183	# 4-31	log: 3	ROC	pg. -	# -	log: -	UL	1581
Submitter: Joseph A. Hertel				Submitter: -				OSHA	1910.304(c)(2)
NFPA 70B	-			NFPA 70E	-			NFPA 79	-

225.18 Clearance for Overhead Conductors and Cables.

Overhead spans of open conductors and open multiconductor cables of not over 600 volts, nominal, shall have a clearance of not less than the following:

(5) 7.5 m (24.5 ft) – over track rails of railroads.

Stallcup's Comment: A new subdivision has been added to address the clearance requirements for overhead conductors and cables installed over track rails of railroads.

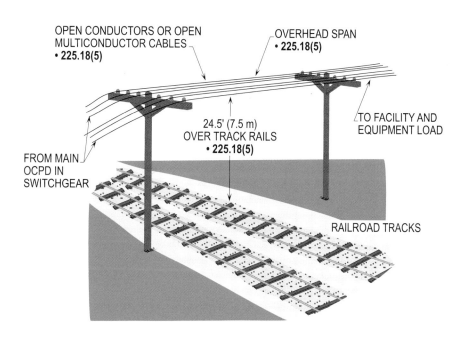

CLEARANCE FOR OVERHEAD CONDUCTORS AND CABLES
NEC 225.18(5)

Purpose of Change: To add clearance requirements for overhead open conductors and open multiconductor cables installed over track rails of railroads.

Type of Change	New Section			Committe Change	Accept			2008 NEC	-
ROP	pg. 184	# 4-35	log: 307	ROC	pg. -	# -	log: -	UL	157
Submitter: Joel A. Rencsok				Submitter: -				OSHA	-
NFPA 70B	-			NFPA 70E	-			NFPA 79	-

225.27 Raceway Seal

Where a raceway enters a building or structure from an underground distribution system, it shall be sealed in accordance with 300.5(G). Spare or unused raceways shall also be sealed. Sealants shall be identified for use with the cable insulation, shield, or other components.

Stallcup's Comment: A new section has been added to require a raceway seal where a raceway enters a building or structure from an underground distribution system.

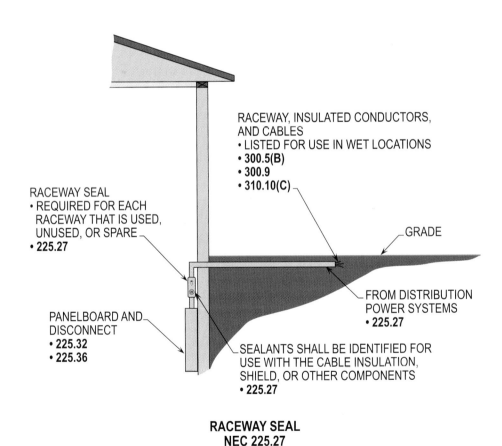

RACEWAY, INSULATED CONDUCTORS, AND CABLES
• LISTED FOR USE IN WET LOCATIONS
• 300.5(B)
• 300.9
• 310.10(C)

RACEWAY SEAL
• REQUIRED FOR EACH RACEWAY THAT IS USED, UNUSED, OR SPARE
• 225.27

GRADE

FROM DISTRIBUTION POWER SYSTEMS
• 225.27

PANELBOARD AND DISCONNECT
• 225.32
• 225.36

SEALANTS SHALL BE IDENTIFIED FOR USE WITH THE CABLE INSULATION, SHIELD, OR OTHER COMPONENTS
• 225.27

RACEWAY SEAL
NEC 225.27

Purpose of Change: To require seals to be installed on used, unused, or spare raceways entering building or structure from underground distribution systems.

Type of Change	New Exception			Committe Change	Accept			2008 NEC	-
ROP	pg. 179	# 4-21	log: 678	ROC	pg. 99	# 4-6	log: 70	UL	-
Submitter: Technical Correlating Committee				Submitter: Technical Correlating Committee				OSHA	1910.304(c)
NFPA 70B	-			NFPA 70E	-			NFPA 79	-

225.52 Disconnecting Means.

(B) Type. Each building or structure disconnect shall simultaneously disconnect all ungrounded supply conductors it controls and shall have a fault-closing rating not less than the maximum available short-circuit current available at its supply terminals.

Exception: Where the individual disconnecting means consists of fused cutouts, the simultaneous disconnection of all ungrounded supply conductors shall not be required if there is a means to disconnect the load before opening the cutouts. A permanent legible sign shall be installed adjacent to the fused cutouts indicating the above requirement.

Stallcup's Comment: A new exception has been added to permit an individual disconnecting means to be installed consisting of fused cutouts. Ungrounded supply conductors shall not be required to be simultaneously disconnected provided that there is a means to disconnect the load before opening the cutouts.

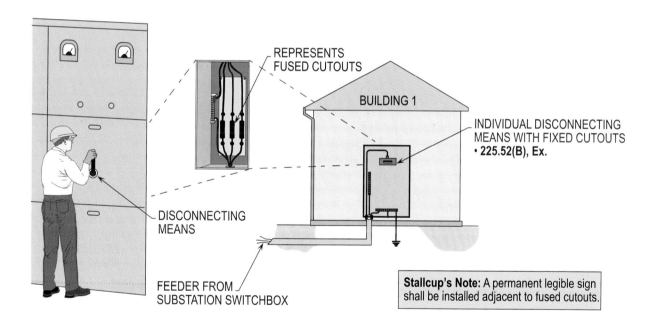

REPRESENTS
FUSED CUTOUTS

BUILDING 1

INDIVIDUAL DISCONNECTING
MEANS WITH FIXED CUTOUTS
• **225.52(B), Ex.**

DISCONNECTING
MEANS

FEEDER FROM
SUBSTATION SWITCHBOX

Stallcup's Note: A permanent legible sign shall be installed adjacent to fused cutouts.

TYPE
225.52(B), Ex.

Purpose of Change: To clarify the use of an individual disconnecting means consisting of cutout fuses.

Type of Change	New Subsections			Committe Change	Accept			2008 NEC	-
ROP	pg. 179	# 4-21	log: 678	ROC	pg. 99	# 4-6	log: 70	UL	-
Submitter: Technical Correlating Committee				Submitter: Technical Correlating Committee				OSHA	1910.304(c)
NFPA 70B	-			NFPA 70E	120.2(F)(2)(c)			NFPA 79	5.5

225.52 Disconnecting Means.

(C) Locking. Disconnecting means shall be capable of being locked in the open position. The provisions for locking shall remain in place with or without the lock installed.

Exception: Where an individual disconnecting means consists of fused cutouts, a suitable enclosure capable of being locked and sized to contain all cutout fuse holders shall be installed at a convenient location to the fused cutouts.

(D) Indicating. Disconnecting means shall clearly indicate whether they are in the open "off" or closed "on" position.

(E) Uniform Position. Where disconnecting means handles are operated vertically, the "up" position of the handle shall be the "on" position.

Exception: A switching device having more than one "on" position, such as a double throw switch, shall not be required to comply with this requirement.

Stallcup's Comment: New subsections have been added to address the locking, indicating, uniform position, and identification requirements for disconnecting means.

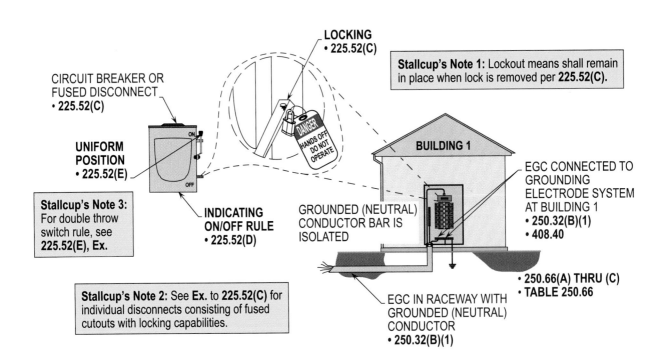

LOCKING, INDICATING, AND UNIFORM POSITION
NEC 225.52(C) THRU (E) AND Ex.

Purpose of Change: To clarify the requirements of the disconnecting means for a building supplied from a substation, main building switchgear, or panelboard.

Type of Change	New Subsection			Committee Change		Accept		2008 NEC	-
ROP	pg. 179	# 4-21	log: 678	ROC	pg. 99	# 4-6	log: 70	UL	-
Submitter: Technical Correlating Committee				Submitter: Technical Correlating Committee				OSHA	1910.304(c)
NFPA 70B	-			NFPA 70E	120.2(F)(2)(o)			NFPA 70	5.5

225.52 Disconnecting Means.

(F) Identification. Where a building or structure has any combination of feeders, branch circuits, or services passing through or supplying it, a permanent plaque or directory shall be installed at each feeder and branch-circuit disconnect location that denotes all other services, feeders, or branch circuits supplying that building or structure or passing through that building or structure and the area served by each.

Stallcup's Comment: New subsections have been added to address the locking, indicating, uniform position, and identification requirements for disconnecting means.

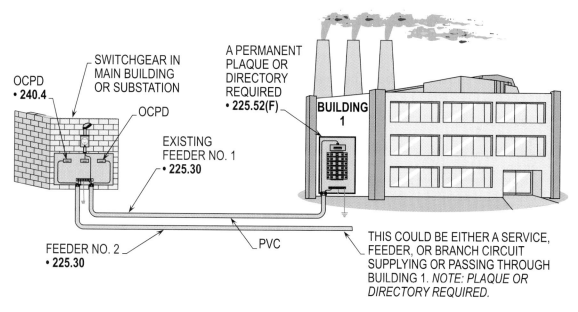

Stallcup's Note: A permanent plaque or directory shall denote at each feeder or branch-circuit disconnect whether power is supplying or passing through building as outlined in **225.52(F)**.

IDENTIFICATION
NEC 225.52(F)

Purpose of Change: To clarify the identification of more than one service, feeder, or branch circuit when supplying or passing through a building.

Type of Change	New Subsections		Committee Change	Accept		2008 NEC	-
ROP	pg. 179	# 4-21	log: 678	ROC	pg. 99	# 4-6	log: 70	UL	-
Submitter: Technical Correlating Committee			Submitter: Technical Correlating Committee			OSHA	1910.304(c)
NFPA 70B	-		NFPA 70E	120.2(F)(2)(c)		NFPA 79	5.5

225.52 Disconnecting Means.

(C) Locking. Disconnecting means shall be capable of being locked in the open position. The provisions for locking shall remain in place with or without the lock installed.

Exception: Where an individual disconnecting means consists of fused cutouts, a suitable enclosure capable of being locked and sized to contain all cutout fuse holders shall be installed at a convenient location to the fused cutouts.

(D) Indicating. Disconnecting means shall clearly indicate whether they are in the open "off" or closed "on" position.

(E) Uniform Position. Where disconnecting means handles are operated vertically, the "up" position of the handle shall be the "on" position.

Exception: A switching device having more than one "on" position, such as a double throw switch, shall not be required to comply with this requirement.

(F) Identification. Where a building or structure has any combination of feeders, branch circuits, or services passing through or supplying it, a permanent plaque or directory shall be installed at each feeder and branch-circuit disconnect location that denotes all other services, feeders, or branch circuits supplying that building or structure or passing through that building or structure and the area served by each.

Stallcup's Comment: New subsections have been added to address the locking, indicating, uniform position, and identification requirements for disconnecting means.

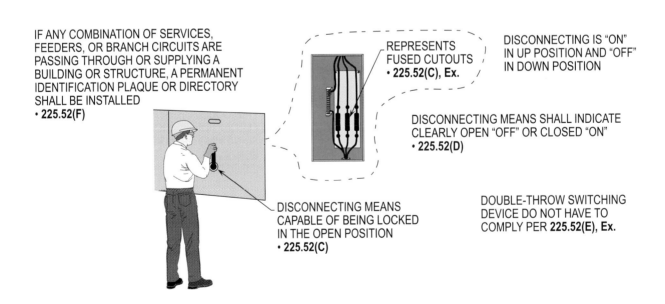

IF ANY COMBINATION OF SERVICES, FEEDERS, OR BRANCH CIRCUITS ARE PASSING THROUGH OR SUPPLYING A BUILDING OR STRUCTURE, A PERMANENT IDENTIFICATION PLAQUE OR DIRECTORY SHALL BE INSTALLED
• **225.52(F)**

REPRESENTS FUSED CUTOUTS
• **225.52(C), Ex.**

DISCONNECTING IS "ON" IN UP POSITION AND "OFF" IN DOWN POSITION

DISCONNECTING MEANS SHALL INDICATE CLEARLY OPEN "OFF" OR CLOSED "ON"
• **225.52(D)**

DISCONNECTING MEANS CAPABLE OF BEING LOCKED IN THE OPEN POSITION
• **225.52(C)**

DOUBLE-THROW SWITCHING DEVICE DO NOT HAVE TO COMPLY PER **225.52(E), Ex.**

LOCKING, INDICATING, UNIFORM POSITION, AND IDENTIFICATION
225.52(C) THRU (F)

Purpose of Change: To clarify the requirements that the disconnecting means has to comply for use with outside branch circuits and feeders.

Type of Change	New Section			Committe Change	Accept			2008 NEC	-
ROP	pg. 179	# 4-21	log: 678	ROC	pg. 99	# 4-6	log: 70	UL	-
Submitter: Technical Correlating Committee				Submitter: Technical Correlating Committee				OSHA	1910.304(c)
NFPA 70B	-			NFPA 70E	-			NFPA 70	C.6

225.56 Inspections and Tests.

(A) Pre-Energization and Operating Tests. The complete electrical system shall be performance tested when first installed on-site. Each protective, switching, and control circuit shall be adjusted in accordance with the recommendations of the protective device study and tested by actual operation using current injection or equivalent methods as necessary to ensure that each and every such circuit operates correctly to the satisfaction of the authority having jurisdiction.

(1) Instrument Transformers. All instrument transformers shall be tested to verify correct polarity and burden.

(2) Protective Relays. Each protective relay shall be demonstrated to operate by injecting current or voltage, or both, at the associated instrument transformer output terminal and observing that the associated switching and signaling functions occur correctly and in proper time and sequence to accomplish the protective function intended.

(3) Switching Circuits. Each switching circuit shall be observed to operate the associated equipment being switched.

(4) Control and Signal Circuits. Each control or signal circuit shall be observed to perform its proper control function or produce a correct signal output.

(5) Metering Circuits. All metering circuits shall be verified to operate correctly from voltage and current sources, similarly to protective relay circuits.

(6) Acceptance Tests. Complete acceptance tests shall be performed, after the station installation is completed, on all assemblies, equipment, conductors, and control and protective systems, as applicable, to verify the integrity of all the systems.

(7) Relays and Metering Utilizing Phase Differences. All relays and metering that use phase differences for operation shall be verified by measuring phase angles at the relay under actual load conditions after operation commences.

(B) Test Report. A test report covering the results of the tests required in 225.56(A) shall be delivered to the authority having jurisdiction prior to energization.

Informational Note: For acceptance specifications, see NETA ATS-2007, *Acceptance Testing Specifications for Electrical Power Distribution Equipment and Systems,* published by the InterNational Electrical Testing Association.

Stallcup's Comment: A new section has been added to address the pre-energization and operating tests for instrument transformers, protective relays, switching circuits, control and signaling circuits, metering circuits, acceptance tests, and relays and metering utilizing phase differences.

PRE-ENERGIZATION AND OPERATING TESTS
• **225.56(A)**

SWITCHING CIRCUITS SHALL OPERATE ASSOCIATED EQUIPMENT
• **225.56(A)(3)**

ACCEPTANCE TESTS SHALL VERIFY THE INTEGRITY OF ALL SYSTEMS
• **225.56(A)(6)**

INSTRUMENT TRANSFORMERS SHALL BE TESTED
• **225.56(A)(1)**

CONTROL AND SIGNAL CIRCUITS SHALL PROPERLY PERFORM FUNCTION
• **225.56(A)(4)**

RELAYS AND METERING UTILIZING PHASE DIFFERENCES SHALL BE VERIFIED
• **225.56(A)(7)**

PROTECTIVE RELAYS SHALL BE TESTED TO VERIFY OPERATION
• **225.56(A)(2)**

METERING CIRCUITS SHALL PERFORM THEIR FUNCTION PROPERLY
• **225.56(A)(5)**

**INSPECTIONS AND TEST
NEC 225.56(A) AND (B)**

Purpose of Change: To address the pre-energization and operating tests for specific items.

Type of Change	New Section			Committee Change	Accept		2008 NEC	-
ROP	pg. 179	# 4-21	log: 678	ROC	pg. 99	# 4-6	log: 70	UL	-
Submitter: Technical Correlating Committee				Submitter: Technical Correlating Committee			OSHA	1910.304(c)
NFPA 70B	-			NFPA 70E	-		NFPA 79	C.6

225.56 Inspections and Tests.

(A) Pre-Energization and Operating Tests. The complete electrical system shall be performance tested when first installed on-site. Each protective, switching, and control circuit shall be adjusted in accordance with the recommendations of the protective device study and tested by actual operation using current injection or equivalent methods as necessary to ensure that each and every such circuit operates correctly to the satisfaction of the authority having jurisdiction.

(1) Instrument Transformers. All instrument transformers shall be tested to verify correct polarity and burden.

(2) Protective Relays. Each protective relay shall be demonstrated to operate by injecting current or voltage, or both, at the associated instrument transformer output terminal and observing that the associated switching and signaling functions occur correctly and in proper time and sequence to accomplish the protective function intended.

(3) Switching Circuits. Each switching circuit shall be observed to operate the associated equipment being switched.

(4) Control and Signal Circuits. Each control or signal circuit shall be observed to perform its proper control function or produce a correct signal output.

(5) Metering Circuits. All metering circuits shall be verified to operate correctly from voltage and current sources, similarly to protective relay circuits.

Stallcup's Comment: A new section has been added to address the pre-energization and operating tests for instrument transformers, protective relays, switching circuits, control and signaling circuits, metering circuits, acceptance tests, and relays and metering utilizing phase differences.

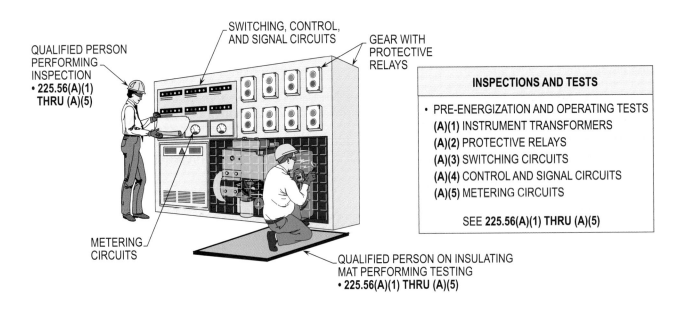

PRE-ENERGIZATION AND OPERATING TESTS
NEC 225.56(A)(1) THRU (A)(5)

Purpose of Change: To verify that inspections and tests shall be performed on specific components that perform particular functions.

Type of Change	New Section			Committee Change	Accept			2008 NEC	-
ROP	pg. 179	# 4-21	log: 678	ROC	pg. 99	# 4-6	log: 70	UL	-
Submitter: Technical Correlating Committee				Submitter: Technical Correlating Committee				OSHA	1910.304(c)
NFPA 70B	-			NFPA 70C				NFPA 70	C.6

225.56 Inspections and Tests.

(6) Acceptance Tests. Complete acceptance tests shall be performed, after the station installation is completed, on all assemblies, equipment, conductors, and control and protective systems, as applicable, to verify the integrity of all the systems.

(7) Relays and Metering Utilizing Phase Differences. All relays and metering that use phase differences for operation shall be verified by measuring phase angles at the relay under actual load conditions after operation commences.

(B) Test Report. A test report covering the results of the tests required in 225.56(A) shall be delivered to the authority having jurisdiction prior to energization.

Informational Note: For acceptance specifications, see NETA ATS-2007, *Acceptance Testing Specifications for Electrical Power Distribution Equipment and Systems,* published by the InterNational Electrical Testing Association.

Stallcup's Comment: A new section has been added to address the pre-energization and operating tests for instrument transformers, protective relays, switching circuits, control and signaling circuits, metering circuits, acceptance tests, and relays and metering utilizing phase differences.

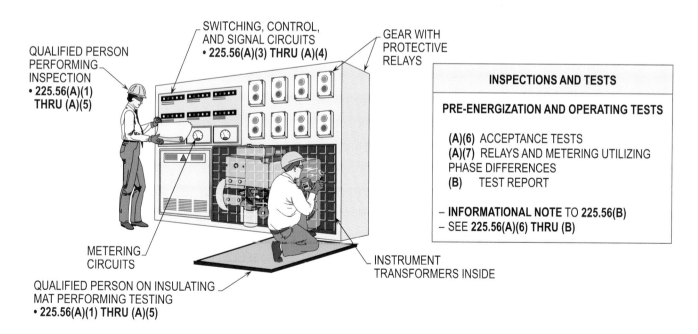

PRE-ENERGIZATION, OPERATING TESTS, AND TEST REPORT
NEC 225.56(A)(6), (A)(7), AND (B)

Purpose of Change: To verify the procedure for the acceptance tests and relays and metering utilizing phase differences as well as the test reports pertaining to the operation of the electrical system.

Type of Change	New Section			Committe Change	Accept			2008 NEC	-
ROP	pg. 179	# 4-21	log: 678	ROC	pg. 99	# 4-6	log: 70	UL	1062
Submitter: Technical Correlating Committee				Submitter: Technical Correlating Committee				OSHA	1910.304(c)
NFPA 70B	15.1.6			NFPA 70E	205.10			NFPA 79	-

225.70 Substations.

(A) Warning Signs.

(1) General. A permanent, legible warning notice carrying the wording "DANGER — HIGH VOLTAGE" shall be placed in a conspicuous position in the following areas:

(a) At all entrances to electrical equipment vaults and electrical equipment rooms, areas, or enclosures

(b) At points of access to conductors on all high-voltage conduit systems and cable systems

(c) On all cable trays containing high-voltage conductors with the maximum spacing of warning notices not to exceed 3 m (10 ft.)

Stallcup's Comment: A new section has been added to address the warning sign requirements for substations.

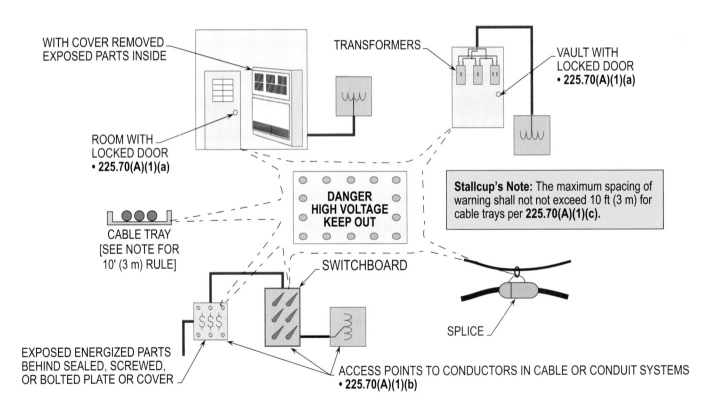

WARNING SIGNS
NED 225.70(A)(1)(a) THRU (A)(1)(c)

Purpose of Change: To clarify when warning signs shall be installed for alerting personnel of high-voltage hazards.

Type of Change	New Section			Committee Change	Accept			2008 NEC	-
ROP	pg. 179	# 4-21	log: 678	ROC	pg. 99	# 4-6	log: 70	UL	1062
Submitter: Technical Correlating Committee				Submitter: Technical Correlating Committee				OSHA	1910.304(c)
NFPA 70B	15.1.6			NFPA 70E	205.10			NFPA 79	-

225.70 Substations.

(A) Warning Signs.

(2) Isolating Equipment. Permanent legible signs shall be installed at isolating equipment warning against operation while carrying current, unless the equipment is interlocked so that it cannot be operated under load.

(3) Fuse Locations. Suitable warning signs shall be erected in a conspicuous place adjacent to fuses, warning operators not to replace fuses while the circuit is energized.

Stallcup's Comment: A new section has been added to address the warning sign requirements for substations.

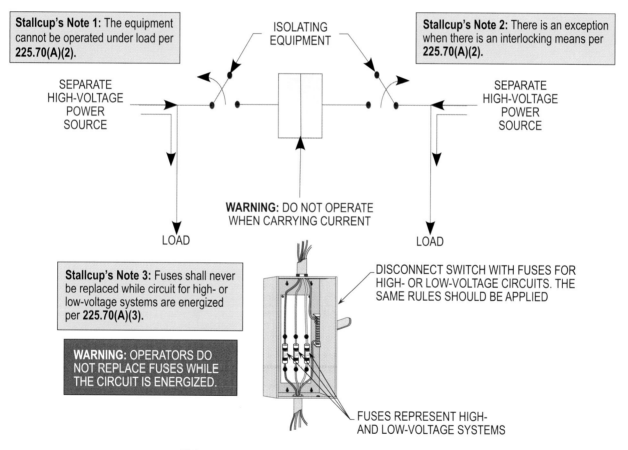

ISOLATING EQUIPMENT AND FUSE LOCATIONS
NEC 225.70(A)(2) AND (A)(3)

Purpose of Change: To clarify that warning signs shall be installed to alert personnel against hazardous switching conditions and fuse replacement procedures.

Type of Change	New Section			Committee Change	Accept			2008 NEC	-
ROP	pg. 179	# 4-21	log: 678	ROC	pg. 99	# 4-6	log: 70	UL	1062
Submitter: Technical Correlating Committee				Submitter: Technical Correlating Committee				OSHA	1910.304(c)
NFPA 70B	15.1.6			NFPA 70E	205.10			NFPA 79	-

225.70 Substations.

(A) Warning Signs.

(4) Backfeed. The following steps shall be taken where the possibility of backfeed exists:

(1) Each group-operated isolating switch or disconnecting means shall bear a warning notice to the effect that contacts on either side of the device might be energized.

(2) A permanent, legible, single-line diagram of the station switching arrangement, clearly identifying each point of connection to the high-voltage section, shall be provided in a conspicuous location within sight of each point of connection.

Stallcup's Comment: A new section has been added to address the warning sign requirements for substations.

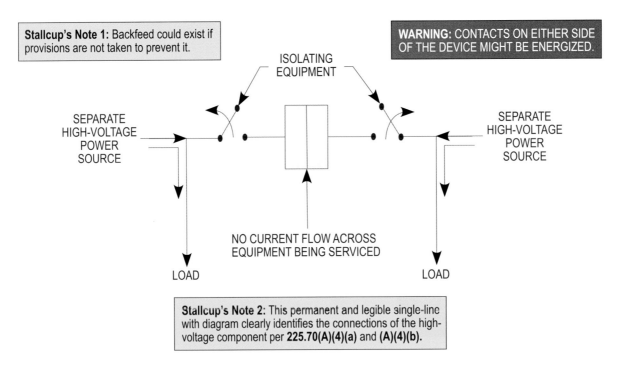

BACKFEED
NEC 225.70(A)(4)(1) AND (A)(4)(2)

Purpose of Change: To clarify that the *National Electrical Code* demands steps to be taken to prevent hazardous backfeed conditions for services.

Type of Change	New Section			Committee Change	Accept			2008 NEC	-
ROP	pg. 179	# 4-21	log: 678	ROC	pg. 99	# 4-6	log: 70	UL	1062
Submitter: Technical Correlating Committee				Submitter: Technical Correlating Committee				OSHA	1910.304(c)
NFPA 70B	15.1.6			NFPA 70E	205.10			NFPA 70	-

225.70 Substations.

(A) Warning Signs.

(5) Metal-Enclosed and Metal-Clad Switchgear. Where metal-enclosed switchgear is installed, the following steps shall be taken:

(a) A permanent, legible, single-line diagram of the switchgear shall be provided in a readily visible location within sight of the switchgear, and this diagram shall clearly identify interlocks, isolation means, and all possible sources of voltage to the installation under normal or emergency conditions, including all equipment contained in each cubicle, and the marking on the switchgear shall cross-reference the diagram.

Exception to (a): Where the equipment consists solely of a single cubicle or metal-enclosed unit substation containing only one set of high-voltage switching devices, diagrams shall not be required.

(b) Permanent, legible signs shall be installed on panels or doors that provide access to live parts over 600 volts and shall carry the wording "DANGER — HIGH VOLTAGE" to warn of the danger of opening while energized.

(c) Where the panel provides access to parts that can only be de-energized and visibly isolated by the serving utility, the warning shall include that access is limited to the serving utility or following an authorization of the serving utility.

Stallcup's Comment: A new section has been added to address the warning sign requirements for substations.

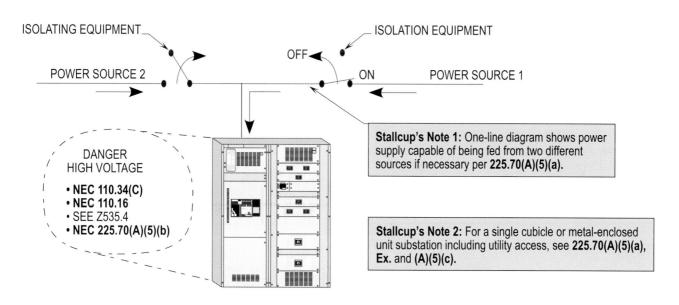

METAL-ENCLOSED AND METAL-CLAD SWITCHGEAR
NEC 225.70(A)(5)(a) THRU (A)(5)(c)

Purpose of Change: To provide requirements for signage and other pertinent items for metal-enclosed and metal-clad switchgear located in substations.

Type of Change	New Exception			Committe Change	Accept			2008 NEC	-
ROP	pg. 195	# 4-82	log: 454	ROC	pg. -	# -	log: -	UL	854
Submitter: Lanny G. McMahill				Submitter: -				OSHA	1910.304(c)(4)
NFPA 70B	-			NFPA 70E	-			NFPA 79	-

230.24 Clearances.

Overhead service conductors shall not be readily accessible and shall comply with 230.24(A) through (E) for services not over 600 volts, nominal.

(A) Above Roofs. Conductors shall have a vertical clearance of not less than 2.5 m (8 ft) above the roof surface. The vertical clearance above the roof level shall be maintained for a distance of not less than 900 mm (3 ft) in all directions from the edge of the roof.

Exception No. 5: Where the voltage between conductors does not exceed 300 and the roof area is guarded or isolated, a reduction in clearance to 900 mm (3 ft) shall be permitted.

Stallcup's Comment: A new exception has been added to correlate with the clearances above roofs in the *National Electrical Safety Code (NESC)*. The *National Electrical Safety Code* permits the service-entrance conductors to be installed a minimum of 3 ft (900 mm) above the roof if the area is guarded or isolated up to 750 volts. The new exception permits this type of installation only up to 300 volts.

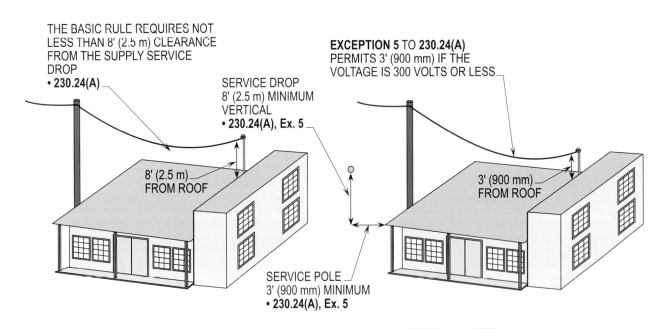

THE BASIC RULE REQUIRES NOT LESS THAN 8' (2.5 m) CLEARANCE FROM THE SUPPLY SERVICE DROP
• 230.24(A)

SERVICE DROP 8' (2.5 m) MINIMUM VERTICAL
• 230.24(A), Ex. 5

8' (2.5 m) FROM ROOF

EXCEPTION 5 TO 230.24(A) PERMITS 3' (900 mm) IF THE VOLTAGE IS 300 VOLTS OR LESS

3' (900 mm) FROM ROOF

SERVICE POLE 3' (900 mm) MINIMUM
• 230.24(A), Ex. 5

Stallcup's Note: Roof is guarded or isolated per **230.24(A), Ex. 5.**

ABOVE ROOFS
NEC 230.24(A), Ex. 5

Purpose of Change: To verify that a 3 ft (900 mm) clearance shall be permitted by the exception instead of the normal 8 ft (2.5 m).

Type of Change	New Exception			Committe Change	Accept			2008 NEC	-
ROP	pg. 202	# 4-107	log: 3043	ROC	pg. -	# -	log: -	UL	854
Submitter: Mike Holt				Submitter: -				OSHA	-
NFPA 70D				NFPA 70E	-			NFPA 79	-

230.42 Minimum Size and Rating.

(A) General. The ampacity of the service-entrance conductors before the application of any adjustment or correction factors shall not be less than either 230.42(A)(1) or (A)(2). Loads shall be determined in accordance with Part III, IV, or V of Article 220, as applicable. Ampacity shall be determined from 310.15. The maximum allowable current of busways shall be that value for which the busway has been listed or labeled.

(1) The sum of the noncontinuous loads plus 125 percent of continuous loads

Exception: Grounded conductors that are not connected to an overcurrent device shall be permitted to be sized at 100 percent of the continuous and noncontinuous load.

Stallcup's Comment: A new exception has been added to address the calculation requirements for the grounded conductor not connected to an overcurrent device for services.

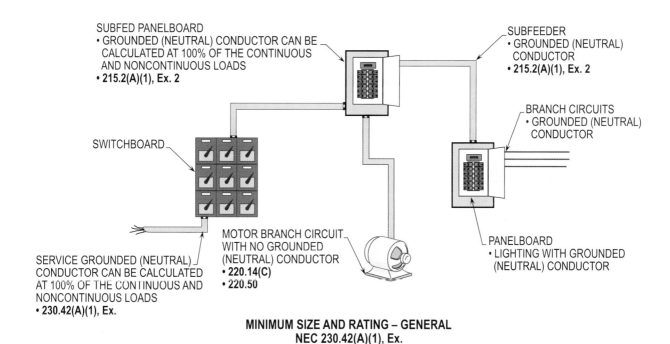

MINIMUM SIZE AND RATING – GENERAL
NEC 230.42(A)(1), Ex.

Purpose of Change: To add an exception to permit the grounded conductor (could be a neutral) to be calculated at 100 percent for continuous or noncontinuous duty loads if not connected to an overcurrent protection device.

Type of Change	Revision			Committe Change	Accept			2008 NEC	230.44
ROP	pg. 203	# 4-113	log: 1740	ROC	pg. 108	# 4-36	log: 73	UL	568
Submitter: Lowell Reith				Submitter: Technical Correlating Committee				OSHA	-
NFPA 70B	20.3			NFPA 70E	-			NFPA 79	-

230.44 Cable Trays.

Cable tray systems shall be permitted to support service-entrance conductors. Cable trays used to support service-entrance conductors shall contain only service-entrance conductors and shall be limited to the following methods:

(1) Type SE cable

(2) Type MC cable

(3) Type MI cable

(4) Type IGS cable

(5) Single thermoplastic-insulated conductors 1/0 and larger with CT rating

Such cable trays shall be identified with permanently affixed labels with the wording "Service-Entrance Conductors." The labels shall be located so as to be visible after installation and placed so that the service-entrance conductors are able to be readily traced through the entire length of the cable tray.

Stallcup's Comment: A revision has been made to clarify the types of cables that shall be permitted to be used for service-entrance conductors in cable trays.

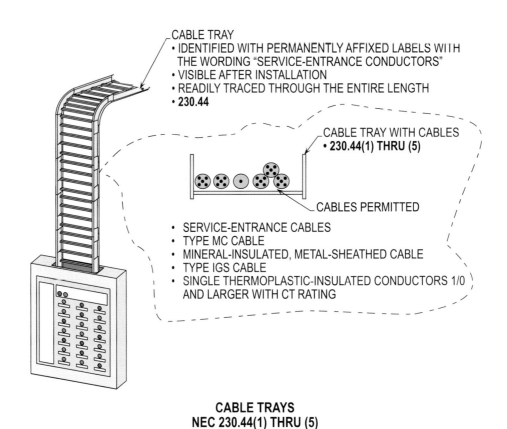

CABLE TRAY
• IDENTIFIED WITH PERMANENTLY AFFIXED LABELS WITH THE WORDING "SERVICE-ENTRANCE CONDUCTORS"
• VISIBLE AFTER INSTALLATION
• READILY TRACED THROUGH THE ENTIRE LENGTH
• **230.44**

CABLE TRAY WITH CABLES
• **230.44(1) THRU (5)**

CABLES PERMITTED

• SERVICE-ENTRANCE CABLES
• TYPE MC CABLE
• MINERAL-INSULATED, METAL-SHEATHED CABLE
• TYPE IGS CABLE
• SINGLE THERMOPLASTIC-INSULATED CONDUCTORS 1/0 AND LARGER WITH CT RATING

CABLE TRAYS
NEC 230.44(1) THRU (5)

Purpose of Change: To clarify the types of cables and identification that shall be permitted to be used for service-entrance conductors in cable trays.

Type of Change	Revision			Committe Change	Accept			2008 NEC	230.66
ROP	pg. 207	# 4-126	log: 828	ROC	pg. 109	# 4-42	log: 1711	UL	869A
Submitter: Dan Leaf				Submitter: Dan Leaf				OSHA	-
NFPA 70B	-			NFPA 70E	-			NFPA 79	16.2

230.66 Marking.

Service equipment rated at 600 volts or less shall be marked to identify it as being suitable for use as service equipment. All service equipment shall be listed. Individual meter socket enclosures shall not be considered service equipment.

Stallcup's Comment: A revision has been made to clarify that service equipment shall be listed and marked to identify it as being suitable for use as service equipment.

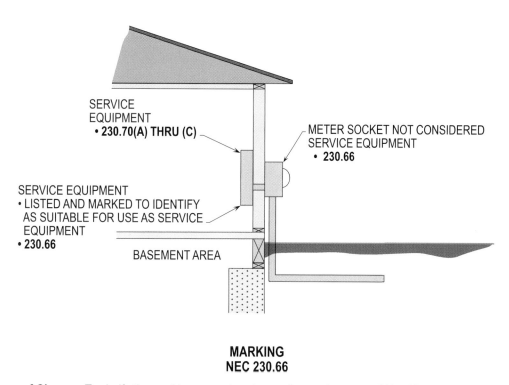

SERVICE
EQUIPMENT
• **230.70(A) THRU (C)**

METER SOCKET NOT CONSIDERED
SERVICE EQUIPMENT
• **230.66**

SERVICE EQUIPMENT
• LISTED AND MARKED TO IDENTIFY
  AS SUITABLE FOR USE AS SERVICE
  EQUIPMENT
• **230.66**

BASEMENT AREA

MARKING
NEC 230.66

Purpose of Change: To clarify the marking procedure for service equipment and identification of meter sockets.

Type of Change	Revision			Committe Change	Accept			2008 NEC	230.72(A), Ex.
ROP	pg. 209	# 4-140	log: 227	ROC	pg. 111	# 4-49	log: 2551	UL	-
Submitter: Don A. Hursey				Submitter: Frederic P. Hartwell				OSHA	1910.304(e)(1)
NFPA 70B	-			NFPA 70E	120.2(F)(2)(c)			NFPA 79	-

230.72 Grouping of Disconnects.

(A) General. The two to six disconnects as permitted in 230.71 shall be grouped. Each disconnect shall be marked to indicate the load served.

Exception: One of the two to six service disconnecting means permitted in 230.71, where used only for a water pump also intended to provide fire protection, shall be permitted to be located remote from the other disconnecting means. If remotely installed in accordance with this exception, a plaque shall be posted at the location of the remaining grouped disconnects denoting its location.

Stallcup's Comment: A revision has been made to clarify the location of the water pump service disconnecting means. If installed remotely, a plaque shall be posted at the location of the remaining grouped disconnects denoting its location.

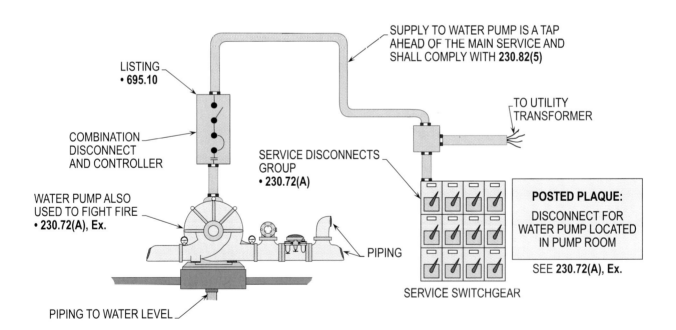

SUPPLY TO WATER PUMP IS A TAP AHEAD OF THE MAIN SERVICE AND SHALL COMPLY WITH **230.82(5)**

LISTING
• **695.10**

COMBINATION DISCONNECT AND CONTROLLER

WATER PUMP ALSO USED TO FIGHT FIRE
• **230.72(A), Ex.**

SERVICE DISCONNECTS GROUP
• **230.72(A)**

TO UTILITY TRANSFORMER

PIPING

PIPING TO WATER LEVEL

POSTED PLAQUE:
DISCONNECT FOR WATER PUMP LOCATED IN PUMP ROOM

SEE **230.72(A), Ex.**

SERVICE SWITCHGEAR

**GROUPING OF DISCONNECTS – GENERAL
NEC 230.72(A), Ex.**

Purpose of Change: To clarify that a posted plaque be installed on one of the two to six disconnects denoting the location of the water pump disconnect switch.

Type of Change	New Subdivision			Committe Change	Accept in Principle			2008 NEC	-
ROP	pg. 211	# 4-148	log: 2310	ROC	pg. 1234	# 4-51	log: 111	UL	-
Submitter: Michael Wright				Submitter: Randolph J. Ivans				OSHA	-
NFPA 70B	-			NFPA 70E	-			NFPA 79	-

230.82 Equipment Connected to the Supply Side of Service Disconnect.

Only the following equipment shall be permitted to be connected to the supply side of the service disconnecting means:

(9) Connections used only to supply listed communications equipment under the exclusive control of the serving electric utility, if suitable overcurrent protection and disconnecting means are provided. For installations of equipment by the serving electric utility, a disconnecting means is not required if the supply is installed as part of a meter socket, such that access can only be gained with the meter removed.

Stallcup's Comment: A new subdivision has been added to address new equipment and installations associated with Smart Grid applications and life-line (such as emergency calling) communications equipment powered at these premises.

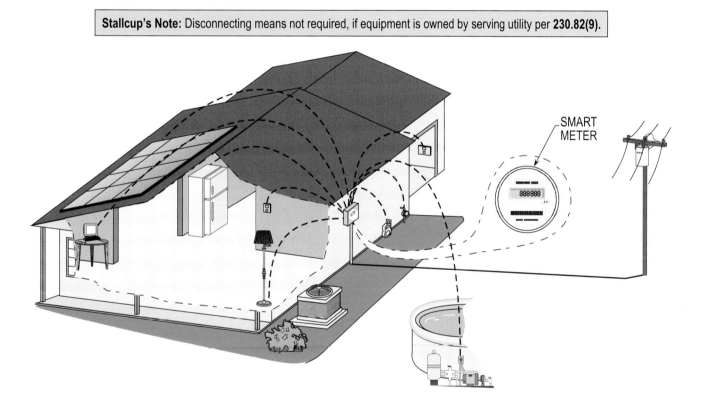

Stallcup's Note: Disconnecting means not required, if equipment is owned by serving utility per **230.82(9)**.

EQUIPMENT CONNECTED TO THE SUPPLY SIDE OF SERVICE DISCONNECT
NEC 230.82(9)

Purpose of Change: To permit a supply side tap for equipment such as emergency calling owned by the servicing utility.

Type of Change	Revision			Committe Change	Accept				2008 NEC	240.21(B)(1)(4)
ROP	pg. 225	# 10-48	log: 4193	ROC	pg. -	# -	log: -		UL	-
Submitter: Paul Dobrowsky				Submitter: -					OSHA	-
NFPA 70B	-			NFPA 70E	-				NFPA 79	7.2.8

240.21 Location in Circuit.

(B) Feeder Taps.

(1) Taps Not over 3 m (10 ft) Long. If the length of the tap conductors does not exceed 3 m (10 ft) and the tap conductors comply with all of the following:

(4) For field installations, if the tap conductors leave the enclosure or vault in which the tap is made, the ampacity of the tap conductors is not less than one-tenth of the rating of the overcurrent device protecting the feeder conductors.

Stallcup's Comment: A revision has been made to clarify the method of describing the ratio of the tap conductor to the feeder conductor is consistent in this section.

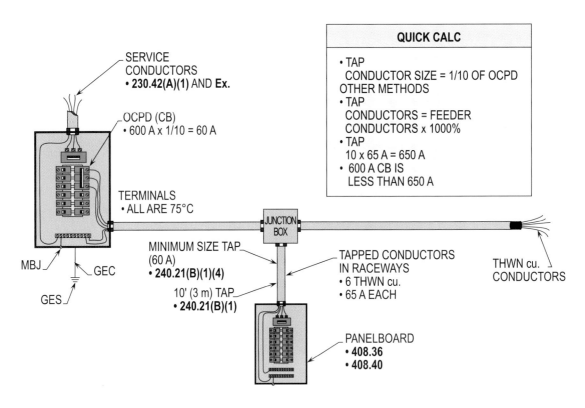

TAPS NOT OVER 3 m (10 FT) LONG
NEC 240.21(B)(1)(4)

Purpose of Change: To clarify the procedure for determining the minimum size conductors for taps not over 10 ft (3 m) in length.

Type of Change	Revision			Committe Change	Accept in Principle in Part			2008 NEC	240.24(E)
ROP	pg. 229	# 10-65	log: 3360	ROC	pg. -	# -	log: -	UL	-
Submitter: Dan Leaf				Submitter: -				OSHA	-
NFPA 70B	-			NFPA 70E	-			NFPA 70	

240.24 Location in or on Premises.

(E) Not Located in Bathrooms. In dwelling units, dormitories, and guest rooms or guest suites, overcurrent devices, other than supplementary overcurrent protection, shall not be located in bathrooms.

Stallcup's Comment: A revision has been made to clarify that overcurrent devices shall not be located in bathrooms of dormitories.

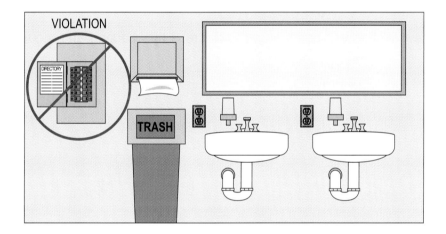

NOT LOCATED IN BATHROOMS
240.24(E)

Purpose of Change: To clarify that overcurrent protection devices are prohibited in the bathrooms of dwelling units, dormortories, and guest rooms or suites of hotels and motels.

Type of Change	New Section			Committe Change		Accept		2008 NEC	-
ROP	pg. 1234	# 10-72	log: 3877	ROC	pg. 1234	# 10-26a	log: CC 1000	UL	-
Submitter: Michael J. Farrell, III				Submitter: Code Making Panel 10				OSHA	-
NFPA 70B	-			NFPA 70E	-			NFPA 79	7.2.9

240.35 Available Fault Current.

(A) Marking. Service equipment in other than dwelling units shall be legibly marked in the field with the maximum available fault current. The field marking(s) shall include the date the fault current calculation was performed and be of sufficient durability to withstand the environment involved.

(B) Modifications. When modifications to the electrical installation occur, that affect the the maximum available fault current at the service, the maximum available fault current shall be verified or recalculated as necessary to ensure the service equipment interrupting ratings are sufficient for the maximum available fault current at the line terminals of the equipment. The required field marking(s) in (A) above shall be adjusted to reflect the new level of maximum available fault current.

Exception: The field marking requirements in (A) and (B) shall not be required in industrial installations where conditions of maintenance and supervision ensure that only qualified persons service the equipment.

Stallcup's Comment: A new section has been added to address the marking and modification requirements for avaiable fault current.

Type of Change	New Section			Committe Change	Accept in Principle			2008 NEC	-
ROP	pg. 233	# 10-82	log: 3562	ROC	pg. 123	# 10-41	log: 2399	UL	489
Submitter: James T. Dollard, Jr.				Submitter: Donald R. Cook				OSHA	-
NFPA 70B	-			NFPA 70E	-			NFPA 79	-

240.87 Noninstantaneous Trip.

Where a circuit breaker is used without an instantaneous trip, documentation shall be available to those authorized to design, install, operate, or inspect the installation as to the location of the circuit breaker(s).

Where a circuit breaker is utilized without an instantaneous trip, one of the following or approved equivalent means shall be provided:

(1) Zone-selective interlocking

(2) Differential relaying

(3) Energy-reducing maintenance switching with local status indicator

Informational Note: An energy-reducing maintenance switch allows a worker to set a circuit breaker trip unit to "no intentional delay" to reduce the clearing time while the worker is working within an arc-flash boundary as defined in NFPA 70E-2009, *Standard for Electrical Safety in the Workplace*, and then to set the trip unit back to a normal setting after the potentially hazardous work is complete.

Stallcup's Comment: A new section has been added to address the requirements for a circuit breaker that is utilized without an instantaneous trip.

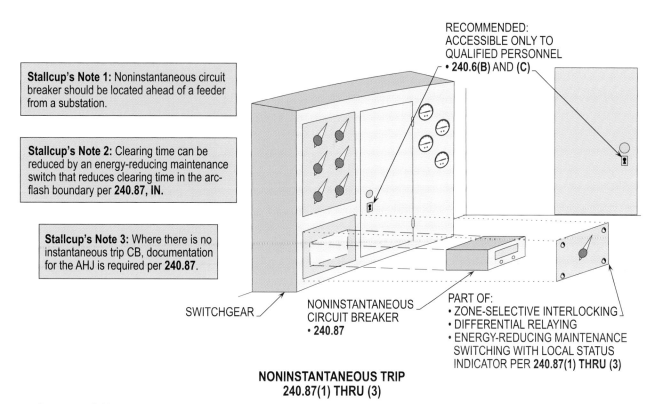

Stallcup's Note 1: Noninstantaneous circuit breaker should be located ahead of a feeder from a substation.

Stallcup's Note 2: Clearing time can be reduced by an energy-reducing maintenance switch that reduces clearing time in the arc-flash boundary per **240.87, IN.**

Stallcup's Note 3: Where there is no instantaneous trip CB, documentation for the AHJ is required per **240.87**.

RECOMMENDED: ACCESSIBLE ONLY TO QUALIFIED PERSONNEL • **240.6(B)** AND **(C)**

SWITCHGEAR

NONINSTANTANEOUS CIRCUIT BREAKER • **240.87**

PART OF: • ZONE-SELECTIVE INTERLOCKING • DIFFERENTIAL RELAYING • ENERGY-REDUCING MAINTENANCE SWITCHING WITH LOCAL STATUS INDICATOR PER **240.87(1) THRU (3)**

NONINSTANTANEOUS TRIP 240.87(1) THRU (3)

Purpose of Change: To outline the requirements pertaining to the use of a circuit breaker without an instantaneous trip.

Type of Change	New Section			Committe Change	Accept in Principle		2008 NEC	-
ROP	pg. 234	# 10-83	log: 214	ROC	pg. 125	# 10-49	log: 2010	UL	-
Submitter: Dorothy Kellogg				Submitter: Dennis M. Darling			OSHA	-
NFPA 70B	-			NFPA 70E	-		NFPA 79	-

240.91 Protection of Conductors.

Conductors shall be protected in accordance with 240.91(A) or (B).

(A) General. Conductors shall be protected in accordance with 240.4.

(B) Devices Rated Over 800 Amperes. Where the overcurrent device is rated over 800 amperes, the ampacity of the conductors it protects shall be equal to or greater than 95 percent of the rating of the overcurrent device specified in 240.6 in accordance with (B)(1) and (2).

(1) The conductors are protected within recognized time vs. current limits for short-circuit currents

(2) All equipment in which the conductors terminate is listed and marked for the application

Stallcup's Comment: A new section has been added to address the requirements for the protection of conductors in supervised industrial installations.

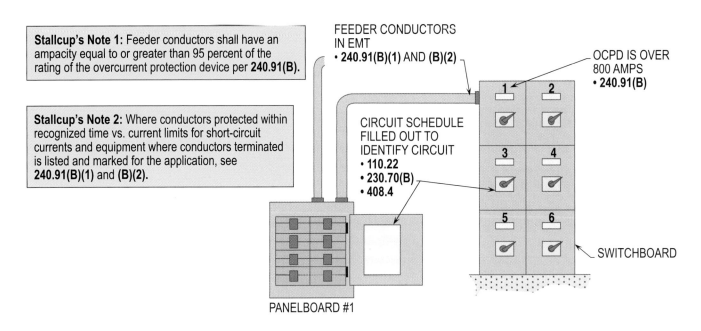

Stallcup's Note 1: Feeder conductors shall have an ampacity equal to or greater than 95 percent of the rating of the overcurrent protection device per **240.91(B).**

Stallcup's Note 2: Where conductors protected within recognized time vs. current limits for short-circuit currents and equipment where conductors terminated is listed and marked for the application, see **240.91(B)(1)** and **(B)(2).**

FEEDER CONDUCTORS IN EMT
• **240.91(B)(1)** AND **(B)(2)**

OCPD IS OVER 800 AMPS
• **240.91(B)**

CIRCUIT SCHEDULE FILLED OUT TO IDENTIFY CIRCUIT
• **110.22**
• **230.70(B)**
• **408.4**

SWITCHBOARD

PANELBOARD #1

**PROTECTION OF CONDUCTORS
NEC 240.91(A), (B)(1), AND (B)(2)**

Purpose of Change: To clarify the requirements for sizing the conductors when the overcurrent protection device is rated over 800 amps and is classified as a supervised industrial installation.

Type of Change	Revision			Committe Change	Accept			2008 NEC	250.8(A)
ROP	pg. 238	# 5-68a	log: CP 506	ROC	pg. -	# -	log: -	UL	467
Submitter: Code Making Panel 5				Submitter: -				OSHA	1910.304(G)(4)(i)
NFPA 70B	-			NFPA 70E	-			NFPA 79	8.2.3

250.8 Connection of Grounding and Bonding Equipment.

(A) Permitted Methods. Equipment grounding conductors, grounding electrode conductors, and bonding jumpers shall be connected by one of the following means:

(1) Listed pressure connectors

(2) Terminal bars

(3) Pressure connectors listed as grounding and bonding equipment

(4) Exothermic welding process

(5) Machine screw-type fasteners that engage not less than two threads or are secured with a nut

(6) Thread-forming machine screws that engage not less than two threads in the enclosure

(7) Connections that are part of a listed assembly

(8) Other listed means

Stallcup's Comment: A revision has been made to clarify the permitted connection methods of equipment grounding conductors, grounding electrode conductors, and bonding jumpers.

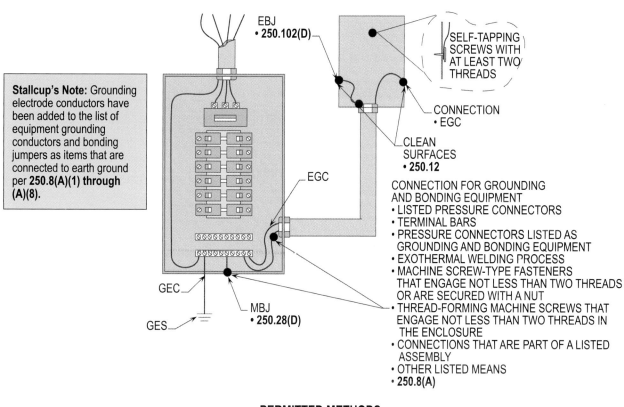

Stallcup's Note: Grounding electrode conductors have been added to the list of equipment grounding conductors and bonding jumpers as items that are connected to earth ground per **250.8(A)(1) through (A)(8).**

EBJ
• 250.102(D)

SELF-TAPPING SCREWS WITH AT LEAST TWO THREADS

CONNECTION
• EGC

CLEAN SURFACES
• 250.12

EGC

CONNECTION FOR GROUNDING AND BONDING EQUIPMENT
• LISTED PRESSURE CONNECTORS
• TERMINAL BARS
• PRESSURE CONNECTORS LISTED AS GROUNDING AND BONDING EQUIPMENT
• EXOTHERMAL WELDING PROCESS
• MACHINE SCREW-TYPE FASTENERS THAT ENGAGE NOT LESS THAN TWO THREADS OR ARE SECURED WITH A NUT
• THREAD-FORMING MACHINE SCREWS THAT ENGAGE NOT LESS THAN TWO THREADS IN THE ENCLOSURE
• CONNECTIONS THAT ARE PART OF A LISTED ASSEMBLY
• OTHER LISTED MEANS
• 250.8(A)

GEC

GES

MBJ
• 250.28(D)

**PERMITTED METHODS
250.8(A)(1) THRU (A)(8)**

Purpose of Change: To clarify the permitted methods for connections of grounding and bonding equipment.

Type of Change	Revision			Committe Change	Accept in Principle			2008 NEC	250.21(B)
ROP	pg. 241	# 5-85	log: 578	ROC	pg. -	# -	log: -	UL	467
Submitter: Michael J. Johnston				Submitter: -				OSHA	1910.304(g)(1)(v)[E]
NFPA 70B	-			NFPA 70E	-			NFPA 79	-

250.21 Alternating-Current Systems of 50 Volts to Less Than 1000 Volts Not Required to Be Grounded.

(B) Ground Detectors. Ground detectors shall be installed in accordance with 250.21(B)(1) and (B)(2).

(1) Ungrounded alternating current systems as permitted in 250.21(A)(1) through (A)(4) operating at not less than 120 volts and not exceeding 1000 volts shall have ground detectors installed on the system.

(2) The ground detection sensing equipment shall be connected as close as practicable to where the system receives its supply.

Stallcup's Comment: A revison has been made to clarify the location of the ground detection sensing equipment.

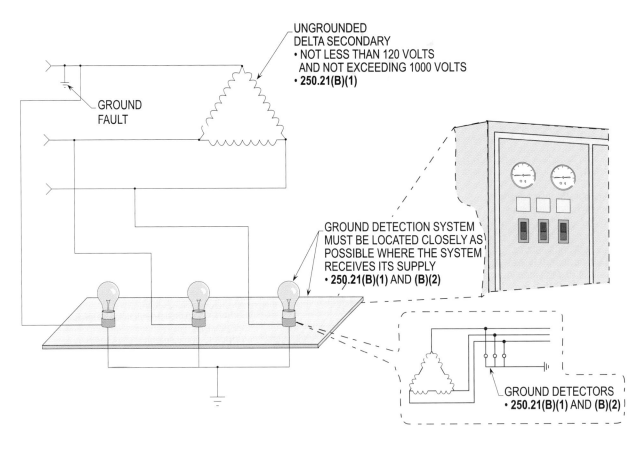

UNGROUNDED
DELTA SECONDARY
• NOT LESS THAN 120 VOLTS
  AND NOT EXCEEDING 1000 VOLTS
• **250.21(B)(1)**

GROUND
FAULT

GROUND DETECTION SYSTEM
MUST BE LOCATED CLOSELY AS
POSSIBLE WHERE THE SYSTEM
RECEIVES ITS SUPPLY
• **250.21(B)(1)** AND **(B)(2)**

GROUND DETECTORS
• **250.21(B)(1)** AND **(B)(2)**

**GROUND DETECTORS
250.21(B)(1) AND (B)(2)**

Purpose of Change: To require the ground detector system to be located as close as practicable to where such systems receive their supply.

Type of Change	New Subsection		Committee Change	Accept in Principle		2008 NEC	-
ROP pg. 241	# 5-86a	log: CP 502	ROC pg. 130	# 5-60	log: 1291	UL	467
Submitter: Code Making Panel 5			Submitter: David E. Shapiro			OSHA	1910.304(g)(1)
NFPA 70B -			NFPA 70E -			NFPA 79	16.1.2

250.21 Alternating-Current Systems of 50 Volts to Less Than 1000 Volts Not Required to be Grounded.

(C) Marking. Ungrounded systems shall be legibly marked "Ungrounded System" at the source or first disconnecting means of the system. The marking shall be of sufficient durability to withstand the environment involved.

Stallcup's Comment: A new subsection has been added to require ungrounded systems to be legibly marked at the source or the first disconnecting means of the system.

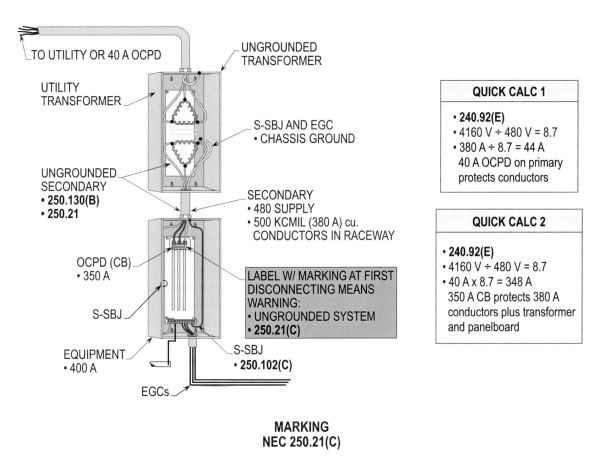

MARKING
NEC 250.21(C)

Purpose of Change: To require a legibly marked label to warn personnel that the system is ungrounded.

Type of Change	Revision			Committe Change		Accept in Principle		2008 NEC	250.30
ROP	pg. 245	# 5-102	log: 3224	ROC	pg. 132	# 5-69	log: 1836	UL	467
Submitter: Mark R. Hilbert				Submitter: Phil Simmons				OSHA	1910.304(g)(1)
NFPA 70B	14.1.6.43			NFPA 70E	-			NFPA 79	4.3.2

250.30 Grounding Separately Derived Alternating-Current Systems.

In addition to complying with 250.30(A) for grounded systems, or as provided in 250.30(B) for ungrounded systems, separately derived systems shall comply with 250.20, 250.21, 250.22, and 250.26.

Informational Note No. 1: An alternate ac power source, such as an on-site generator, is not a separately derived system if the grounded conductor is solidly interconnected to a service-supplied system grounded conductor. An example of such a situation is where alternate source transfer equipment does not include a switching action in the grounded conductor and allows it to remain solidly connected to the service-supplied grounded conductor when the alternate source is operational and supplying the load served.

Informational Note No. 2: See 445.13 for the minimum size of conductors that carry fault current.

Stallcup's Comment: A revision has been made to clarify that separately derived systems shall comply with the following system grounding requirements:
• Alternating-current systems to be grounded
• Alternating-current systems of 50 volts to 1000 volts not required to be grounded
• Circuits not to be grounded
• Conductor to be grounded – alternating-current systems

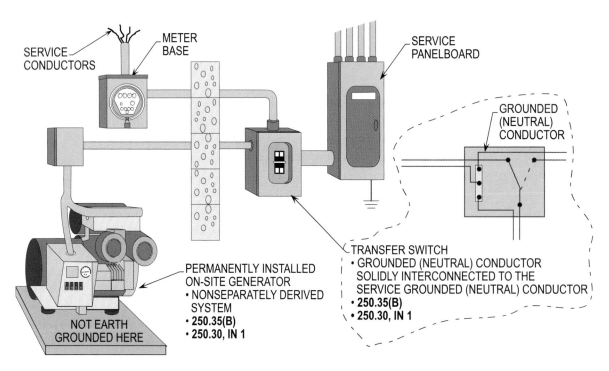

GROUNDING SEPARATELY DERIVED ALTERNATING-CURRENT SYSTEMS
NEC 250.30, IN 1, AND IN 2

Purpose of Change: This revision clarifies that a separately derived system shall not be grounded where transfer equipment does not switch the grounded (neutral) conductor and where the grounded (neutral) conductor is solidly interconnected to the service-supplied grounded (neutral) conductor.

Type of Change	Revision			Committee Change	Accept in Principle			2008 NEC	250.30(A)(1)
ROP	pg. 245	# 5-102	log: 3224	ROC	pg. 132	# 5-69	log: 1836	UL	467
Submitter: Mark R. Hilbert				Submitter: Phil SImmons				OSHA	-
NFPA 70B	14.1.6.43			NFPA 70E	-			NFPA 79	-

250.30 Grounding Separately Derived Alternating-Current Systems.

(A) Grounded Systems.

(1) System Bonding Jumper. An unspliced system bonding jumper shall comply with 250.28(A) through (D). This connection shall be made at any single point on the separately derived system from the source to the first system disconnecting means or overcurrent device, or it shall be made at the source of a separately derived system that has no disconnecting means or overcurrent devices, in accordance with 250.30(A)(1)(a) or (b). The system bonding jumper shall remain within the enclosure where it originates. If the source is located outside the building or structure supplied, a system bonding jumper shall be installed at the grounding electrode connection in compliance with 250.30(C).

(a) *Installed at the Source*. The system bonding jumper shall connect the grounded conductor to the supply-side bonding jumper and the normally non-current-carrying metal enclosure.

(b) *Installed at the First Disconnecting Means*. The system bonding jumper shall connect the grounded conductor to the supply-side bonding jumper, the disconnecting means enclosure, and the equipment grounding conductor(s).

Stallcup's Comment: A revision has been made to clarify that the system bonding jumper shall remain within the enclosure where it originates. The grounded (neutral) conductor to the supply-side bonding jumper and the normally non-current-carrying metal enclosure shall be connected to the system bonding jumper where installed at the source. The grounded (neutral) conductor to the supply-side bonding jumper, the disconnecting means enclosure, and the equipment grounding conductor(s) shall be connected to the system bonding jumper where installed at the first disconnecting means.

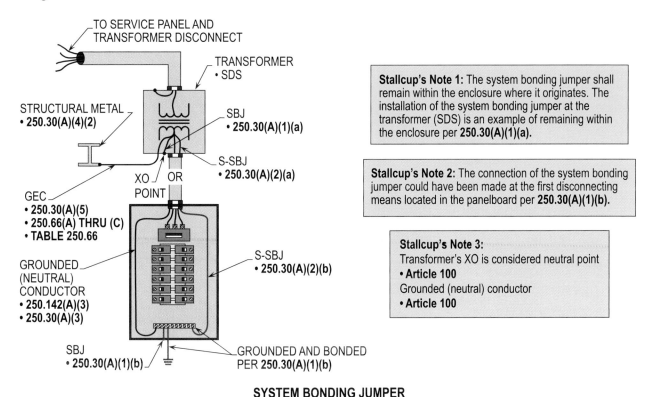

SYSTEM BONDING JUMPER
NEC 250.30(A)(1)(a) AND (A)(1)(b)

Purpose of Change: To clarify the requirements for the placement of the system bonding jumper.

Type of Change	Revision			Committe Change		Accept in Principle		2008 NEC	250.30(A)(2)
ROP	pg. 245	# 5-102	log: 3224	ROC	pg. 132	# 5-69	log: 1836	UL	467
Submitter: Mark R. Hilbert				Submitter: Phil Simmons				OSHA	-
NFPA 70B	14.1.6.43			NFPA 70E	-			NFPA 79	-

250.30 Grounding Separately Derived Alternating-Current Systems.

(A) Grounded Systems.

(2) Supply-Side Bonding Jumper. If the source of a separately derived system and the first disconnecting means are located in separate enclosures, a supply-side bonding jumper shall be installed with the circuit conductors from the source enclosure to the first disconnecting means. A supply-side bonding jumper shall not be required to be larger than the derived ungrounded conductors. The supply-side bonding jumper shall be permitted to be of nonflexible metal raceway type or of the wire or bus type as follows:

(a) A supply-side bonding jumper of the wire type shall comply with 250.102(C), based on the size of the derived ungrounded conductors.

(b) A supply-side bonding jumper of the bus type shall have a cross-sectional area not smaller than a supply-side bonding jumper of the wire type as determined in 250.102(C).

Stallcup's Comment: A revision has been made to clarify that a supply-side bonding jumper shall be installed with the circuit conductors where the source of a separately derived system and the first disconnecting means are located in separate enclosures. A supply-side bonding jumper shall not be required to be larger than ungrounded (phase) conductors. A supply-side bonding jumper of the wire type shall be sized based on the ungrounded (phase) conductors, and a bus type shall be sized to have a cross-sectional area not smaller than a supply-side bonding jumper.

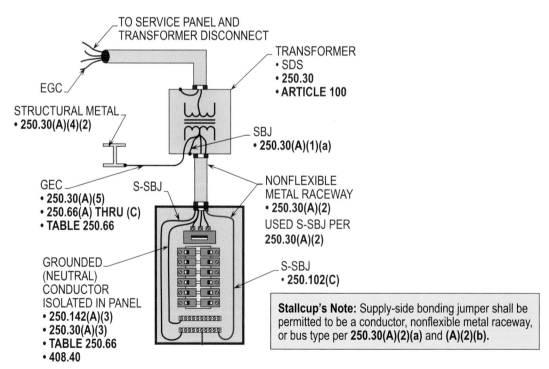

SUPPLY-SIDE BONDING JUMPER
NEC 250.30(A)(2)(a) AND (A)(2)(b)

Purpose of Change: To clarify that wire type and bus type shall be permitted to be used as a supply-side bonding jumper in a separately derived system installation.

Type of Change	Revision			Committee Change	Accept in Princple			2008 NEC	250.30(A)(8)
ROP	pg. 245	# 5-102	log: 3224	ROC	pg. 132	# 5-69	log: 1836	UL	467
Submitter: Mark R. Hilbert				Submitter: Phil Simmons				OSHA	-
NFPA 70B	14.1.6.43			NFPA 70E	-			NFPA 79	-

250.30 Grounding Separately Derived Alternating-Current Systems.

(A) Grounded Systems.

(3) Grounded Conductor. If a grounded conductor is installed and the system bonding jumper connection is not located at the source, 250.30(A)(3)(a) through (A)(3)(d) shall apply.

(a) *Sizing for a Single Raceway.* The grounded conductor shall not be smaller than the required grounding electrode conductor specified in Table 250.66 but shall not be required to be larger than the largest derived ungrounded conductor(s). In addition, for sets of derived ungrounded conductors larger than 1100 kcmil copper or 1750 kcmil aluminum, the grounded conductor shall not be smaller than 12 1/2 percent of the circular mil area of the largest set of derived ungrounded conductors.

Stallcup's Comment: Section **250.30(A)(8)** has been relocated to **250.30(A)(3)** to clarify the requirements for sizing for a single raceway, parallel conductors in two or more raceways, delta-connected systems, and impedance grounded systems where a grounded (neutral) conductor is installed and the system bonding jumper connection is not located at the source.

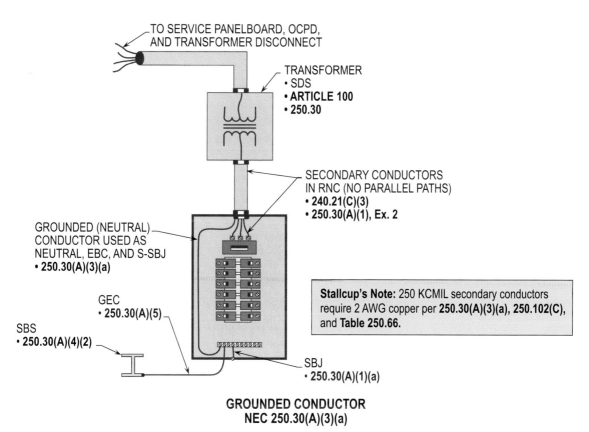

GROUNDED CONDUCTOR
NEC 250.30(A)(3)(a)

Purpose of Change: To clarify the procedure for sizing the grounded (neutral) conductor from the separately derived system to the first disconnecting means.

Type of Change	Revision			Committee Change	Accept in Princple			2008 NEC	250.30(A)(8)
ROP	pg. 245	# 5-102	log: 3224	ROC	pg. 132	# 5-69	log: 1836	UL	467
Submitter: Mark R. Hilbert				Submitter: Phil Simmons				OSHA	-
NFPA 70B	14.1.6.43			NFPA 70E	-			NFPA 79	-

250.30 Grounding Separately Derived Alternating-Current Systems.

(A) Grounded Systems.

(3) Grounded Conductor. If a grounded conductor is installed and the system bonding jumper connection is not located at the source, 250.30(A)(3)(a) through (A)(3)(d) shall apply.

(b) *Parallel Conductors in Two or More Raceways.* If the ungrounded conductors are installed in parallel in two or more raceways, the grounded conductor shall also be installed in parallel. The size of the grounded conductor in each raceway shall be based on the total circular mil area of the parallel derived ungrounded conductors in the raceway as indicated in 250.30(A)(3)(a), but not smaller than 1/0 AWG.

Informational Note: See 310.10(H) for grounded conductors connected in parallel.

Stallcup's Comment: Section **250.30(A)(8)** has been relocated to **250.30(A)(3)** to clarify the requirements for sizing for a single raceway, parallel conductors in two or more raceways, delta-connected systems, and impedance grounded systems where a grounded (neutral) conductor is installed and the system bonding jumper connection is not located at the source.

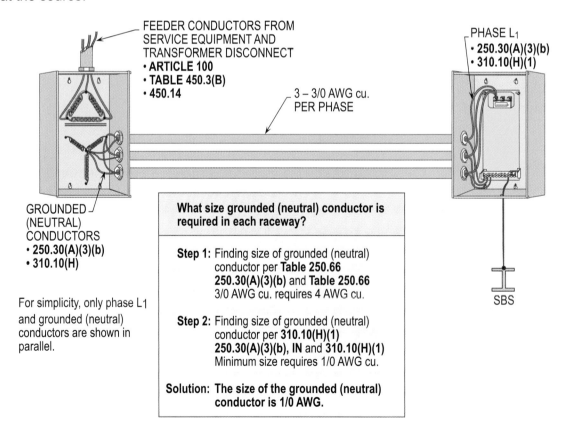

FEEDER CONDUCTORS FROM SERVICE EQUIPMENT AND TRANSFORMER DISCONNECT
- **ARTICLE 100**
- **TABLE 450.3(B)**
- **450.14**

3 – 3/0 AWG cu. PER PHASE

PHASE L1
- **250.30(A)(3)(b)**
- **310.10(H)(1)**

GROUNDED (NEUTRAL) CONDUCTORS
- **250.30(A)(3)(b)**
- **310.10(H)**

For simplicity, only phase L1 and grounded (neutral) conductors are shown in parallel.

What size grounded (neutral) conductor is required in each raceway?

Step 1: Finding size of grounded (neutral) conductor per **Table 250.66** **250.30(A)(3)(b)** and **Table 250.66** 3/0 AWG cu. requires 4 AWG cu.

Step 2: Finding size of grounded (neutral) conductor per **310.10(H)(1)** **250.30(A)(3)(b), IN** and **310.10(H)(1)** Minimum size requires 1/0 AWG cu.

Solution: The size of the grounded (neutral) conductor is 1/0 AWG.

SBS

GROUNDED CONDUCTOR
NEC 250.30(A)(3)(b)

Purpose of Change: To relocate the requirements for sizing the grounded conductor in a parallel installation.

Type of Change	Revision			Committee Change	Accept in Princple		2008 NEC	250.30(A)(8)
ROP	pg. 245	# 5-102	log: 3224	ROC	pg. 132	# 5-69	log: 1836	UL	467
Submitter: Mark R. Hilbert				Submitter: Phil Simmons			OSHA	-
NFPA 70B	14.1.0.40			NFPA 70E	▪		NFPA 79	▪

250.30 Grounding Separately Derived Alternating-Current Systems.

(A) Grounded Systems.

(3) Grounded Conductor. If a grounded conductor is installed and the system bonding jumper connection is not located at the source, 250.30(A)(3)(a) through (A)(3)(d) shall apply.

(c) *Delta-Connected System.* The grounded conductor of a 3-phase, 3-wire delta system shall have an ampacity not less than that of the ungrounded conductors.

Stallcup's Comment: Section **250.30(A)(8)** has been relocated to **250.30(A)(3)** to clarify the requirements for sizing for a single raceway, parallel conductors in two or more raceways, delta-connected systems, and impedance grounded systems where a grounded (neutral) conductor is installed and the system bonding jumper connection is not located at the source.

FOR TRANSFORMER DISCONNECTING MEANS, SEE **450.14**.

CURRENT ON EACH CONDUCTOR
• 200 AMPS

TRANSFORMER (SDS)
• COILS 480 V

Stallcup's Note 1: Coil current is (58% of 200 A) 116 A.

Stallcup's Note 3: Grounded (neutral) conductor shall have an ampacity not less than the grounded (neutral) conductors per **250.30(A)(3)(c)**.

CORNER GROUNDED DELTA SYSTEM
• **250.26(4)**

RNC

SYSTEM BONDING JUMPER

Stallcup's Note 2: Phase-to-phase voltage is equal to coil voltage (480 V).

CB
• MARKED 1Ø - 3Ø
• **240.85**

THIS SYSTEM SHALL BE PERMITTED TO BE GROUNDED
• **250.20** AND **250.26**

GE

GEC

GROUNDED CONDUCTOR
NEC 250.30(A)(3)(c)

Purpose of Change: To clarify that the requirements for corner grounded systems have been relocated and renumbered

Type of Change	Revision			Committee Change	Accept in Princple		2008 NEC	250.30(A)(8)
ROP	pg. 245	# 5-102	log: 3224	ROC	pg. 132	# 5-69	log: 1836	UL	467
Submitter: Mark R. Hilbert				Submitter: Phil Simmons			OSHA	-
NFPA 70B	14.1.6.43			NFPA 70E	-		NFPA 79	-

250.30 Grounding Separately Derived Alternating-Current Systems.

(A) Grounded Systems.

(3) Grounded Conductor. If a grounded conductor is installed and the system bonding jumper connection is not located at the source, 250.30(A)(3)(a) through (A)(3)(d) shall apply.

(d) *Impedance Grounded System.* The grounded conductor of an impedance grounded neutral system shall be installed in accordance with 250.36 or 250.186, as applicable.

Stallcup's Comment: Section **250.30(A)(8)** has been relocated to **250.30(A)(3)** to clarify the requirements for sizing for a single raceway, parallel conductors in two or more raceways, delta-connected systems, and impedance grounded systems where a grounded (neutral) conductor is installed and the system bonding jumper connection is not located at the source.

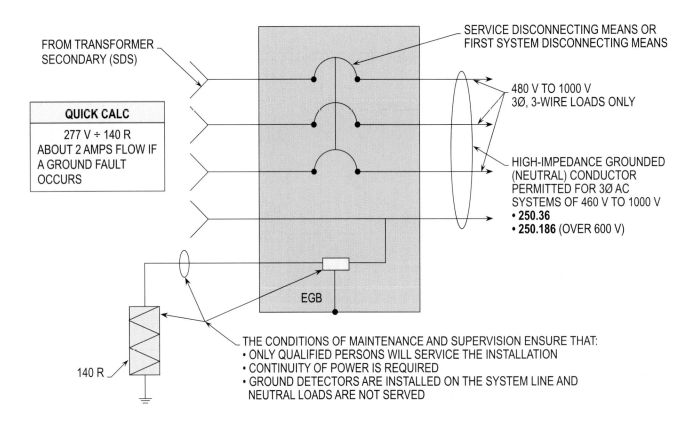

GROUNDED CONDUCTOR
NEC 250.30(A)(3)(d)

Purpose of Change: To locate rules for high-impedance grounding of **Article 250.**

Type of Change	Revision			Committe Change	Accept in Principle			2008 NEC	250.30(A)(4)(a)
ROP	pg. 245	# 5-102	log: 3224	ROC	pg. 132	# 5-69	log: 1836	UL	467
Submitter: Mark R. Hilbert				Submitter: Phil Simmons				OSHA	-
NFPA 70B	14.1.6.43			NFPA 70E	-			NFPA 79	-

250.30 Grounding Separately Derived Alternating-Current Systems.

(A) Grounded Systems.

(6) Grounding Electrode Conductor, Multiple Separately Derived Systems. A common grounding electrode conductor for multiple separately derived systems shall be permitted. If installed, the common grounding electrode conductor shall be used to connect the grounded conductor of the separately derived systems to the grounding electrode as specified in 250.30(A)(4). A grounding electrode conductor tap shall then be installed from each separately derived system to the common grounding electrode conductor. Each tap conductor shall connect the grounded conductor of the separately derived system to the common grounding electrode conductor. This connection shall be made at the same point on the separately derived system where the system bonding jumper is connected.

(a) *Common Grounding Electrode Conductor.* The common grounding electrode conductor shall be permitted to be one of the following:

(1) A conductor of the wire type not smaller than 3/0 AWG copper or 250 kcmil aluminum

(2) The metal frame of the building or structure that complies with 250.52(A)(2) or is connected to the grounding electrode system by a conductor that shall not be smaller than 3/0 AWG copper or 250 kcmil aluminum

Stallcup's Comment: A revision has been made to clarify that the common grounding electrode conductor shall be permitted to be installed to connect the grounded (neutral) conductor of the separately derived systems to the grounding electrode by a wire type not smaller than 3/0 AWG copper or 250 kcmil aluminum or by the metal frame of the building or structure by a conductor not smaller than 3/0 AWG copper or 250 KCMIL aluminum.

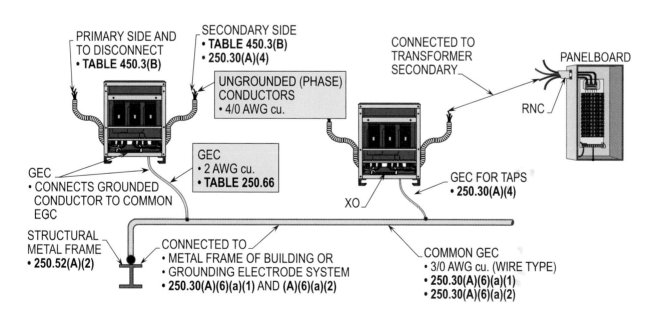

GROUNDING ELECTRODE CONDUCTOR, MULTIPLE SEPARATELY DERIVED SYSTEMS
250.30(A)(6)(a)(1) AND (A)(6)(a)(2)

Purpose of Change: To provide requirements for connecting the common grounding electrode conductor for bonding and grounding separately derived systems to earth ground.

Type of Change	New Subdivision			Committee Change		Accept in Princple		2008 NEC	-
ROP	pg. 245	# 5-102	log: 3224	ROC	pg. 132	# 5-69	log: 1836	UL	467
Submitter: Mark R. Hilbert				Submitter: Phil Simmons				OSHA	-
NFPA 70B	14.1.6.43			NFPA 70E	-			NFPA 79	-

250.30 Grounding Separately Derived Alternating-Current Systems.

(B) Ungrounded Systems. The equipment of an ungrounded separately derived system shall be grounded and bonded as specified in 250.30(B)(1) through (B)(3).

(3) Bonding Path and Conductor. A supply-side bonding jumper shall be installed from the source of a separately derived system to the first disconnecting means in compliance with 250.30(A)(2).

Stallcup's Comment: A new subdivision has been added to require a supply-side bonding jumper to be installed from the source of a separately derived system to the first disconnecting means.

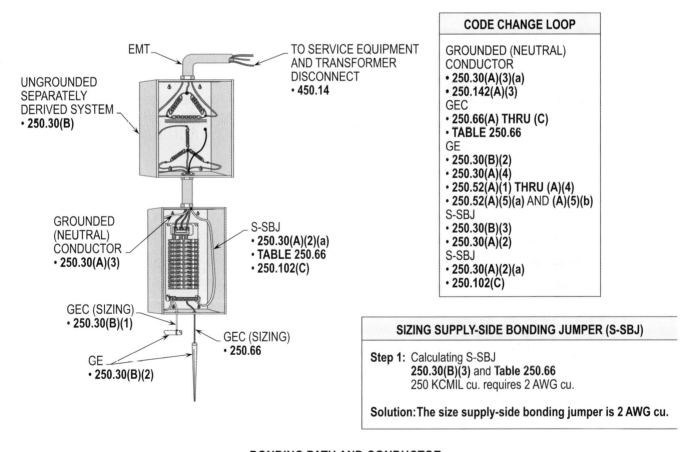

BONDING PATH AND CONDUCTOR
NEC 250.30(B)(3)

Purpose of Change: To clarify the sizing of the supply-side bonding jumper on an ungrounded separately derived system.

Type of Change	New Subsection			Committe Change	Accept in Principle			2008 NEC	-
ROP	pg. 245	# 5-102	log: 3224	ROC	pg. 132	# 5-69	log: 1836	UL	467
Submitter: Mark R. Hilbert				Submitter: Phil Simmons				OSHA	-
NFPA 70B	14.1.6.43			NFPA 70E	-			NFPA 79	-

250.30 Grounding Separately Derived Alternating-Current Systems.

(C) Outdoor Source. If the source of the separately derived system is located outside the building or structure supplied, a grounding electrode connection shall be made at the source location to one or more grounding electrodes in compliance with 250.50. In addition, the installation shall comply with 250.30(A) for grounded systems or with 250.30(B) for ungrounded systems.

Exception: The grounding electrode conductor connection for impedance grounded neutral systems shall comply with 250.36 or 250.186, as applicable.

Stallcup's Comment: A new subsection has been added to require a grounding electrode connection to be made at the source location to one or more grounding electrodes if the source of the separately derived system is located outside the building or structure.

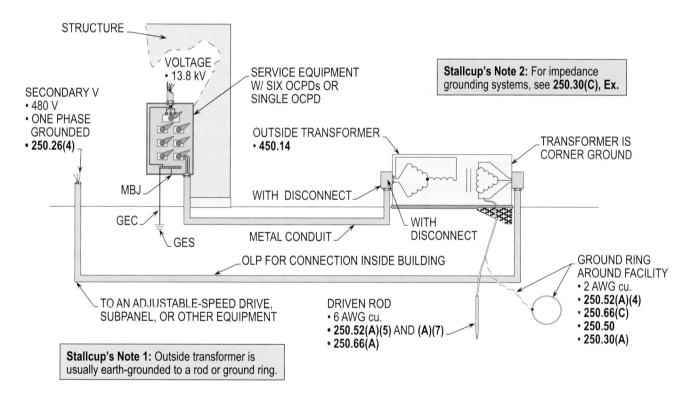

OUTDOOR SOURCE
NEC 250.30(C)

Purpose of Change: To clarify the earth grounding of separately derived systems located outside of structure.

Type of Change	New Subdivision			Committee Change	Accept			2008 NEC	-
ROP	pg. 1234	# 5-126	log: 3225	ROC	pg. 1234	# 5-89	log: 82	UL	467
Submitter: Mark R. Hilbert				Submitter: Technical Correlating Committee				OSHA	-
NFPA 70B	14.1.6.48			NFPA 70E	-			NFPA 79	-

250.32 Buildings or Structures Supplied by a Feeder(s) or Branch Circuit(s).

(B) Grounded Systems.

(2) Supplied by Separately Derived System.

(a) *With Overcurrent Protection.* If overcurrent protection is provided where the conductors originate, the installation shall comply with 250.32(B)(1).

Stallcup's Comment: A new subdivision has been added to address the grounding requirements for electrical systems in buildings or structures supplied by separately derived systems having overcurrent protection or not having overcurrent protection.

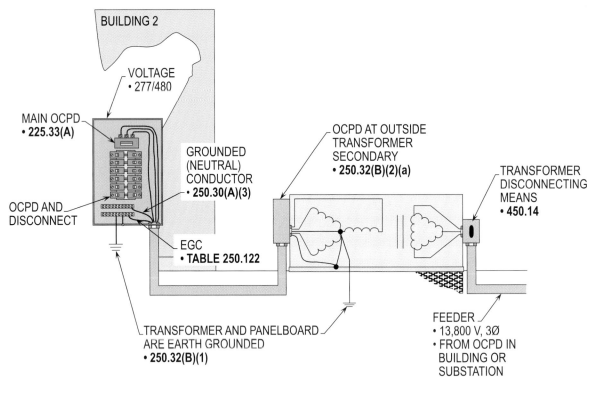

WITH OVERCURRENT PROTECTION
NEC 250.32(B)(2)(a)

Purpose of Change: To clarify the grounding and bonding techniques for separately derived systems supplying another building or structure.

Type of Change	New Subdivision			Committee Change	Accept			2008 NEC	-
ROP	pg. 1234	# 5-126	log: 3225	ROC	pg. 1234	# 5-89	log: 82	UL	467
Submitter: Mark R. Hilbert				Submitter: Technical Correlating Committee				OSHA	-
NFPA 70B	14.1.6.48			NFPA 70E	-			NFPA 79	-

250.32 Buildings or Structures Supplied by a Feeder(s) or Branch Circuit(s).

(B) Grounded Systems.

(2) Supplied by Separately Derived System.

(b) *Without Overcurrent Protection.* If overcurrent protection is not provided where the conductors originate, the installation shall comply with 250.30(A) . If installed, the supply-side bonding jumper shall be connected to the building or structure disconnecting means and to the grounding electrode(s).

Stallcup's Comment: A new subdivision has been added to address the grounding requirements for electrical systems in buildings or structures supplied by separately derived systems having overcurrent protection or not having overcurrent protection.

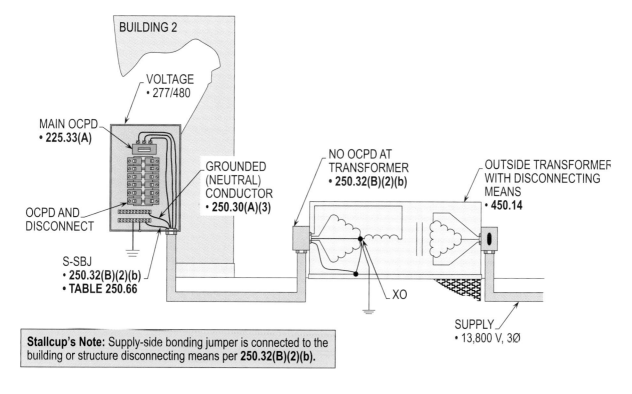

WITHOUT OVERCURRENT PROTECTION
NEC 250.32(B)(2)(b)

Purpose of Change: To clarify the grounding and bonding techniques for separately derived systems supplying another building or structure.

Type of Change	Revision			Committe Change	Accept in Principle			2008 NEC	250.32(C)
ROP	pg. 253	# 5-129	log: 3531	ROC	pg. 137	# 5-93	log: 2561	UL	467
Submitter: Phil Simmons				Submitter: Frederic P. Hartwell				OSHA	-
NFPA 70B	14.1.6.48			NFPA 70E	-			NFPA 79	-

250.32 Buildings or Structures Supplied by a Feeder(s) or Branch Circuit(s).

(C) Ungrounded Systems.

(1) Supplied by a Feeder or Branch Circuit. An equipment grounding conductor, as described in 250.118, shall be installed with the supply conductors and be connected to the building or structure disconnecting means and to the grounding electrode(s). The grounding electrode(s) shall also be connected to the building or structure disconnecting means.

Stallcup's Comment: A revision has been made to clarify that an equipment grounding conductor shall be installed with the supply conductors and be connected to the building or structure disconnecting means and to the grounding electrode(s).

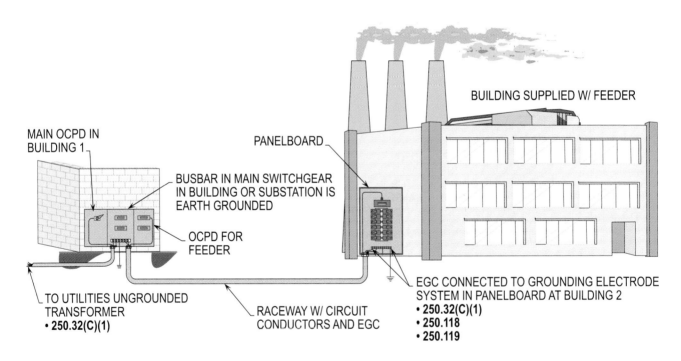

MAIN OCPD IN BUILDING 1

PANELBOARD

BUILDING SUPPLIED W/ FEEDER

BUSBAR IN MAIN SWITCHGEAR IN BUILDING OR SUBSTATION IS EARTH GROUNDED

OCPD FOR FEEDER

TO UTILITIES UNGROUNDED TRANSFORMER
• 250.32(C)(1)

RACEWAY W/ CIRCUIT CONDUCTORS AND EGC

EGC CONNECTED TO GROUNDING ELECTRODE SYSTEM IN PANELBOARD AT BUILDING 2
• 250.32(C)(1)
• 250.118
• 250.119

SUPPLIED BY A FEEDER OR BRANCH CIRCUIT
NEC 250.32(C)(1)

Purpose of Change: To clarify that an equipment grounding conductor shall be routed with the feeder when the supply is ungrounded.

Type of Change		Revision			Committe Change		Accept in Principle		2008 NEC	250.32(C)
ROP	pg. 253	# 5-129	log: 3531		ROC	pg. 137	# 5-93	log: 2561	UL	467
Submitter: Phil Simmons					Submitter: Frederic P. Hartwell				OSHA	-
NFPA 70B	14.1.6.48				NFPA 70E	-			NFPA 79	-

250.32 Buildings or Structures Supplied by a Feeder(s) or Branch Circuit(s).

(C) Ungrounded Systems.

(1) Supplied by a Feeder or Branch Circuit. An equipment grounding conductor, as described in 250.118, shall be installed with the supply conductors and be connected to the building or structure disconnecting means and to the grounding electrode(s). The grounding electrode(s) shall also be connected to the building or structure disconnecting means.

Stallcup's Comment: A revision has been made to clarify that an equipment grounding conductor shall be installed with the supply conductors and be connected to the building or structure disconnecting means and to the grounding electrode(s).

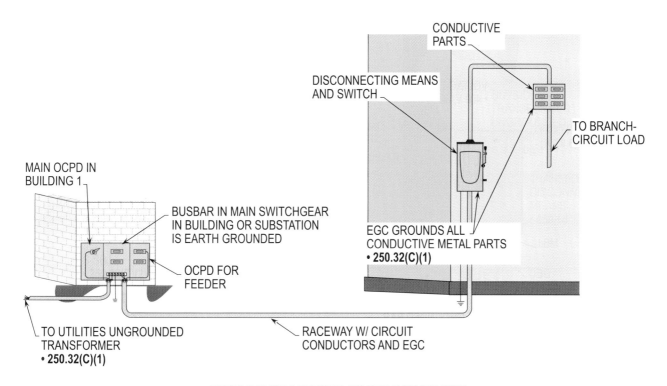

SUPPLIED BY A FEEDER OR BRANCH CIRCUIT
NEC 250.32(C)(1)

Purpose of Change: To clarify that an equipment grounding conductor shall be routed with the branch circuit and bond and ground all ungrounded noncurrent-carrying parts of the electrical system when the supply is ungrounded.

Type of Change	Revision			Committee Change	Accept in Principle			2008 NEC	250.30(A)(2)
ROP	pg. 264	# 5-150	log: 3705	ROC	pg. -	# -	log: -	UL	467
Submitter: Vince Baclawski				Submitter: -				OSHA	-
NFPA 70B	14.1.6.27			NFPA 70E	-			NFPA 79	-

250.52 Grounding Electrodes.

(A) Electrodes Permitted for Grounding.

(2) Metal Frame of the Building or Structure. The metal frame of the building or structure that is connected to the earth by one or more of the following methods:

(1) At least one structural metal member that is in direct contact with the earth for 3.0 m (10 ft) or more, with or without concrete encasement.

(2) Hold-down bolts securing the structural steel column that are connected to a concrete-encased electrode that complies with 250.52(A)(3) and is located in the support footing or foundation. The hold-down bolts shall be connected to the concrete-encased electrode by welding, exothermic welding, the usual steel tie wires, or other approved means.

Stallcup's Comment: A revision has been made to clarify the permitted electrode methods to be used for the metal frame of the building or structure that is connected to the earth.

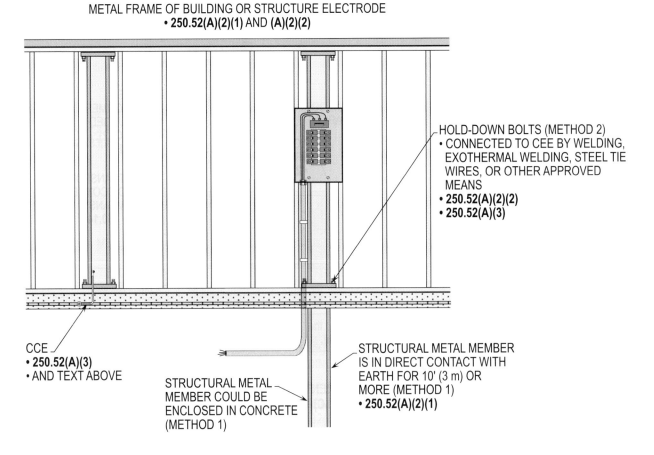

METAL FRAME OF BUILDING OR STRUCTURE ELECTRODE
• **250.52(A)(2)(1)** AND **(A)(2)(2)**

HOLD-DOWN BOLTS (METHOD 2)
• CONNECTED TO CEE BY WELDING, EXOTHERMAL WELDING, STEEL TIE WIRES, OR OTHER APPROVED MEANS
• **250.52(A)(2)(2)**
• **250.52(A)(3)**

CCE
• **250.52(A)(3)**
• AND TEXT ABOVE

STRUCTURAL METAL MEMBER COULD BE ENCLOSED IN CONCRETE (METHOD 1)

STRUCTURAL METAL MEMBER IS IN DIRECT CONTACT WITH EARTH FOR 10' (3 m) OR MORE (METHOD 1)
• **250.52(A)(2)(1)**

METAL FRAME OF THE BUILDING OR STRUCTURE
NEC 250.52(A)(2)(1) AND (A)(2)(2)

Purpose of Change: To clarify the requirements for using the metal frame of a building or structure as a grounding electrode.

Type of Change	Revision			Committe Change	Accept in Principle		2008 NEC	250.52(A)(3)
ROP	pg. 265	# 5-158	log: 3536	ROC	pg. 139	# 5-103	log: 1840	UL	467
Submitter: Phil Simmons				Submitter: Phil Simmons			OSHA	-
NFPA 70B	14.1.6.27			NFPA 70E	-		NFPA 79	-

250.52 Grounding Electrodes.

(A) Electrodes Permitted for Grounding.

(3) Concrete-Encased Electrode.
A concrete-encased electrode shall consist of at least 6.0 m (20 ft) of either (1) or (2):

(1) One or more bare or zinc galvanized or other electrically conductive coated steel reinforcing bars or rods of not less than 13 mm (1/2 in.) in diameter, installed in one continuous 6.0 m (20 ft) length, or if in multiple pieces connected together by the usual steel tie wires, exothermic welding, welding, or other effective means to create a 6.0 m (20 ft) or greater length; or

(2) Bare copper conductor not smaller than 4 AWG

Metallic components shall be encased by at least 50 mm (2 in.) of concrete and shall be located horizontally within that portion of a concrete foundation or footing that is in direct contact with the earth or within vertical foundations or structural components or members that are in direct contact with the earth. If multiple concrete-encased electrodes are present at a building or structure, it shall be permissible to bond only one into the grounding electrode system.

Informational Note: Concrete installed with insulation, vapor barriers, films or similar items separating the concrete from the earth is not considered to be in "direct contact" with the earth.

Stallcup's Comment: A revision has been made to clarify the permitted electrode methods to be used for concrete-encased electrodes.

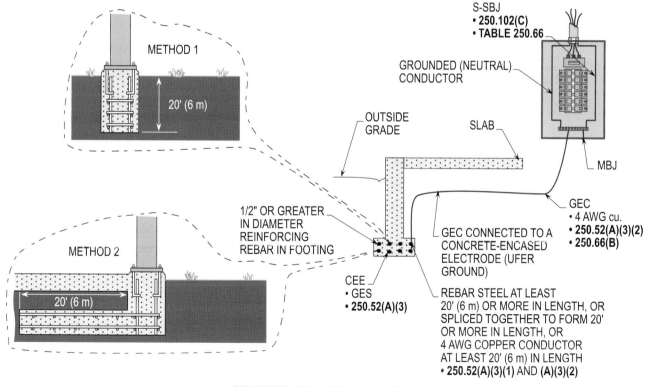

CONCRETE-ENCASED ELECTRODE
NEC 250.52(A)(3)(1) AND (A)(3)(2)

Purpose of Change: To clarify the requirements for installing and using a concrete-encased electrode.

Type of Change	Revision		Committee Change	Accept in Principle		2008 NEC	250.53(A)
ROP	pg. 268	# 5-169a	log: CP 500	ROC	pg. 142	# 5-115	log: 1842	UL	467
Submitter: Code-Making Panel 5			Submitter: Phil Simmons			OSHA	-
NFPA 70B	14.1.6.27		NFPA 70E	-		NFPA 79	-

250.53 Grounding Electrode System Installation.

(A) Rod, Pipe, and Plate Electrodes. Rod, pipe, and plate electrodes shall meet the requirements of 250.53(A)(1) through (A)(3).

(2) Supplemental Electrode Required. A single rod, pipe, or plate electrode shall be supplemented by an additional electrode of a type specified in 250.52(A)(2) through (A)(8). The supplemental electrode shall be permitted to be bonded to one of the following:

(1) Rod, pipe, or plate electrode

(2) Grounding electrode conductor

(3) Grounded service-entrance conductor

(4) Nonflexible grounded service raceway

(5) Any grounded service enclosure

Exception: If a single rod, pipe, or plate grounding electrode has a resistance to earth of 25 ohms or less, the supplemental electrode shall not be required.

Stallcup's Comment: A revision has been made to clarify the installation requirements for rod, pipe, and plate electrodes.

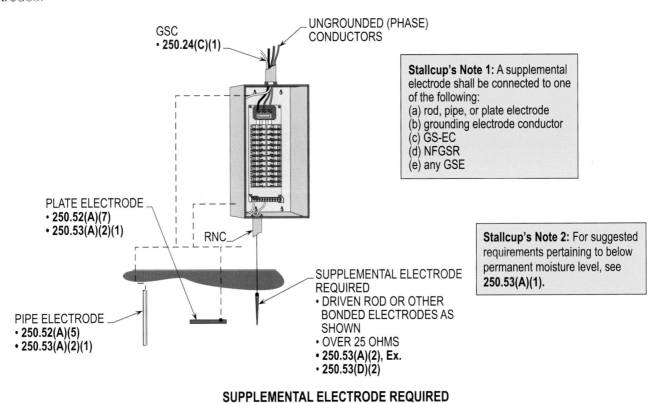

GSC
• **250.24(C)(1)**

UNGROUNDED (PHASE) CONDUCTORS

Stallcup's Note 1: A supplemental electrode shall be connected to one of the following:
(a) rod, pipe, or plate electrode
(b) grounding electrode conductor
(c) GS-EC
(d) NFGSR
(e) any GSE

PLATE ELECTRODE
• **250.52(A)(7)**
• **250.53(A)(2)(1)**

RNC

PIPE ELECTRODE
• **250.52(A)(5)**
• **250.53(A)(2)(1)**

SUPPLEMENTAL ELECTRODE REQUIRED
• DRIVEN ROD OR OTHER BONDED ELECTRODES AS SHOWN
• OVER 25 OHMS
• **250.53(A)(2), Ex.**
• **250.53(D)(2)**

Stallcup's Note 2: For suggested requirements pertaining to below permanent moisture level, see **250.53(A)(1).**

**SUPPLEMENTAL ELECTRODE REQUIRED
NEC 250.53(A)(2) AND Ex.**

Purpose of Change: To clarify when a required supplemental electrode is to be supplemented by another electrode.

Type of Change	Revision			Committee Change	Accept in Principle		2008 NEC	250.53(A)
ROP	pg. 268	# 5-169a	log: CP 500	ROC	pg. 142	# 5-115	log: 1842	UL	467
Submitter: Code-Making Panel 5				Submitter: Phil Simmons			OSHA	-
NFPA 70B	14.1.6.27			NFPA 70E	-		NFPA 70	-

250.53 Grounding Electrode System Installation.

(A) Rod, Pipe, and Plate Electrodes. Rod, pipe, and plate electrodes shall meet the requirements of 250.53(A)(1) through (A)(3).

(3) Supplemental Electrode. If multiple rod, pipe, or plate electrodes are installed to meet the requirements of this section, they shall not be less than 1.8 m (6 ft) apart.

Informational Note: The paralleling efficiency of rods is increased by spacing them twice the length of the longest rod.

Stallcup's Comment: A revision has been made to clarify the installation requirements for rod, pipe, and plate electrodes.

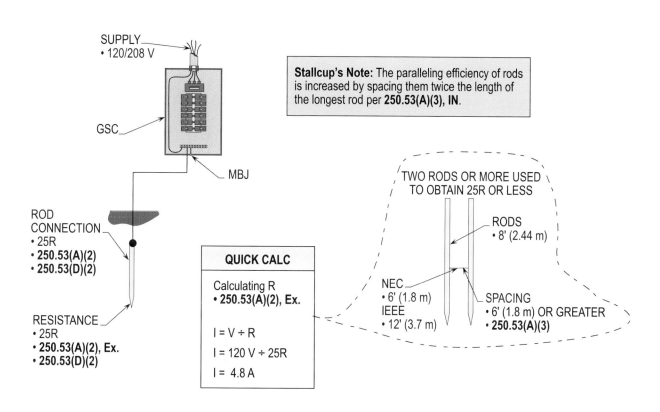

SUPPLEMENTAL ELECTRODE
NEC 250.53(A)(3)

Purpose of Change: To clarify the resistance-to-ground and the spacing of driven rods, pipes, and plates.

Type of Change	Revision			Committe Change	Accept			2008 NEC	250.60
ROP	pg. 1234	# 5-179	log: 1516	ROC	pg. -	# -	log: -	UL	467
Submitter: Mark S. Harger				Submitter: -				OSHA	-
NFPA 70B	14.1.6.6			NFPA 70E	-			NFPA 79	-

250.60 Use of Strike Termination Devices.

Conductors and driven pipes, rods, or plate electrodes used for grounding strike termination devices shall not be used in lieu of the grounding electrodes required by 250.50 for grounding wiring systems and equipment. This provision shall not prohibit the required bonding together of grounding electrodes of different systems.

Informational Note No. 1: See 250.106 for spacing from strike termination devices. See 800.100(D), 810.21(J), and 820.100(D) for bonding of electrodes.

Informational Note No. 2: Bonding together of all separate grounding electrodes will limit potential differences between them and between their associated wiring systems.

Stallcup's Comment: A revision has been made to clarify the language used in NFPA 780, *Standard for the Installation of Lightning Protection Systems,* which defines a strike termination device as a component of a lightning protection system that intercepts lightning flashes and connects them to a path to ground. Strike termination devices include air terminals, metal masts, permanent metal parts of structures, and overhead ground wires installed in catenary lightning protection systems.

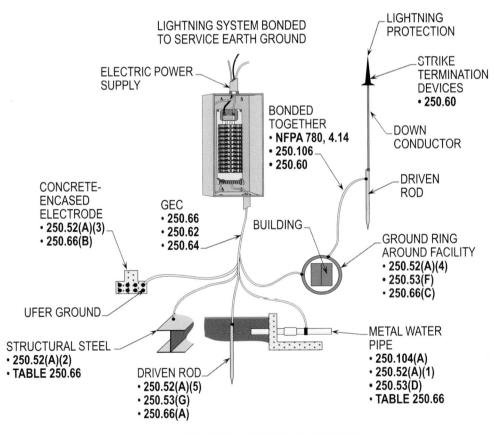

USE OF STRIKE TERMINATION DEVICES
NEC 250.60

Purpose of Change: To change the "title head" of the section and clarify the use of language used in NFPA 780.

Type of Change	Revision			Committee Change	Accept			2008 NEC	250.64(B)
ROP	pg. 273	# 5-195	log: 3539	ROC	pg. 145	# 5-123	log: 1844	UL	467
Submitter: Phil Simmons				Submitter: Phil Simmons				OSHA	-
NFPA 70B	-			NFPA 70E	-			NFPA 79	-

250.64 Grounding Electrode Conductor Installation.

(B) Securing and Protection Against Physical Damage. Where exposed, a grounding electrode conductor or its enclosure shall be securely fastened to the surface on which it is carried. Grounding electrode conductors shall be permitted to be installed on or through framing members. A 4 AWG or larger copper or aluminum grounding electrode conductor shall be protected if exposed to physical damage. A 6 AWG grounding electrode conductor that is free from exposure to physical damage shall be permitted to be run along the surface of the building construction without metal covering or protection if it is securely fastened to the construction; otherwise, it shall be protected in rigid metal conduit (RMC), intermediate metal conduit (IMC), rigid polyvinyl chloride conduit (PVC), reinforced thermosetting resin conduit (RTRC), electrical metallic tubing (EMT), or cable armor. Grounding electrode conductors smaller than 6 AWG shall be protected in RMC, IMC, PVC, RTRC, EMT, or cable armor.

Stallcup's Comment: A revision has been made to clarify that grounding electrode conductors shall be permitted to be installed on or through framing members.

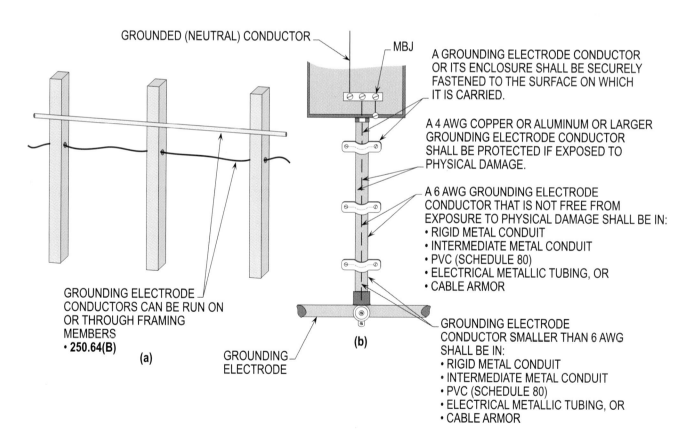

GROUNDED (NEUTRAL) CONDUCTOR

MBJ

A GROUNDING ELECTRODE CONDUCTOR OR ITS ENCLOSURE SHALL BE SECURELY FASTENED TO THE SURFACE ON WHICH IT IS CARRIED.

A 4 AWG COPPER OR ALUMINUM OR LARGER GROUNDING ELECTRODE CONDUCTOR SHALL BE PROTECTED IF EXPOSED TO PHYSICAL DAMAGE.

A 6 AWG GROUNDING ELECTRODE CONDUCTOR THAT IS NOT FREE FROM EXPOSURE TO PHYSICAL DAMAGE SHALL BE IN:
• RIGID METAL CONDUIT
• INTERMEDIATE METAL CONDUIT
• PVC (SCHEDULE 80)
• ELECTRICAL METALLIC TUBING, OR
• CABLE ARMOR

GROUNDING ELECTRODE CONDUCTORS CAN BE RUN ON OR THROUGH FRAMING MEMBERS
• 250.64(B)

(a)

GROUNDING ELECTRODE

(b)

GROUNDING ELECTRODE CONDUCTOR SMALLER THAN 6 AWG SHALL BE IN:
• RIGID METAL CONDUIT
• INTERMEDIATE METAL CONDUIT
• PVC (SCHEDULE 80)
• ELECTRICAL METALLIC TUBING, OR
• CABLE ARMOR

**SECURING AND PROTECTION AGAINST PHYSICAL DAMAGE
NEC 250.64(B)**

Purpose of Change: To provide requirements for permitting grounding electrode conductors to be routed through framing members.

Type of Change	Revision			Committe Change		Accept in Principle		2008 NEC	250.64(D)
ROP	pg. 274	# 5-196	log: 3540	ROC	pg -	# -	log: -	UL	467
Submitter: Phil Simmons				Submitter: -				OSHA	-
NFPA 70B	-			NFPA 70E	-			NFPA 79	-

250.64 Grounding Electrode Conductor Installation.

(D) Service with Multiple Disconnecting Means Enclosures. If a service consists of more than a single enclosure as permitted in 230.71(A), grounding electrode connections shall be made in accordance with 250.64(D)(1), (D)(2), or (D)(3).

(1) Common Grounding Electrode Conductor and Taps. A common grounding electrode conductor and grounding electrode conductor taps shall be installed. The common grounding electrode conductor shall be sized in accordance with 250.66, based on the sum of the circular mil area of the largest ungrounded service-entrance conductor(s). If the service-entrance conductors connect directly to a service drop or service lateral, the common grounding electrode conductor shall be sized in accordance with Table 250.66, Note 1.

Stallcup's Comment: A revision has been made to clarify the permitted connection methods of a common grounding electrode conductor.

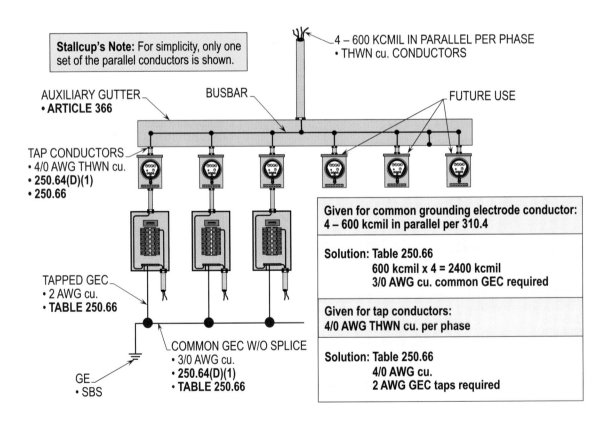

COMMON GROUNDING ELECTRODE CONDUCTOR AND TAPS
NEC 250.64(D)(1)

Purpose of Change: To clarify the common grounding electrode conductor, including sizing requirements.

Type of Change	Revision			Committe Change	Accept in Principle			2008 NEC	250.64(D)
ROP	pg. 274	# 5-196	log: 3540	ROC	pg. -	# -	log: -	UL	467
Submitter: Phil Simmons				Submitter: -				OSHA	-
NFPA 70B	-			NFPA 70E	-			NFPA 79	-

250.64 Grounding Electrode Conductor Installation.

(D) Service with Multiple Disconnecting Means Enclosures. If a service consists of more than a single enclosure as permitted in 230.71(A), grounding electrode connections shall be made in accordance with 250.64(D)(1), (D)(2), or (D)(3).

(1) Common Grounding Electrode Conductor and Taps.

A grounding electrode conductor tap shall extend to the inside of each service disconnecting means enclosure. The grounding electrode conductor taps shall be sized in accordance with 250.66 for the largest service-entrance conductor serving the individual enclosure. The tap conductors shall be connected to the common grounding electrode conductor by one of the following methods in such a manner that the common grounding electrode conductor remains without a splice or joint:

(1) Exothermic welding.

(2) Connectors listed as grounding and bonding equipment.

(3) Connections to an aluminum or copper busbar not less than 6 mm x 50 mm (1⁄4 in. x 2 in.). The busbar shall be securely fastened and shall be installed in an accessible location. Connections shall be made by a listed connector or by the exothermic welding process. If aluminum busbars are used, the installation shall comply with 250.64(A).

Stallcup's Comment: A revision has been made to clarify the permitted connection methods of common grounding electrode conductor.

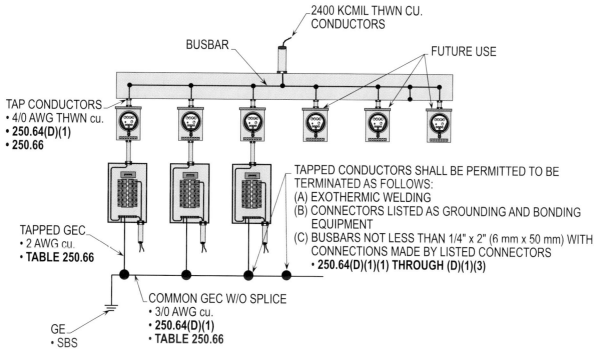

COMMON GROUNDING ELECTRODE CONDUCTOR AND TAPS
NEC 250.64(D)(1)

Purpose of Change: To clarify the appropriate connection methods for the common grounding electrode conductor.

Type of Change	New Subsection			Committee Change	Accept in Principle			2008 NEC	-
ROP	pg. 278	# 5-212	log: 4013	ROC	pg. 144	# 5-131	log: 1845	UL	467
Submitter: Paul Dobrowsky				Submitter: Phil Simmons				OSHA	-
NFPA 70B	14.1.6.27			NFPA 70E	-			NFPA 79	-

250.68 Grounding Electrode Conductor and Bonding Jumper Connection to Grounding Electrodes.

(C) Metallic Water Pipe and Structural Metal. Grounding electrode conductors and bonding jumpers shall be permitted to be connected at the following locations and used to extend the connection to an electrode(s):

(1) Interior metal water piping located not more than 1.52 m (5 ft) from the point of entrance to the building shall be permitted to be used as a conductor to interconnect electrodes that are part of the grounding electrode system.

(2) The structural frame of a building that is directly connected to a grounding electrode as specified in 250.52(A)(2) or 250.68(C)(2)(a), (b), or (c) shall be permitted as a bonding conductor to interconnect electrodes that are part of the grounding electrode system, or as a grounding electrode conductor.

a. By connecting the structural metal frame to the reinforcing bars of a concrete-encased electrode, as provided in 250.52(A)(3), or ground ring as provided in 250.52(A)(4)

b. By bonding the structural metal frame to one or more of the grounding electrodes, as specified in 250.52(A)(5) or (A)(7), that comply with (2)

c. By other approved means of establishing a connection to earth

Stallcup's Comment: A new subsection has been added to address the requirements for grounding electrode conductors and bonding jumpers connected to interior metal water piping or the structural frame of a building and used to extend the connection to an electrode(s).

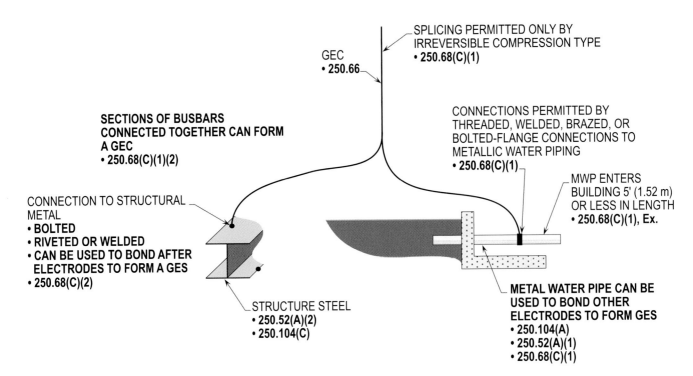

**METALLIC WATER PIPE AND STRUCTURAL METAL
NEC 250.68(C)(1) AND (C)(2)**

Purpose of Change: To clarify how to splice and connect the grounding electrode conductor to metallic water pipe and structural metal.

Type of Change	New Section			Committee Change	Accept		2008 NEC	-
ROP	pg. 290	# 5-259	log: 4526	ROC	pg. 151	# 5-162 log: 2340	UL	467
Submitter: Phil Simmons				Submitter: Mike Holt			OSHA	-
NFPA 70B	14.1.6.15			NFPA 70E	-		NFPA 79	-

250.121 Use of Equipment Grounding Conductors.

An equipment grounding conductor shall not be used as a grounding electrode conductor.

Stallcup's Comment: A new section has been added to clarify that an equipment grounding conductor shall not be used as a grounding electrode conductor.

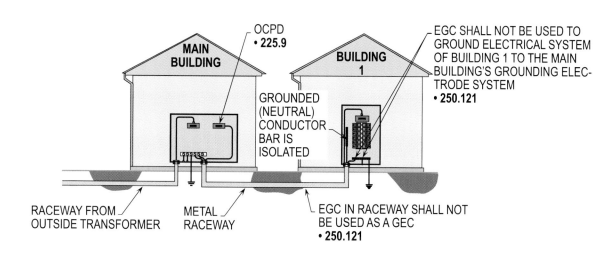

USE OF EQUIPMENT GROUNDING CONDUCTORS
NEC 250.121

Purpose of Change: To clarify and designate the use of the equipment grounding conductor.

Type of Change	Revision			Committee Change	Accept in Principle			2008 NEC	250.122(F)
ROP	pg. 298	# 5-295	log: 1479	ROC	pg. 158	# 5-184	log: 1855	UL	467
Submitter: Richard A. Janoski				Submitter: Phil Simmons				OSHA	-
NFPA 70B	14.1.6.15			NFPA 70E	-			NFPA 79	8.2.2.3

250.122 Size of Equipment Grounding Conductors.

(F) Conductors in Parallel. Where conductors are installed in parallel in multiple raceways or cables as permitted in 310.10(H), the equipment grounding conductors, where used, shall be installed in parallel in each raceway or cable. Where conductors are installed in parallel in the same raceway, cable, or cable tray as permitted in 310.10(H), a single equipment grounding conductor shall be permitted. Equipment grounding conductors installed in cable tray shall meet the minimum requirements of 392.10(B)(1)(c).

Each equipment grounding conductor shall be sized in compliance with 250.122.

Stallcup's Comment: A revision has been made to clarify that if ungrounded (phase) conductors are installed in parallel in the same raceway, cable, or cable tray, only one equipment grounding conductor shall be required to be installed for the entire parallel set.

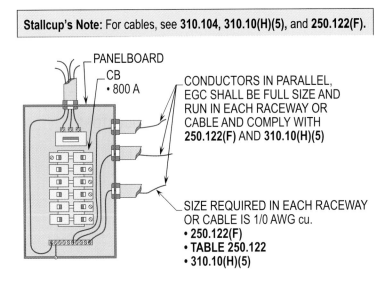

Stallcup's Note: For cables, see **310.104, 310.10(H)(5),** and **250.122(F).**

**CONDUCTORS IN PARALLEL
NEC 250.122(F)**

Purpose of Change: To clarify that the equipment grounding conductor(s) shall be installed in parallel in each raceway or cable, if used.

Type of Change	Revision			Committee Change	Accept in Principle		2008 NEC	250.122(F)
ROP	pg. 298	# 5-295	log: 1479	ROC	pg. 158	# 5-184	log: 1855	UL	467
Submitter: Richard A. Janoski				Submitter: Phil Simmons			OSHA	-
NFPA 70B	14.1.6.15			NFPA 70E	-		NFPA 79	8.2.3

250.122 Size of Equipment Grounding Conductors.

(F) Conductors in Parallel. Where conductors are installed in parallel in multiple raceways or cables as permitted in 310.10(H), the equipment grounding conductors, where used, shall be installed in parallel in each raceway or cable. Where conductors are installed in parallel in the same raceway, cable, or cable tray as permitted in 310.10(H), a single equipment grounding conductor shall be permitted. Equipment grounding conductors installed in cable tray shall meet the minimum requirements of 392.10(B)(1)(c).

Each equipment grounding conductor shall be sized in compliance with 250.122.

Stallcup's Comment: A revision has been made to clarify that if ungrounded (phase) conductors are installed in parallel in the same raceway, cable, or cable tray, only one equipment grounding conductor shall be required to be installed for the entire parallel set.

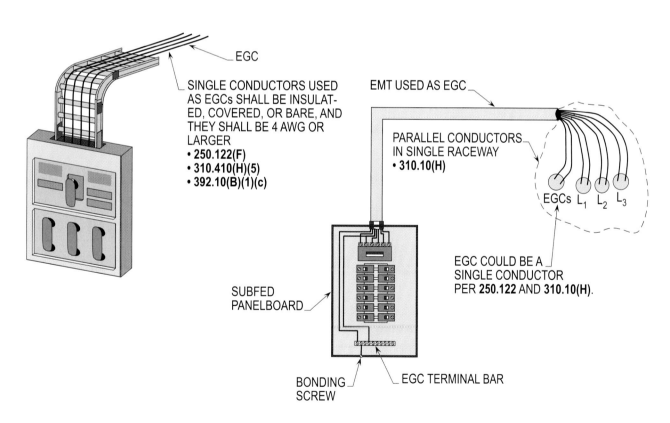

SINGLE CONDUCTORS USED AS EGCs SHALL BE INSULATED, COVERED, OR BARE, AND THEY SHALL BE 4 AWG OR LARGER
- 250.122(F)
- 310.410(H)(5)
- 392.10(B)(1)(c)

EMT USED AS EGC

PARALLEL CONDUCTORS IN SINGLE RACEWAY
- 310.10(H)

EGCs L₁ L₂ L₃

EGC COULD BE A SINGLE CONDUCTOR PER **250.122** AND **310.10(H)**.

SUBFED PANELBOARD

BONDING SCREW

EGC TERMINAL BAR

**SIZE OF EQUIPMENT GROUNDING CONDUCTORS
NEC 250.122(F)**

Purpose of Change: To clarify the requirements of routing equipment grounding conductors in parallel runs or single runs.

Type of Change	Revision			Committee Change	Accept in Principle		2008 NEC	250.190
ROP	pg. 1234	# 5-313	log: 615	ROC	pg. 1234	# 5-193	log: 570	UL	467
Submitter: Paul Guidry				Submitter: Code-Making Panel 6			OSHA	-
NFPA 70B	14.1.6.14			NFPA 70E	-		NFPA 79	-

250.190 Grounding of Equipment.

(A) Equipment Grounding. All non-current-carrying metal parts of fixed, portable, and mobile equipment and associated fences, housings, enclosures, and supporting structures shall be grounded.

Exception: Where isolated from ground and located such that any person in contact with ground cannot contact such metal parts when the equipment is energized, the metal parts shall not be required to be grounded.

Informational Note: See 250.110, Exception No. 2, for pole-mounted distribution apparatus.

Stallcup's Comment: A revision has been made to separate equipment connected to an equipment grounding conductor from equipment connected to a grounding electrode conductor, and this revision provides requirements for connection and conductor sizing.

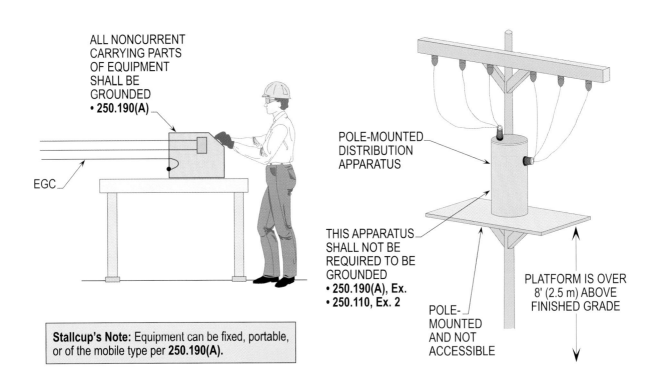

GROUNDING OF EQUIPMENT
NEC 250.190(A) AND Ex.

Purpose of Change: To clarify the requirements when noncurrent carrying parts of equipment have to be grounded for personnel safety.

Type of Change	Revision			Committee Change	Accept in Principle		2008 NEC	250.190
ROP	pg. 1234	# 5-313	log: 615	ROC	pg. 1234	# 5-193	log: 570	UL	467
Submitter: Paul Guidry				Submitter: Code-Making Panel 6			OSHA	-
NFPA 70B	14.1.6.27			NFPA 70E	-		NFPA 79	-

250.190 Grounding of Equipment.

(B) Grounding Electrode Conductor. If a grounding electrode conductor connects non-current-carrying metal parts to ground, the grounding electrode conductor shall be sized in accordance with Table 250.66, based on the size of the largest ungrounded service, feeder, or branch-circuit conductors supplying the equipment. The grounding electrode conductor shall not be smaller than 6 AWG copper or 4 AWG aluminum.

Stallcup's Comment: A revision has been made to separate equipment connected to an equipment grounding conductor from equipment connected to a grounding electrode conductor, and this revision provides requirements for connection and conductor sizing.

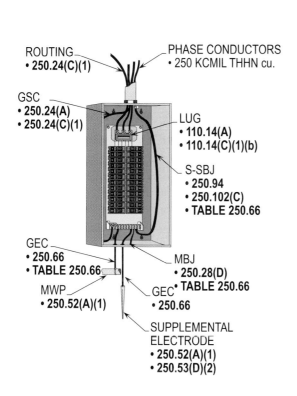

STALLCUP'S CODE LOOP

GROUNDED CONDUCTOR
- 250.24(C)(1)
- 250.142(A)(1)

MBJ
- 250.28(A) THRU (D)

S-SBJ
- 250.102(C)

GEC
- 250.66(A) THRU (C)
- TABLE 250.66

GES
- 250.52(A)(1) THRU (A)(4)
- 250.52(A)(5)

Sizing Grounding Electrode Conductor 250.190(B)

Step 1: Calculating GEC
250.190(B) and Table 250.66
250 KCMIL cu. requires 2 AWG cu.

Solution: The size of the grounding electrode conductor is 2 AWG cu.

Note: Grounding electrode conductor shall not be smaller than 6 AWG cu. or 4 AWG alu.

GROUNDING ELECTRODE CONDUCTOR
NEC 250.190(B)

Purpose of Change: To verify the sizing of the grounding electrode conductor.

Type of Change	Revision			Committee Change	Accept in Principle			2008 NEC	250.190
ROP	pg. 1234	# 5-313	log: 615	ROC	pg. 1234	# 5-193	log: 570	UL	467
Submitter: Paul Guidry				Submitter: Code-Making Panel 6				OSHA	-
NFPA 70B	14.1.6.15			NFPA 70E	-			NFPA 79	-

250.190 Grounding of Equipment.

(C) Equipment Grounding Conductor. Equipment grounding conductors shall comply with 250.190(C)(1) through (C)(3).

(1) General. Equipment grounding conductors that are not an integral part of a cable assembly shall not be smaller than 6 AWG copper or 4 AWG aluminum.

(2) Shielded Cables. The metallic insulation shield encircling the current carrying conductors shall be permitted to be used as an equipment grounding conductor, if it is rated for clearing time of ground fault current protective device operation without damaging the metallic shield. The metallic tape insulation shield and drain wire insulation shield shall not be used as an equipment grounding conductor for solidly grounded systems.

(3) Sizing. Equipment grounding conductors shall be sized in accordance with Table 250.122 based on the current rating of the fuse or the overcurrent setting of the protective relay.

Informational Note: The overcurrent rating for a circuit breaker is the combination of the current transformer ratio and the current pickup setting of the protective relay.

Stallcup's Comment: A revision has been made to separate equipment connected to an equipment grounding conductor from equipment connected to a grounding electrode conductor, and this revision provides requirements for connection and conductor sizing.

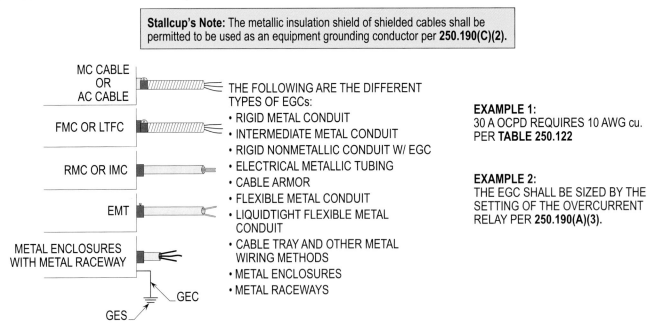

Stallcup's Note: The metallic insulation shield of shielded cables shall be permitted to be used as an equipment grounding conductor per **250.190(C)(2)**.

MC CABLE OR AC CABLE

FMC OR LTFC

RMC OR IMC

EMT

METAL ENCLOSURES WITH METAL RACEWAY

GEC

GES

THE FOLLOWING ARE THE DIFFERENT TYPES OF EGCs:
- RIGID METAL CONDUIT
- INTERMEDIATE METAL CONDUIT
- RIGID NONMETALLIC CONDUIT W/ EGC
- ELECTRICAL METALLIC TUBING
- CABLE ARMOR
- FLEXIBLE METAL CONDUIT
- LIQUIDTIGHT FLEXIBLE METAL CONDUIT
- CABLE TRAY AND OTHER METAL WIRING METHODS
- METAL ENCLOSURES
- METAL RACEWAYS

EXAMPLE 1:
30 A OCPD REQUIRES 10 AWG cu.
PER **TABLE 250.122**

EXAMPLE 2:
THE EGC SHALL BE SIZED BY THE SETTING OF THE OVERCURRENT RELAY PER **250.190(A)(3)**.

EQUIPMENT GROUNDING CONDUCTOR
NEC 250.190(C)(1) THRU (C)(3)

Purpose of Change: To clarify the procedure for using metallic shield as an equipment grounding conductor and sizing the equipment grounding conductor.

Type of Change	New Section			Committee Change	Accept in Principle			2008 NEC	-
ROP	pg. 303	# 5-315	log: 616	ROC	pg. -	# -	log: -	UL	467
Submitter: Paul Guidry				Submitter: -				OSHA	-
NFPA 70B	14.1.6.48			NFPA 70E	-			NFPA 70	-

250.191 Grounding System at Alternating-Current Substations.

For ac substations, the grounding system shall be in accordance with Part III of Article 250.

Informational Note: For further information on outdoor ac substation grounding, see ANSI/IEEE 80-2000, *IEEE Guide for Safety in AC Substation Grounding.*

Stallcup's Comment: A new section has been added to address the requirements for the grounding system at AC substations.

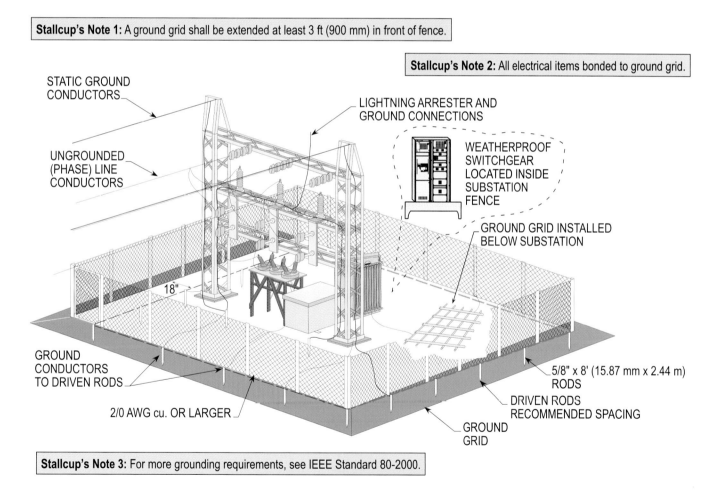

Stallcup's Note 1: A ground grid shall be extended at least 3 ft (900 mm) in front of fence.

Stallcup's Note 2: All electrical items bonded to ground grid.

STATIC GROUND CONDUCTORS

LIGHTNING ARRESTER AND GROUND CONNECTIONS

WEATHERPROOF SWITCHGEAR LOCATED INSIDE SUBSTATION FENCE

UNGROUNDED (PHASE) LINE CONDUCTORS

GROUND GRID INSTALLED BELOW SUBSTATION

18"

GROUND CONDUCTORS TO DRIVEN RODS

5/8" x 8' (15.87 mm x 2.44 m) RODS

DRIVEN RODS RECOMMENDED SPACING

2/0 AWG cu. OR LARGER

GROUND GRID

Stallcup's Note 3: For more grounding requirements, see IEEE Standard 80-2000.

**GROUNDING SYSTEM AT ALTERNATING-CURRENT SUBSTATIONS
NEC 250.191**

Purpose of Change: To clarify the grounding system at an AC substation shall comply with **Part III** of **Article 250.**

Name   Date

Chapter 2
Wiring and Protection

Section   Answer

1. GFCI protection shall be provided for receptacles installed within _____ ft of the outside edge
of the sink located in areas other than kitchens.

(A) 3	**(B)** 5
(C) 6	**(D)** 10

2. GFCI protection shall be provided for receptacles in the following locations:

(A) Sinks located in areas other than kitchens
(B) Indoor wet locations
(C) Locker rooms with associated showering facilities
(D) All of the above

3. Foyers that are not part of a hallway and that have an area that is greater than 60 ft² shall have a
receptacle(s) located in each wall space _____ ft or more in width and unbroken by doorways,
floor-to-ceiling, and similar openings.

(A) 3	**(B)** 5
(C) 6	**(D)** 10

4. Overhead spans of open conductors and open multiconductor cables of not over 600 volts shall have
a clearance of not less than _____ ft over track rails of railroads.

(A) 22.5	**(B)** 24.5
(C) 25	**(D)** 30

5. Where a raceway enters a building or structure from an underground distribution system, it shall be
_____.

(A) sealed	**(B)** marked
(C) purged and pressurized	**(D)** provided with a drain hole

6. Where an individual disconnecting means consists of fused cutouts, the simultaneous disconnection
of all _____ supply conductors shall not be required if there is a means to disconnect the load
before opening the cutouts.

(A) grounded	**(B)** ungrounded
(C) equipment	**(D)** bonding

7. Disconnecting means shall be capable of being locked in the open position. The provisions for
locking shall remain in place:

(A) with the lock installed	**(B)** without the lock installed
(C) all of the above	**(D)** none of the above

Section   Answer

8. A permanent, legible warning notice carrying the wording "DANGER – HIGH VOLTAGE" shall be placed in a conspicuous position on all cable trays containing high-voltage conductors with the maximum spacing of warning notices not to exceed _____ ft.

(A) 3
(C) 6

(B) 5
(D) 1

9. Overhead service conductors above roofs shall be permitted to have a reduction in clearance to _____ ft where the voltage between conductors does not exceed 300 and the roof is guarded or isolated.

(A) 3
(C) 10

(B) 6
(D) 15

10. Grounded conductors that are not connected to an overcurrent device shall be permitted to be sized at _____ percent of the continuous and noncontinuous load.

(A) 80
(C) 125

(B) 100
(D) 135

11. For field installations, if the tap conductors leave the enclosure or vault in which the tap is made, the ampacity of the tap conductors is not less than _____ of the rating of the overcurrent device protecting the feeder conductors.

(A) 1/6
(C) 1/10

(B) 1/8
(D) 1/16

12. Overcurrent devices shall not be located in bathrooms of the following locations:

(A) Dwelling units
(C) Guest rooms or guest suites

(B) Dormitories
(D) All of the above

13. Where the overcurrent device is rated over 800 amperes, the ampacity of the conductors it protects shall be equal to or greater than _____ percent of the rating of the overcurrent device specified in **240.6**, in accordance with the conductors being protected within recognized time vs. current limits for short-circuit currents, and all equipment in which the conductors terminate is listed and marked for the application.

(A) 80
(C) 95

(B) 90
(D) 100

14. The grounded conductor shall not be smaller than the required _____ conductor but shall not be required to be larger than the largest derived ungrounded conductor(s) for a separately derived system.

(A) grounding electrode
(C) equipment bonding

(B) equipment grounding
(D) bonding

15. For sets of derived ungrounded conductors larger than 1100 KCMIL copper or 1750 KCMIL aluminum, the grounded conductor shall not be smaller than _____ percent of the circular mil area of the largest set of derived ungrounded conductors.

(A) 6-1/2
(C) 13-1/2

(B) 12-1/2
(D) 15

Section   Answer

16. The common grounding electrode conductor for a separately derived system shall be permitted to
be a conductor of a wire type not smaller than _____ AWG copper.

(A) 1/0   **(B)** 2/0
(C) 3/0   **(D)** 4/0

17. The common grounding electrode conductor for a separately derived system shall be permitted to
be the metal frame of the building or structure or be connected to the grounding electrode system by a
conductor that shall not be smaller than _____ AWG copper.

(A) 6   **(B)** 2
(C) 1/0   **(D)** 3/0

18. At least one structural member that is in direct contact with the earth for _____ or more, with
or without concrete encasement, shall be permitted to be used as a grounding electrode.

(A) 6   **(B)** 10
(C) 12   **(D)** 2

19. A concrete-encased electrode shall be permitted to consist of 20 ft of bare copper conductor not
smaller than _____ AWG.

(A) 1   **(B)** 2
(C) 3   **(D)** 4

20. Equipment grounding conductors that are not an integral part of a cable assembly shall not be
smaller than _____ AWG copper.

(A) 6   **(B)** 8
(C) 4   **(D)** 2

Wiring Methods and Materials

Chapter 3 of the *National Electrical Code* has always been referred to as the "Installation Chapter" and is used by electricians to install electrical systems in a safe, dependable, and reliable manner. **Chapter 3** covers the requirements needed by on-the-job electricians who are installing service, feeder, and branch circuits and thus providing electrical power to the end of branch circuits and on to the point of use.

Chapter 3 is also called the "Rough-In Chapter" by electricians. All articles in **Chapter 3** are of the 300 series, and each contains the rules that pertain to installing electrical wiring methods and accessories. On-the-job work procedures of the average electrician bring them into almost daily contact with an article in **Chapter 3** concerning these installation rules.

For example, when installing wiring in cable trays, electricians cannot install and fill cable trays with different wiring, cables, and systems without knowing the requirements of **Article 392**. The same is true for electricians installing rigid metal conduit. Installers would not know how many 90° bends are permitted or how often supports are required for rigid metal conduit without first studying **Article 344**.

Type of Change	Revision			Committee Change	Accept in Principle			2008 NEC	300.4(E)
ROP	pg. 312	# 3-34	log: 2763	ROC	pg. -	# -	log: -	UL	-
Submitter: Donald R. Offerdahl				Submitter: -				OSHA	-
NFPA 70B	-			NFPA 70E	-			NFPA 79	-

300.4 Protection Against Physical Damage.

(E) Cables, Raceways, or Boxes Installed In or Under Roof Decking. A cable, raceway, or box, installed in exposed or concealed locations under metal-corrugated sheet roof decking, shall be installed and supported so there is not less than 38 mm (1 1/2 in.) measured from the lowest surface of the roof decking to the top of the cable, raceway, or box. A cable, raceway, or box shall not be installed in concealed locations in metal-corrugated, sheet decking-type roof.

Stallcup's Comment: A revision has been made to clarify that cables, raceways, or boxes shall be installed and supported so as to be protected where installed in or under roof decking.

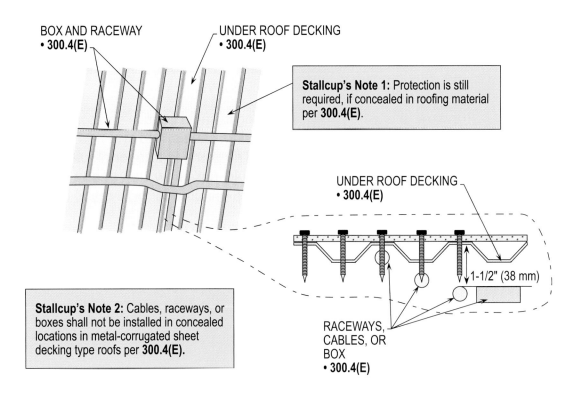

BOX AND RACEWAY
• 300.4(E)

UNDER ROOF DECKING
• 300.4(E)

Stallcup's Note 1: Protection is still required, if concealed in roofing material per **300.4(E)**.

UNDER ROOF DECKING
• 300.4(E)

1-1/2" (38 mm)

Stallcup's Note 2: Cables, raceways, or boxes shall not be installed in concealed locations in metal-corrugated sheet decking type roofs per **300.4(E)**.

RACEWAYS, CABLES, OR BOX
• 300.4(E)

**CABLES, RACEWAYS, OR BOXES INSTALLED
IN OR UNDER ROOF DECKING
NEC 300.4(E)**

Purpose of Change: To clarify the procedure and methods of protecting wiring methods installed in or under roof decking.

Type of Change	New Exceptions			Committee Change	Accept			2008 NEC	-
ROP	pg. 316	# 3-52	log: 4531	ROC	pg. 168	# 3-17	log: 574	UL	1990
Submitter: Phil Simmons				Submitter: Vince Baclawski				OSHA	-
NFPA 70B	-			NFPA 70E	-			NFPA 70	

√

300.5 Underground Installations.

(C) Underground Cables Under Buildings. Underground cable installed under a building shall be in a raceway.

Exception No. 1: Type MI Cable shall be permitted under a building without installation in a raceway where embedded in concrete, fill, or other masonry in accordance with 332.10(6) or in underground runs where suitably protected against physical damage and corrosive conditions in accordance with 332.10(10).

Exception No. 2: Type MC Cable listed for direct burial or concrete encasement shall be permitted under a building without installation in a raceway in accordance with 330.10(A)(5) and in wet locations in accordance with 330.10(11).

Stallcup's Comment: New exceptions have been added to clarify the permitted methods for installation of Type MI and Type MC cables in an underground installation.

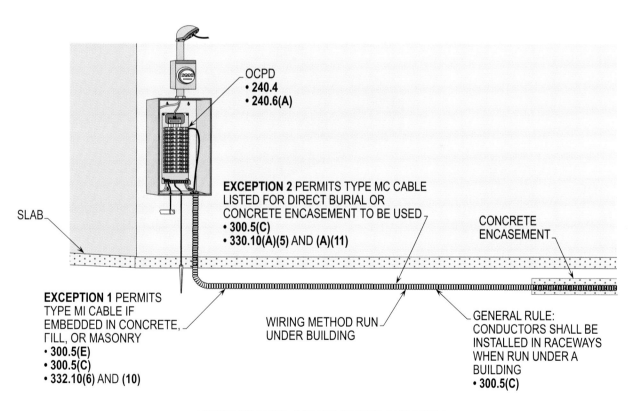

OCPD
• **240.4**
• **240.6(A)**

EXCEPTION 2 PERMITS TYPE MC CABLE
LISTED FOR DIRECT BURIAL OR
CONCRETE ENCASEMENT TO BE USED
• **300.5(C)**
• **330.10(A)(5) AND (A)(11)**

SLAB

CONCRETE
ENCASEMENT

EXCEPTION 1 PERMITS
TYPE MI CABLE IF
EMBEDDED IN CONCRETE,
FILL, OR MASONRY
• **300.5(E)**
• **300.5(C)**
• **332.10(6) AND (10)**

WIRING METHOD RUN
UNDER BUILDING

GENERAL RULE:
CONDUCTORS SHALL BE
INSTALLED IN RACEWAYS
WHEN RUN UNDER A
BUILDING
• **300.5(C)**

UNDERGROUND CABLES UNDER BUILDINGS
NEC 300.5(C), Ex. 1 AND Ex. 2

Purpose of Change: To permit Type MI and MC cables to be installed under buildings when certain conditions are applied.

Type of Change	Revision			Committee Change	Accept in Principle			2008 NEC	300.5(I), Ex. 1
ROP	pg. 318	# 3-61	log: 4607	ROC	pg. -	# -	log: -	UL	1990
Submitter: Frederic P. Hartwell				Submitter: -				OSHA	-
NFPA 70B	-			NFPA 70E	-			NFPA 79	-

300.5 Underground Installations.

(I) Conductors of the Same Circuit. All conductors of the same circuit and, where used, the grounded conductor and all equipment grounding conductors shall be installed in the same raceway or cable or shall be installed in close proximity in the same trench.

Exception No. 1: Conductors shall be permitted to be installed in parallel in raceways, multiconductor cables, or direct-buried single conductor cables. Each raceway or multiconductor cable shall contain all conductors of the same circuit, including equipment grounding conductors. Each direct-buried single conductor cable shall be located in close proximity in the trench to the other single conductor cables in the same parallel set of conductors in the circuit, including equipment grounding conductors.

Stallcup's Comment: A revision has been made to clarify that direct-buried single conductor cable shall be permitted to be installed in parallel for underground installations.

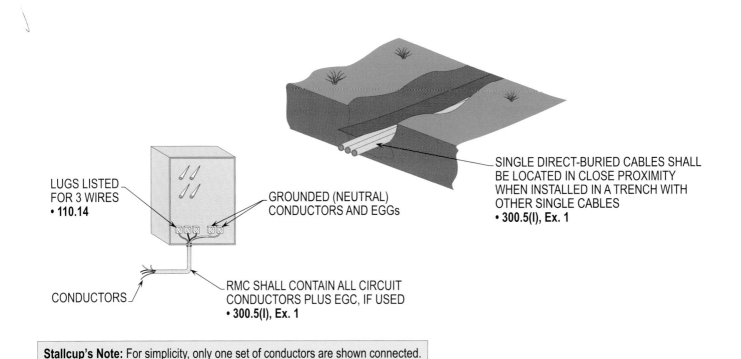

LUGS LISTED FOR 3 WIRES
• 110.14

GROUNDED (NEUTRAL) CONDUCTORS AND EGGs

SINGLE DIRECT-BURIED CABLES SHALL BE LOCATED IN CLOSE PROXIMITY WHEN INSTALLED IN A TRENCH WITH OTHER SINGLE CABLES
• 300.5(I), Ex. 1

CONDUCTORS

RMC SHALL CONTAIN ALL CIRCUIT CONDUCTORS PLUS EGC, IF USED
• 300.5(I), Ex. 1

Stallcup's Note: For simplicity, only one set of conductors are shown connected.

CONDUCTORS OF THE SAME CIRCUIT
NEC 300.5(I), Ex. 1

Purpose of Change: To verify that ungrounded conductors, neutrals, and equipment grounding conductors, if present, shall be routed together.

Type of Change	Revision			Committee Change	Accept in Principle			2008 NEC	300.6(A)
ROP	pg. 319	# 3-64	log: 3646	ROC	pg. 169	# 3-22	log: 1984	UL	-
Submitter: Donald A. Ganiere				Submitter: Donald A. Ganiere				OSHA	-
NFPA 70B	-			NFPA 70E				NFPA 79	-

300.6 Protection Against Corrosion and Deterioration.

(A) Ferrous Metal Equipment. Ferrous metal raceways, cable trays, cablebus, auxiliary gutters, cable armor, boxes, cable sheathing, cabinets, metal elbows, couplings, nipples, fittings, supports, and support hardware shall be suitably protected against corrosion inside and outside (except threads at joints) by a coating of approved corrosion-resistant material. Where corrosion protection is necessary and the conduit is threaded in the field, the threads shall be coated with an approved electrically conductive, corrosion-resistant compound.

Stallcup's Comment: A revision has been made to clarify that where corrosion protection is necessary, threads shall be coated with an approved electrically conductive, corrosion-resistant compound if threads do not have corrosion protection.

> **Stallcup's Note 1:** Raceways, cable trays, cablebus, auxiliary gutter, cable armor boxes, cable sheathing, cabinets, elbows, couplings, supports, etc. shall be suitable for where they are installed per **300.6(A) through (D).**

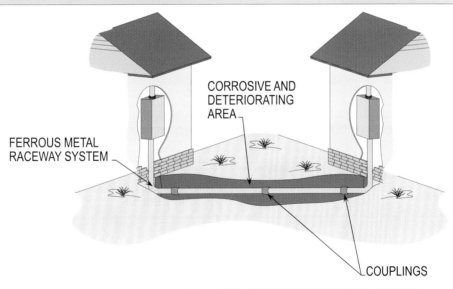

CORROSIVE AND DETERIORATING AREA

FERROUS METAL RACEWAY SYSTEM

COUPLINGS

> **Stallcup's Note 2:** Threads without corrosion protection shall be coated with an approved electrically conductive, corrosion-resistant compound per **300.6(A).**

FERROUS METAL EQUIPMENT
NEC 300.6(A)

Purpose of Change: To revise and make clear that threads of couplings for metallic raceways, enclosures, and associated equipment are properly protected from corrosion and deterioration.

Type of Change	New Subsection			Committee Change	Accept			2008 NEC	-
ROP	pg. 329	# 3-105	log: 3701	ROC	pg. -	# -	log: -	UL	1990
Submitter: Vince Baclawski				Submitter: -				OSHA	-
NFPA 70B	-			NFPA 70E	-			NFPA 79	-

300.50 Underground Installations.

(B) Wet Locations. The interior of enclosures or raceways installed underground shall be considered to be a wet location. Insulated conductors and cables installed in these enclosures or raceways in underground installations shall be listed for use in wet locations and shall comply with 310.10(C). Any connections or splices in an underground installation shall be approved for wet locations.

Stallcup's Comment: A new subsection has been added to address that the interior of enclosures or raceways shall be considered to be in a wet location where installed underground. The conductors and cables shall be insulated and be listed for use in wet locations. Any connections or splices shall be approved for wet locations.

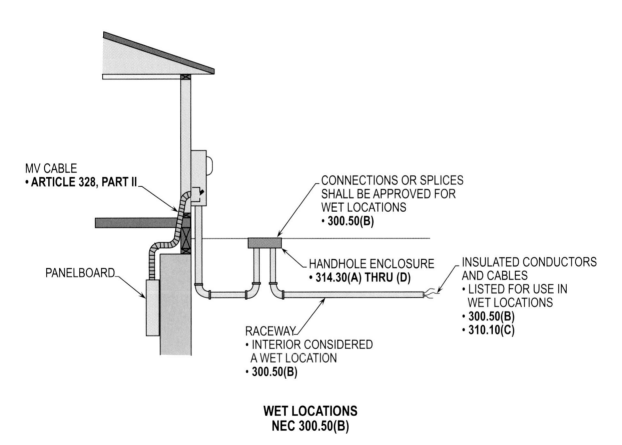

MV CABLE
• **ARTICLE 328, PART II**

PANELBOARD

CONNECTIONS OR SPLICES
SHALL BE APPROVED FOR
WET LOCATIONS
• **300.50(B)**

HANDHOLE ENCLOSURE
• **314.30(A) THRU (D)**

INSULATED CONDUCTORS
AND CABLES
• LISTED FOR USE IN
WET LOCATIONS
• **300.50(B)**
• **310.10(C)**

RACEWAY
• INTERIOR CONSIDERED
A WET LOCATION
• **300.50(B)**

WET LOCATIONS
NEC 300.50(B)

Purpose of Change: A new subsection has been added to clarify that the interior of enclosures or raceways shall be considered a wet location in an underground installation. All insulated conductors and cables shall be listed for use in wet locations and shall comply with **310.10(C)**.

Type of Change	Article Renumbered			Committee Change	Accept in Principle in Part			2008 NEC	Article 310
ROP	pg. 329	# 6-8	log: 2395	ROC	pg. 174	# 6-4	log: 2571	UL	-
Submitter: James M. Daly				Submitter: Frederic P. Hartwell				OSHA	1910.305
NFPA 70B	-			NFPA 70E	210.4			NFPA 79	Chapter 12

Article 310 Conductors for General Wiring.

Part I General

Part II Installation

Part III Construction Specifications

Stallcup's Comment: Article 310 has been renumbered as **310.1 through 310.120** and placed into three parts for general requirements, installation requirements, and construction specifications.

ARTICLE 310 – OUTLINED

PART I – GENERAL
• **310.1** AND **310.2**

PART II – INSTALLATION
• **310.10 THRU 310.60**

PART III – CONSTRUCTION SPECIFICATIONS
• **310.104 THRU 310.120**

Stallcup's Note: Electrical personnel must very carefully review this new layout of **Article 310,** which includes the renumbering and relocation of many existing sections and tables from the 2008 *National Electrical Code*.

APPLYING RENUMBERED SECTIONS AND LAYOUT OF **ARTICLE 310**

ENGINEER, CONTRACTOR, ELECTRICIAN, OR INSPECTOR

CONDUCTORS FOR GENERAL WIRING
ARTICLE 310

Purpose of Change: To provide a new layout of **Article 310** in an effort to make it more user-friendly and line up with the other articles in the 300 series.

CROSS REFERENCE	
2011 NEC	**SECTION HEADING**
PART I – GENERAL	
310.1	SCOPE
310.2	DEFINITIONS
PART II – INSTALLATIONS	
310.10	USES PERMITTED
310.10(A)	DRY LOCATIONS
310.10(B)	DRY AND DAMP LOCATIONS
310.10(C)	WET LOCATIONS
310.10(D)	LOCATIONS EXPOSED TO DIRECT SUNLIGHT
310.10(E)	SHIELDING
310.10(F)	DIRECT-BURIAL CONDUCTORS
310.10(G)	CORROSIVE CONDITIONS
310.10(H)	CONDUCTORS IN PARALLEL
310.15	AMPACITIES FOR CONDUCTORS RATED 0-2000 VOLTS
310.15(A)(1)	TABLES OR ENGINEERING SUPERVISION
310.15(A)(2)	SELECTION OF AMPACITY
310.15(A)(3)	TEMPERATURE LIMITATIONS OF CONDUCTORS
310.15(B)	TABLES
310.15(B)(2)	AMBIENT TEMPERATURE CORRECTION FACTORS [**TABLE 310.15(B)(2)(b)**]
310.15(B)(3)	ADJUSTMENT FACTORS [**TABLE 310.15(B)(3)(a)**]
310.15(B)(4)	BARE OR COVERED CONDUCTORS
310.15(B)(5)	NEUTRAL CONDUCTOR
310.15(B)(6)	GROUNDING OR BONDING CONDUCTOR
310.15(B)(7)	120/240 V, 3-WIRE, SINGLE-PHASE DWELLING SERVICE AND FEEDERS
310.15(C)	ENGINEERING SUPERVISION
TABLE 310.15(B)(16)	AMPACITY TABLE (EXISTING **TABLE 310.16**)
TABLE 310.15(B)(17) THRU (B)(20)	AMPACITY TABLES (EXISTING **TABLES 310.17 THRU 310.20**)
310.60(A)	DEFINITIONS
310.60(B)	AMPACITIES OF CONDUCTORS RATED 2001 TO 35,000 VOLTS
310.60(B)(1)	SELECTION OF AMPACITIES
310.60(C)	TABLES
310.60(C)(1)	GROUNDED SHIELDS
310.60(C)(2)	BURIAL DEPTH OF UNDERGROUND CIRCUITS
310.60(C)(3)	ELECTRICAL DUCTS IN FIGURE 310.60
310.60(C)(4)	AMBIENT TEMPERATURE CORRECTION
310.60(D)	ENGINEERING SUPERVISION
PART III – CONSTRUCTION SPECIFICATIONS	
310.104	CONDUCTOR, CONSTRUCTION, AND APPLICATION [SEE **TABLES 310.104(A) THRU 310.104(E)**]
310.106(A)	MINIMUM SIZE OF CONDUCTORS
310.106(B)	CONDUCTOR MATERIAL
310.106(C)	STRANDED CONDUCTORS
310.106(D)	INSULATED
310.110	CONDUCTOR IDENTIFICATION
310.110(A)	GROUNDED CONDUCTOR
310.110(B)	EQUIPMENT GROUNDING CONDUCTORS
310.110(C)	UNGROUNDED CONDUCTORS
310.120(A)	MARKINGS (REQUIRED INFORMATION)
310.120(B)(1)	SURFACE MARKING
310.120(B)(2)	MARKER TAPE
310.120(B)(3)	TAG MARKING
310.120(B)(4)	OPTIONAL MARKING OF WIRE SIZE
310.120(C)	SUFFIXES TO DESIGNATE NUMBER OF CONDUCTORS
310.120(D)	OPTIONAL MARKINGS

CONDUCTORS FOR GENERAL WIRING
ARTICLE 310

Purpose of Change: To provide a new layout of **Article 310** in an effort to make it more user friendly.

Type of Change	Revision			Committee Change	Accept in Principle			2008 NEC	310.6
ROP	pg. 338	# 6-24	log: 4089	ROC	pg. 177	# 6-8	log: 1790	UL	1569
Submitter: Michael P. Walls				Submitter: Michael P. Walls				OSHA	-
NFPA 70B	-			NFPA 70E	-			NFPA 79	-

310.10 Uses Permitted.

(E) Shielding. Non-shielded, ozone-resistant insulated conductors with a maximum phase-to-phase voltage of 5000 volts shall be permitted in Type MC cables in industrial establishments where the conditions of maintenance and supervision ensure that only qualified persons service the installation. For other establishments, solid dielectric insulated conductors operated above 2000 volts in permanent installations shall have ozone-resistant insulation and shall be shielded. All metallic insulation shields shall be connected to a grounding electrode conductor, a grounding busbar, and equipment grounding conductor, or a grounding electrode.

Stallcup's Comment: A revision has been made to recognize the enhanced reliability of the construction of metal-armored cable containing nonshielded conductors. The nonshielded conductors have a concentric lay-orientation and their insulation is protected from damage during installation. The installation of nonshielded conductors shall be permitted in Type MC cables in industrial establishments where the conditions of maintenance and supervision ensure that only qualified persons service the installation.

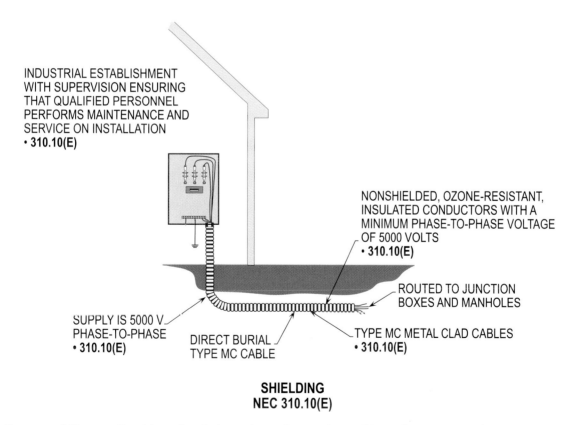

INDUSTRIAL ESTABLISHMENT WITH SUPERVISION ENSURING THAT QUALIFIED PERSONNEL PERFORMS MAINTENANCE AND SERVICE ON INSTALLATION
• **310.10(E)**

NONSHIELDED, OZONE-RESISTANT, INSULATED CONDUCTORS WITH A MINIMUM PHASE-TO-PHASE VOLTAGE OF 5000 VOLTS
• **310.10(E)**

ROUTED TO JUNCTION BOXES AND MANHOLES

SUPPLY IS 5000 V PHASE-TO-PHASE
• **310.10(E)**

DIRECT BURIAL TYPE MC CABLE

TYPE MC METAL CLAD CABLES
• **310.10(E)**

SHIELDING
NEC 310.10(E)

Purpose of Change: To add wording that permits, under certain conditions of use, the use of Type MC cable on electrical systems of 5000 volts.

Type of Change	Revision			Committee Change	Accept			2008 NEC	310.6, Ex. 2
ROP	pg. 337	# 6-21b	log: CP 601	ROC	pg. 176	# 6-7	log: 95	UL	-
Submitter: Code Making Panel 6				Submitter: Technical Correlating Committee				OSHA	-
NFPA 70B	-			NFPA 70E	-			NFPA 79	-

310.10 Uses Permitted.

(E) Shielding.

Exception No. 2: Nonshielded insulated conductors listed by a qualified testing laboratory shall be permitted for use up to 5000 volts to replace existing nonshielded conductors, on existing equipment in industrial establishments only, under the following conditions:

(a) Where the condition of maintenance and supervision ensures that only qualified personnel install and service the installation.

(b) Conductors shall have insulation resistant to electric discharge and surface tracking, or the insulated conductor(s) shall be covered with a material resistant to ozone, electric discharge, and surface tracking.

(c) Where used in wet locations, the insulated conductor(s) shall have an overall nonmetallic jacket or a continuous metallic sheath.

(d) Insulation and jacket thicknesses shall be in accordance with Table 310.13(D).

Stallcup's Comment: A revision has been made to recognize that nonshielded insulated conductors listed by a qualified testing laboratory shall be permitted to replace existing nonshielded conductors up to 5000 volts on existing equipment in industrial establishments only if certain conditions are complied with.

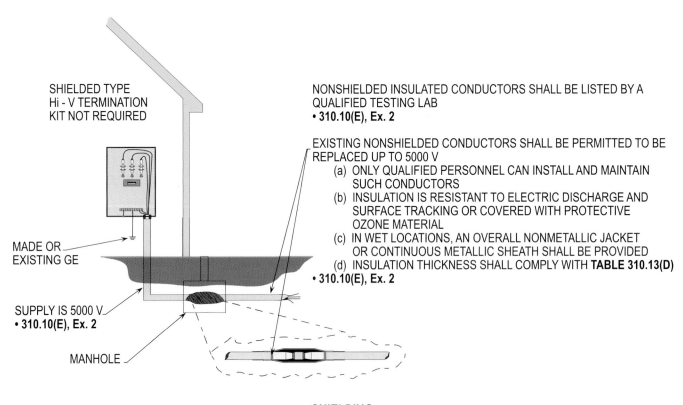

SHIELDED TYPE
Hi - V TERMINATION
KIT NOT REQUIRED

MADE OR
EXISTING GE

SUPPLY IS 5000 V
• **310.10(E), Ex. 2**

MANHOLE

NONSHIELDED INSULATED CONDUCTORS SHALL BE LISTED BY A
QUALIFIED TESTING LAB
• **310.10(E), Ex. 2**

EXISTING NONSHIELDED CONDUCTORS SHALL BE PERMITTED TO BE
REPLACED UP TO 5000 V
   (a) ONLY QUALIFIED PERSONNEL CAN INSTALL AND MAINTAIN
      SUCH CONDUCTORS
   (b) INSULATION IS RESISTANT TO ELECTRIC DISCHARGE AND
      SURFACE TRACKING OR COVERED WITH PROTECTIVE
      OZONE MATERIAL
   (c) IN WET LOCATIONS, AN OVERALL NONMETALLIC JACKET
      OR CONTINUOUS METALLIC SHEATH SHALL BE PROVIDED
   (d) INSULATION THICKNESS SHALL COMPLY WITH **TABLE 310.13(D)**
• **310.10(E), Ex. 2**

SHIELDING
NEC 310.10(E), Ex. 2

Purpose of Change: To permit the use of 5000 volt nonshielded insulated conductors.

Type of Change	New Subdivision			Committee Change	Accept			2008 NEC	-
ROP	pg. 335	# 6-17	log: 3754	ROC	pg. -	# -	log: -	UL	-
Submitter: Jim Pauley				Submitter: -				OSHA	-
NFPA 70B	14.1.0.13			NFPA 70E	205.5			NFPA 79	8.2.2

310.10 Uses Permitted.

(H) Conductors in Parallel.

(6) Equipment Bonding Jumpers. Where parallel equipment bonding jumpers are installed in raceways, they shall be sized and installed in accordance with 250.102.

Stallcup's Comment: A new subdivison has been added to address the installation and sizing of equipment bonding jumpers installed in parallel.

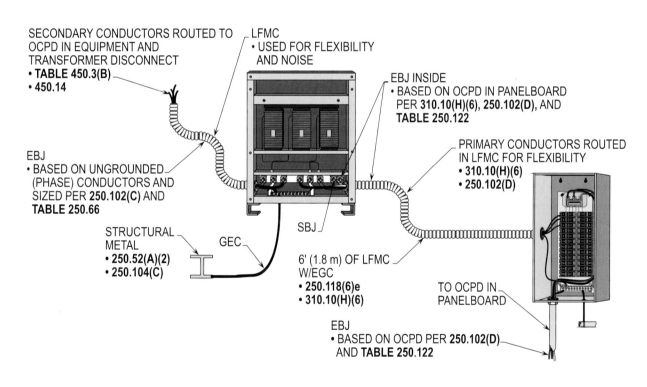

SECONDARY CONDUCTORS ROUTED TO OCPD IN EQUIPMENT AND TRANSFORMER DISCONNECT
• **TABLE 450.3(B)**
• **450.14**

LFMC
• USED FOR FLEXIBILITY AND NOISE

EBJ INSIDE
• BASED ON OCPD IN PANELBOARD PER **310.10(H)(6)**, **250.102(D)**, AND **TABLE 250.122**

PRIMARY CONDUCTORS ROUTED IN LFMC FOR FLEXIBILITY
• **310.10(H)(6)**
• **250.102(D)**

EBJ
• BASED ON UNGROUNDED (PHASE) CONDUCTORS AND SIZED PER **250.102(C)** AND **TABLE 250.66**

STRUCTURAL METAL
• **250.52(A)(2)**
• **250.104(C)**

GEC

SBJ

6' (1.8 m) OF LFMC W/EGC
• **250.118(6)e**
• **310.10(H)(6)**

TO OCPD IN PANELBOARD

EBJ
• BASED ON OCPD PER **250.102(D)** AND **TABLE 250.122**

EQUIPMENT BONDING JUMPERS
NEC 310.10(H)(6)

Purpose of Change: To clarify the appropriate method for sizing the supply-side and load-side equipment bonding jumpers.

Type of Change	Revision		Committee Change	Accept in Principle	2008 NEC	310.15(B)
ROP pg. 344	# 6-51	log: 1646	ROC pg. -	# -	log: -	UL	-
Submitter: James M. Daly			Submitter: -		OSHA	-
NFPA 70B -			NFPA 70E -		NFPA 79	-

310.15 Ampacities for Conductors Rated 0-2000 Volts.

(B) Tables. Ampacities for conductors rated 0 to 2000 volts shall be as specified in the Allowable Ampacity Table 310.15(B)(16) through Table 310.15(B)(19), and Ampacity Table 310.15(B)(20) and Table 310.15(B)(21) as modified by 310.15(B)(1) through (B)(7).

The temperature correction and adjustment factors shall be permitted to be applied to the ampacity for the temperature rating of the conductor, if the corrected and adjusted ampacity does not exceed the ampacity for the temperature rating of the termination in accordance with the provisions of 110.14(C).

Stallcup's Comment: A revision has been made to clarify that the actual temperature rating of the conductor shall be permitted to be used for ampacity adjustment or correction factors. After calculations have been made, the ampacity shall not exceed the temperature limitations of the termination.

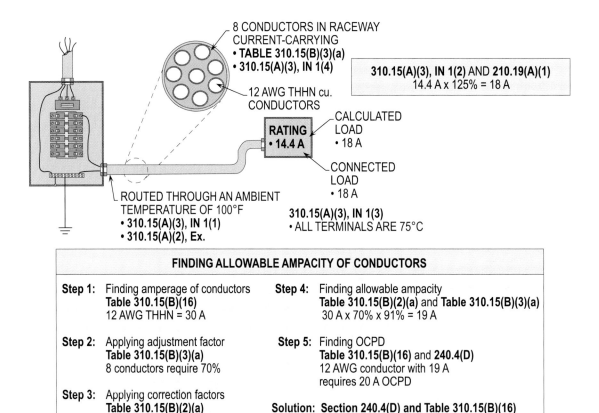

Purpose of Change: Renumbering of **Table 310.16** to **Table 310.15(B)(16)**, **Table 310.17** to **Table 310.15(B)(17)**, **Table 310.18** to **Table 310.16(B)(18)**, and **Table 310.19** to **Table 310.15(B)(19)** in an effort to make the *National Electrical Code* more user-friendly.

Type of Change	New Subsection			Committee Change	Accept			2008 NEC	Table 310.16
ROP	pg. 344	# 6-53	log: 628	ROC	pg. -	# -	log: -	UL	-
Submitter: James M. Daly				Submitter:				OSHA	-
NFPA 70B	-			NFPA 70E	-			NFPA 79	-

310.15 Ampacities for Conductors Rated 0-2000 Volts.

(B) Tables.

(2) Ambient Temperature Correction Factors. Ampacities for ambient temperatures other than those shown in the ampacity tables shall be corrected in accordance with Table 310.15(B)(2)(a) or Table 310.15(B)(2)(b), or shall be permitted to be calculated using the following equation:

Stallcup's Comment: A new subsection has been added to address the ambient temperature correction factors for conductors. The ambient correction factors used for derating purposes are found in **Tables 310.15(B)(2)(a)** and **310.15(B)(2)(b).** These ambient correction factors have been relocated from existing **Table 310.16** in the 2008 *National Electrical Code.*

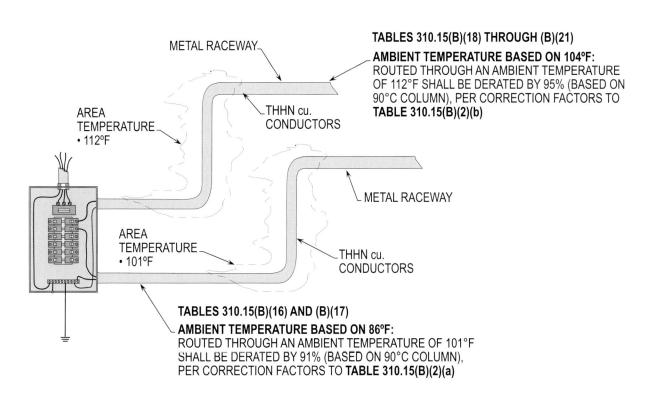

AMBIENT TEMPERATURE CORRECTION FACTORS
NEC 310.15(B)(2)

Purpose of Change: To introduce two tables that have been renumbered to address the requirements for ambient temperature correction factors.

Type of Change	Renumbered		Committee Change	Accept in Part		2008 NEC	Table 310.15(B)(2)(a)
ROP	pg. 349	# 6-57 / log: 635	ROC	pg. 179	# 6-24 / log: 471	UL	-
Submitter: James M. Daly			Submitter: Steven R. Terry			OSHA	-
NFPA 70B	-		NFPA 70E	-		NFPA 79	-

310.15 Ampacities for Conductors Rated 0-2000 Volts.

(B) Tables.

Table 310.15(B)(3)(a) Adjustment Factors for More Than Three Current-Carrying Conductors in a Raceway or Cable

[1]Number of conductors is the total number of conductors in the raceway or cable adjusted in accordance with 310.15(B)(5) and (6).

Stallcup's Comment: A new note has been added below **Table 310.15(B)(3)(a)** to correlate with the heading change of "Number of Current-Carrying Conductors" to "Number of Conductors." This new note clarifies that the total number of conductors in the cable or raceway shall be counted and subtract the grounded (neutral) conductors not required to be counted per **310.15(B)(4)** and the grounding or bonding conductors not required to be counted per **310.15(B)(5).**

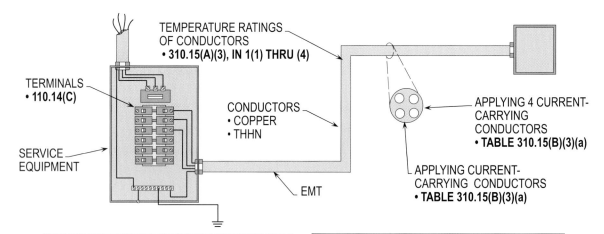

ADJUSTMENT FACTORS FOR MORE THAN THREE CURRENT-CARRYING CONDUCTORS IN A RACEWAY OR CABLE
TABLE 310.15(B)(3)(a), NOTE 1

Type of Change	Revision			Committee Change	Accept in Principle			2008 NEC	312.8
ROP	pg. 368	# 9-34	log: 3758	ROC	pg. -	# -	log: -	UL	50
Submitter: Jim Pauley				Submitter: -				OSHA	-
NFPA 70B	24.7			NFPA 70E	-			NFPA 79	13.3

312.8 Switch and Overcurrent Device Enclosures with Splices, Taps, and Feed-Through Conductors.

The wiring space of enclosures for switches or overcurrent devices shall be permitted for conductors feeding through, spliced, or tapping off to other enclosures, switches, or overcurrent devices where all of the following conditions are met:

(1) The total of all conductors installed at any cross section of the wiring space does not exceed 40 percent of the cross-sectional area of that space.

(2) The total area of all conductors, splices, and taps installed at any cross section of the wiring space does not exceed 75 percent of the cross-sectional area of that space.

(3) A warning label is applied to the enclosure that identifies the closest disconnecting means for any feed-through conductors.

Stallcup's Comment: A revision has been made to address the requirements for switch and overcurrent device enclosures with splices, taps, and feed-through conductors.

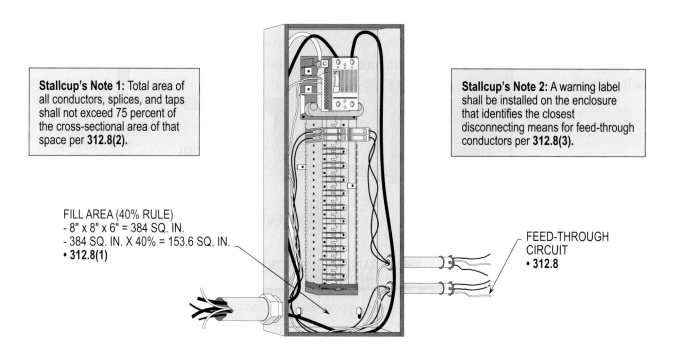

Stallcup's Note 1: Total area of all conductors, splices, and taps shall not exceed 75 percent of the cross-sectional area of that space per **312.8(2).**

Stallcup's Note 2: A warning label shall be installed on the enclosure that identifies the closest disconnecting means for feed-through conductors per **312.8(3).**

FILL AREA (40% RULE)
- 8" x 8" x 6" = 384 SQ. IN.
- 384 SQ. IN. X 40% = 153.6 SQ. IN.
• **312.8(1)**

FEED-THROUGH CIRCUIT
• **312.8**

SWITCH AND OVERCURRENT DEVICE ENCLOSURES WITH SPLICES, TAPS, AND FEED-THROUGH CONDUCTORS
NEC 312.8(1) THRU (3)

Purpose of Change: To add requirements for placing a warning label to identify closest disconnecting means for any feed-through conductors.

Type of Change	Revision			Committee Change	Accept			2008 NEC	314.27(A)
ROP	pg. 376	# 9-77	log: 3905	ROC	pg. 200	# 9-29	log: 108	UL	514A
Submitter: Bradford D. Rupp				Submitter: Technical Correlating Committee				OSHA	-
NFPA 70B	-			NFPA 70E	-			NFPA 79	-

314.27 Outlet Boxes.

(A) Boxes at Luminaire or Lampholder Outlets. Outlet boxes or fittings designed for the support of luminaires and lampholders, and installed as required by 314.23, shall be permitted to support a luminaire or lampholder.

Stallcup's Comment: A revision has been made to clarify that outlet boxes and fittings designed for the support of luminaires and lampholders shall be permitted to support a luminaire or lampholder.

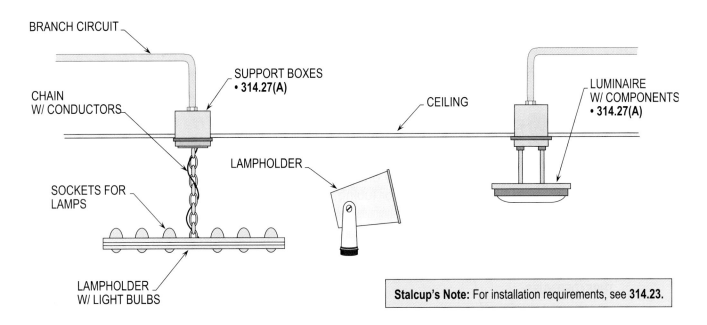

BOXES AT LUMINAIRE OR LAMPHOLDER OUTLETS
NEC 314.27(A)

Purpose of Change: To clarify that outlet boxes and fittings designed for such purpose shall be permitted to support luminaires and lampholders.

Type of Change	New Subdivision			Committee Change	Accept			2008 NEC	-
ROP	pg. 376	# 9-77	log: 3905	ROC	pg. 200	# 9-29	log: 108	UL	514A
Submitter: Bradford D. Rupp				Submitter: Technical Correlating Committee				OSHA	-
NFPA 70B	-			NFPA 70E				NFPA 79	-

314.27 Outlet Boxes.

(A) Boxes at Luminaire or Lampholder Outlets. Outlet boxes or fittings designed for the support of luminaires and lampholders, and installed as required by 314.23, shall be permitted to support a luminaire or lampholder.

(1) Wall Outlets. Boxes used at luminaire or lampholder outlets in a wall shall be marked on the interior of the box to indicate the maximum weight of the luminaire that is permitted to be supported by the box in the wall, if other than 23 kg (50 lb).

Stallcup's Comment: A new subdivison has been added to clarify the box requirements for supporting a luminaire at wall outlets. Boxes shall be required to be marked on the interior of the box to indicate the maximum weight that shall be permitted to be supported at the wall outlet if greater than 50 lb (23 kg).

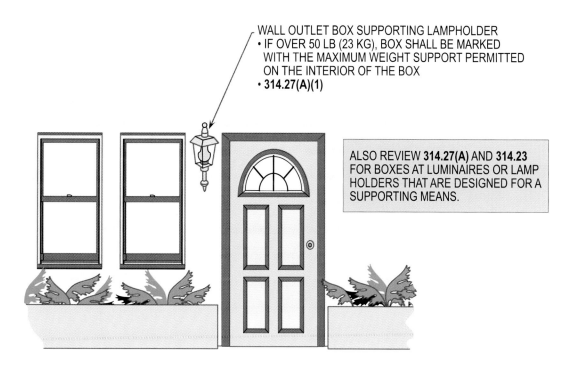

WALL OUTLET BOX SUPPORTING LAMPHOLDER
• IF OVER 50 LB (23 KG), BOX SHALL BE MARKED
  WITH THE MAXIMUM WEIGHT SUPPORT PERMITTED
  ON THE INTERIOR OF THE BOX
• 314.27(A)(1)

ALSO REVIEW **314.27(A)** AND **314.23** FOR BOXES AT LUMINAIRES OR LAMP HOLDERS THAT ARE DESIGNED FOR A SUPPORTING MEANS.

**WALL OUTLETS
NEC 314.27(A)(1)**

Purpose of Change: To address the requirements for identifying the maximum supporting weight for wall outlets.

Type of Change	New Subdivision			Committee Change		Accept		2008 NEC	-
ROP	pg. 376	# 9-77	log: 3905	ROC	pg. 200	# 9-29	log: 108	UL	514A
Submitter: Bradford D. Rupp				Submitter: Technical Correlating Committee				OSHA	-
NFPA 70B	-			NFPA 70E		-		NFPA 79	-

314.27 Outlet Boxes.

(A) Boxes at Luminaire or Lampholder Outlets. Outlet boxes or fittings designed for the support of luminaires and lampholders, and installed as required by 314.23, shall be permitted to support a luminaire or lampholder.

(2) Ceiling Outlets. At every outlet used exclusively for lighting, the box shall be designed or installed so that a luminaire or lampholder may be attached. Boxes shall be required to support a luminaire weighing a minimum of 23 kg (50 lb). A luminaire that weighs more than 23 kg (50 lb) shall be supported independently of the outlet box, unless the outlet box is listed and marked for the maximum weight to be supported.

Stallcup's Comment: A new subdivison has been added to clarify the box requirements for supporting a luminaire at ceiling outlets. Boxes shall be required to support a luminaire weighing a minimum of 50 lb (23 kg) or shall be supported independently of the outlet box if the luminaire weighs more than 50 lb (23 kg), unless the outlet box is listed and marked to support the maximum weight.

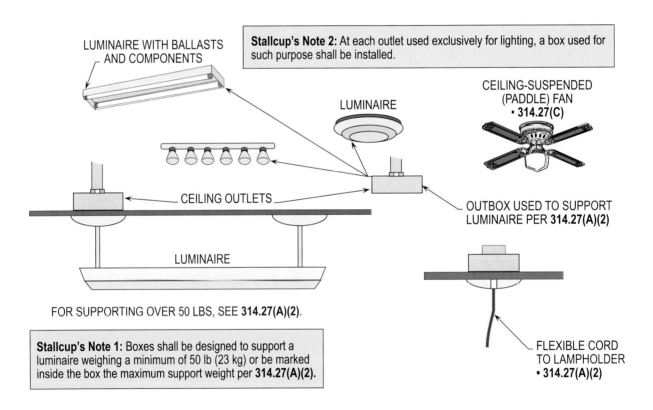

LUMINAIRE WITH BALLASTS AND COMPONENTS

Stallcup's Note 2: At each outlet used exclusively for lighting, a box used for such purpose shall be installed.

LUMINAIRE

CEILING-SUSPENDED (PADDLE) FAN • **314.27(C)**

CEILING OUTLETS

OUTBOX USED TO SUPPORT LUMINAIRE PER **314.27(A)(2)**

LUMINAIRE

FOR SUPPORTING OVER 50 LBS, SEE **314.27(A)(2)**.

Stallcup's Note 1: Boxes shall be designed to support a luminaire weighing a minimum of 50 lb (23 kg) or be marked inside the box the maximum support weight per **314.27(A)(2)**.

FLEXIBLE CORD TO LAMPHOLDER • **314.27(A)(2)**

CEILING OUTLETS
NEC 314.27(A)(2)

Purpose of Change: To address the requirements for ceiling outlets and supporting boxes.

Type of Change	New Paragraph			Committee Change	Accept			2008 NEC	314.27(C)
ROP	pg. 377	# 9-81	log: 1157	**ROC**	pg. 201	# 9-34	log: 1237	**UL**	514A
Submitter: Vince Baclawski				Submitter: Vince Baclawski				**OSHA**	-
NFPA 70B	-			**NFPA 70E**	-			**NFPA 70**	

314.27 Outlet Boxes.

(C) Boxes at Ceiling-Suspended (Paddle) Fan Outlets. Outlet boxes or outlet box systems used as the sole support of a ceiling-suspended (paddle) fan shall be listed, shall be marked by their manufacturer as suitable for this purpose, and shall not support ceiling-suspended (paddle) fans that weigh more than 32 kg (70 lb). For outlet boxes or outlet box systems designed to support ceiling-suspended (paddle) fans that weigh more than 16 kg (35 lb), the required marking shall include the maximum weight to be supported.

Where spare, separately switched, ungrounded conductors are provided to a ceiling mounted outlet box, in a location acceptable for a ceiling-suspended (paddle) fan in single or multi-family dwellings, the outlet box or outlet box system shall be listed for sole support of a ceiling-suspended (paddle) fan.

Stallcup's Comment: A new paragraph has been added to require an outlet box or outlet box system to be listed for the sole support of a ceiling-suspended (paddle) fan where installed in a location acceptable for a ceiling-suspended (paddle) fan in single- or multi-family dwellings.

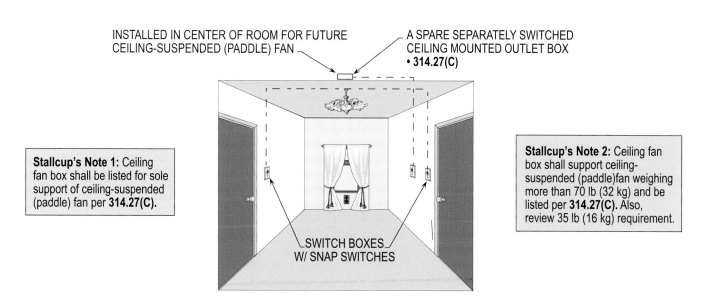

INSTALLED IN CENTER OF ROOM FOR FUTURE CEILING-SUSPENDED (PADDLE) FAN

A SPARE SEPARATELY SWITCHED CEILING MOUNTED OUTLET BOX
• 314.27(C)

Stallcup's Note 1: Ceiling fan box shall be listed for sole support of ceiling-suspended (paddle) fan per **314.27(C).**

Stallcup's Note 2: Ceiling fan box shall support ceiling-suspended (paddle)fan weighing more than 70 lb (32 kg) and be listed per **314.27(C).** Also, review 35 lb (16 kg) requirement.

SWITCH BOXES W/ SNAP SWITCHES

**BOXES AT CEILING-SUSPENDED (PADDLE) FAN OUTLETS
NEC 314.27(C)**

Purpose of Change: To provide the requirements for mounting a spare ceiling box for future installation of a ceiling-suspended (paddle) fan.

Type of Change	New Subsection		Committee Change	Accept in Principle		2008 NEC	-
ROP	pg. 379	# 9-87	log: 577	ROC	pg. 202	# 9-38	log: 1210	UL	1059
Submitter: Michael J. Johnston			Submitter: David H. Kendall			OSHA	-
NFPA 70B	-		NFPA 70E	-		NFPA 79	-

314.28 Pull and Junction Boxes and Conduit Bodies.

(E) Power Distribution Blocks. Power distribution blocks shall be permitted in pull and junction boxes over 1650 cm³ (100 in.³) for connections of conductors where installed in boxes and where the installation complies with (1) through (5).

Exception: Equipment grounding terminal bars shall be permitted in smaller enclosures.

(1) Installation. Power distribution blocks installed in boxes shall be listed.

(2) Size. In addition to the overall size requirement in the first sentence of 314.28(A)(2), the power distribution block shall be installed in a box with dimensions not smaller than specified in the installation instructions of the power distribution block.

(3) Wire Bending Space. Wire bending space at the terminals of power distribution blocks shall comply with 312.6.

(4) Live Parts. Power distribution blocks shall not have uninsulated live parts exposed within a box, whether or not the box cover is installed.

(5) Through Conductors. Where the pull or junction boxes are used for conductors that do not terminate on the power distribution block(s), the through conductors shall be arranged so the power distribution block terminals are unobstructed following installation.

Stallcup's Comment: A new subsection has been added to address the permitted uses for a power distribution block in pull and junction boxes.

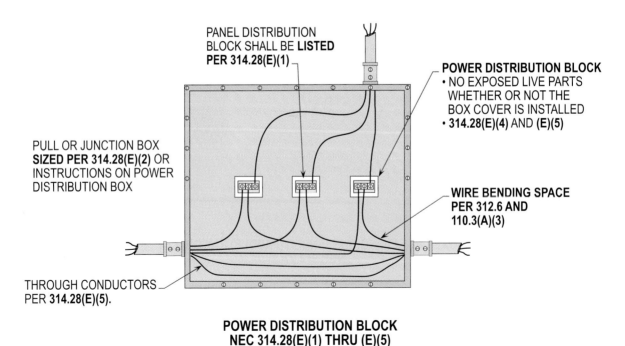

PANEL DISTRIBUTION BLOCK SHALL BE **LISTED** PER **314.28(E)(1)**

POWER DISTRIBUTION BLOCK
• NO EXPOSED LIVE PARTS WHETHER OR NOT THE BOX COVER IS INSTALLED
• **314.28(E)(4)** AND **(E)(5)**

PULL OR JUNCTION BOX **SIZED PER 314.28(E)(2)** OR INSTRUCTIONS ON POWER DISTRIBUTION BOX

WIRE BENDING SPACE PER **312.6 AND 110.3(A)(3)**

THROUGH CONDUCTORS PER **314.28(E)(5).**

**POWER DISTRIBUTION BLOCK
NEC 314.28(E)(1) THRU (E)(5)**

Purpose of Change: To provide requirements for the installation of a power distribution block in a pull or junction box.

Type of Change	New Subsection			Committee Change	Accept in Principle		2008 NEC	-
ROP	pg. 380	# 9-92	log: 3347	ROC	pg. -	# -	log: -	UL	-
Submitter: Dan Leaf				Submitter: -			OSHA	-
NFPA 70B	-			NFPA 70E	-		NFPA 79	-

314.70 General.

(B) Conduit Bodies. Where conduit bodies are used on systems over 600 volts, the installation shall comply with the provisions of Part IV and with the following general provisions of this article:

(1) Part I, 314.4

(2) Part II, 314.15; 314.17; 314.23(A), (E), or (G); and 314.29

(3) Part III, 314.40(A); and 314.41

Stallcup's Comment: A new subsection has been added to address the requirements of conduit bodies for pull and junction boxes for use on systems over 600 volts, nominal.

Stallcup's Note 1: Conduit bodies used with systems over 600 volts shall comply with **314.4, 314.15, 314.17, 314.23(A), (E), or (G), 314.29** as well as with **314.40(A)** and **314.41.**

Stallcup's Note 2: Shielding shall be grounded at each cable joint if exposed to personnel contact per **250.184(C)(5).**

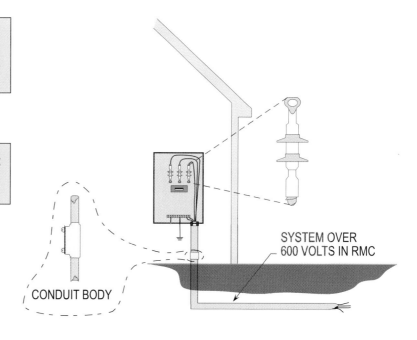

SYSTEM OVER
600 VOLTS IN RMC

CONDUIT BODY

CONDUIT BODIES
NEC 314.70(B)(1) THRU (B)(3)

Purpose of Change: To address the requirements and sections for installing conduit bodies in systems over 600 volts.

Type of Change	New Subsection			Committee Change	-				2008 NEC	-
ROP	pg. 380	# 9-92	log: 3347	ROC	pg. -		# -	log: -	UL	-
Submitter: Dan Leaf				Submitter: -					OSHA	-
NFPA 70B	-			NFPA 70E	-				NFPA 79	-

314.70 General.

(C) Handhole Enclosures. Where handhole enclosures are used on systems over 600 volts, the installation shall comply with the provisions of Part IV and with the following general provisions of this article:

(1) Part I, 314.3; and 314.4

(2) Part II, 314.15; 314.17; 314.23(G); 314.28(B); 314.29; and 314.30

Stallcup's Comment: A new subsection has been added to address the requirements of handhole enclosure for pull and junction boxes for use on systems over 600 volts, nominal.

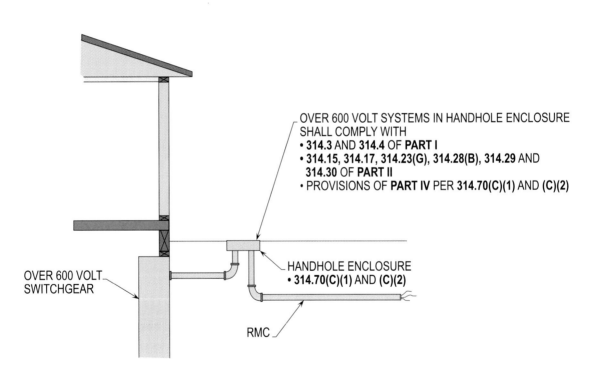

HANDHOLE ENCLOSURES
NEC 314.70(C)(1) AND (C)(2)

Purpose of Change: To address the use of handholes in systems over 600 volts.

Type of Change	Revision			Committee Change	Accept in Principle			2008 NEC	334.10(1)
ROP	pg. 391	# 7-77	log: 3789	ROC	pg. 208	# 7-26	log: 1859	UL	719
Submitter: James Grant				Submitter: Phil Simmons				OSHA	-
NFPA 70B	-			NFPA 70E	-			NFPA 79	-

334.10 Uses Permitted.

Type NM, Type NMC, and Type NMS cables shall be permitted to be used in the following:

(1) One- and two-family dwellings and their attached or detached garages, and their storage buildings.

Stallcup's Comment: A revision has been made to clarify that nonmetallic-sheathed cable (Types NM, NMC, and NMS) shall be permitted in one- and two-family dwellings, their attached or detached garages, and their storage buildings.

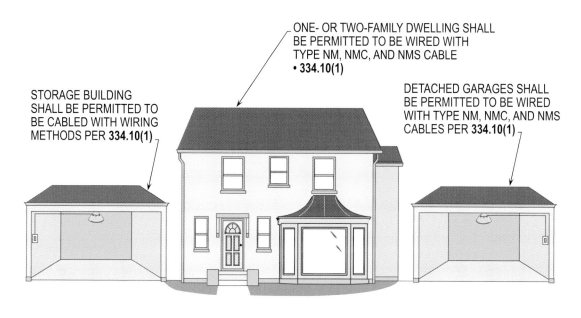

STORAGE BUILDING SHALL BE PERMITTED TO BE CABLED WITH WIRING METHODS PER **334.10(1)**

ONE- OR TWO-FAMILY DWELLING SHALL BE PERMITTED TO BE WIRED WITH TYPE NM, NMC, AND NMS CABLE
• **334.10(1)**

DETACHED GARAGES SHALL BE PERMITTED TO BE WIRED WITH TYPE NM, NMC, AND NMS CABLES PER **334.10(1)**

**USES PERMITTED
NEC 334.10(1)**

Purpose of Change: To permit the use of nonmetallic-sheathed cable, such as Type NM, NMC, and NMS in one- and two-family dwellings, their attached or detached garages, and their storage buildings.

Type of Change	New Subdivision			Committee Change	Accept			2008 NEC	334.12(A), Ex.
ROP	pg. 392	# 7-79	log: 1697	ROC	pg. 208	# 7-30	log: 112	UL	719
Submitter: Mike Theisen				Submitter: Technical Correlating Committee				OSHA	-
NFPA 70B	-			NFPA 70E	-			NFPA 79	-

334.10 Uses Permitted.

Type NM, Type NMC, and Type NMS cables shall be permitted to be used in the following:

(5) Types I and II construction where installed within raceways permitted to be installed in Types I and II construction.

Stallcup's Comment: A new subdivison has been added to place Types I and II construction into positive language instead of being used as an exception.

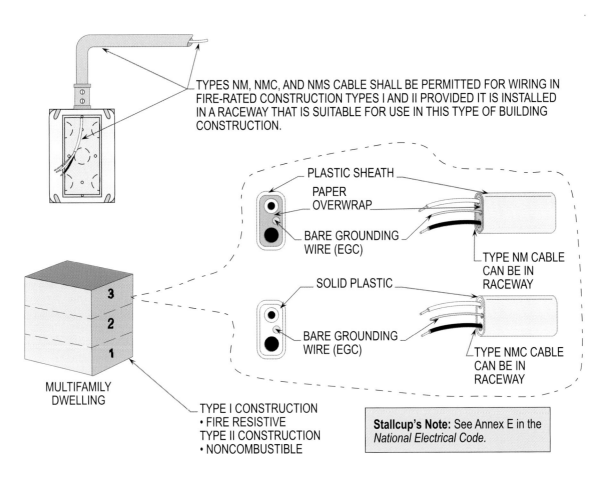

TYPES NM, NMC, AND NMS CABLE SHALL BE PERMITTED FOR WIRING IN FIRE-RATED CONSTRUCTION TYPES I AND II PROVIDED IT IS INSTALLED IN A RACEWAY THAT IS SUITABLE FOR USE IN THIS TYPE OF BUILDING CONSTRUCTION.

PLASTIC SHEATH

PAPER OVERWRAP

BARE GROUNDING WIRE (EGC)

TYPE NM CABLE CAN BE IN RACEWAY

SOLID PLASTIC

BARE GROUNDING WIRE (EGC)

TYPE NMC CABLE CAN BE IN RACEWAY

MULTIFAMILY DWELLING

TYPE I CONSTRUCTION
• FIRE RESISTIVE
TYPE II CONSTRUCTION
• NONCOMBUSTIBLE

Stallcup's Note: See Annex E in the *National Electrical Code.*

USES PERMITTED
NEC 334.10(5)

Purpose of Change: Revised to permit Types NM, NMC, and NMS cable to be installed in Type I and II construction when installed within raceways permitted in that type of construction.

Type of Change	New Paragraph			Committee Change	Accept in Principle			2008 NEC	338.10(B)(4)(a)
ROP	pg. -	# 7-133	log: 451	ROC	pg. -	# 7-48	log: 1824	UL	854
Submitter: Richard W. Likes				Submitter: Richard W. Likes				OSHA	-
NFPA 70B	-			NFPA 70E	-			NFPA 79	-

338.10 Uses Permitted.

(B) Branch Circuits or Feeders.

(4) Installation Methods for Branch Circuits and Feeders.

(a) *Interior Installations*. In addition to the provisions of this article, Type SE service-entrance cable used for interior wiring shall comply with the installation requirements of Part II of Article 334, excluding 334.80

Where installed in thermal insulation, the ampacity shall be in accordance with the 60°C (140°F) conductor temperature rating. The maximum conductor temperature rating shall be permitted to be used for ampacity adjustment and correction purposes, if the final derated ampacity does not exceed that for a 60°C (140°F) rated conductor.

Stallcup's Comment: A new paragraph has been added to address concerns that service-entrance cable (Type SE) is listed to 75°C. The ampacity of the Type SE cable shall be permitted to be decreased if installed in thermal insulation.

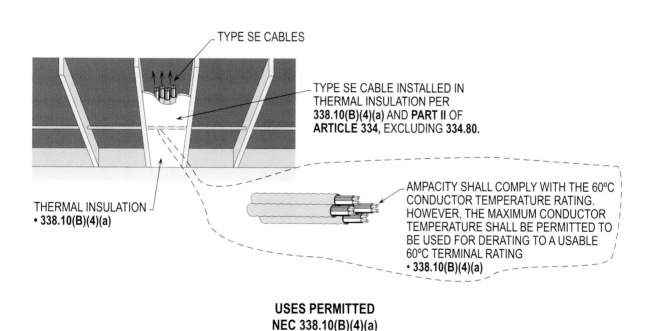

TYPE SE CABLES

TYPE SE CABLE INSTALLED IN THERMAL INSULATION PER **338.10(B)(4)(a)** AND **PART II** OF **ARTICLE 334**, EXCLUDING **334.80**.

AMPACITY SHALL COMPLY WITH THE 60°C CONDUCTOR TEMPERATURE RATING. HOWEVER, THE MAXIMUM CONDUCTOR TEMPERATURE SHALL BE PERMITTED TO BE USED FOR DERATING TO A USABLE 60°C TERMINAL RATING
• **338.10(B)(4)(a)**

THERMAL INSULATION
• **338.10(B)(4)(a)**

USES PERMITTED
NEC 338.10(B)(4)(a)

Purpose of Change: To clarify the derating procedure for determining ampacity of conductors in Type SE cable installed in thermal insulation.

Type of Change	Revision			Committee Change	Accept in Principle			2008 NEC	342.30(A)
ROP	pg. 403	# 8-17	log: 4745	ROC	pg. -	# -	log: -	UL	1242
Submitter: James M. Imlah				Submitter: -				OSHA	-
NFPA 70B	-			NFPA 70E	-			NFPA 79	-

342.30 Securing and Supporting.

(A) Securely Fastened. IMC shall be secured in accordance with one of the following:

(1) IMC shall be securely fastened within 900 mm (3 ft) of each outlet box, junction box, device box, cabinet, conduit body, or other conduit termination.

(2) Where structural members do not readily permit fastening within 900 mm (3 ft), fastening shall be permitted to be increased to a distance of 1.5 m (5 ft).

(3) Where approved, conduit shall not be required to be securely fastened within 900 mm (3 ft) of the service head for above-the-roof termination of a mast.

Stallcup's Comment: A revision has been made to clarify the requirements for securely fastening intermediate metal conduit (Type IMC) in a list format.

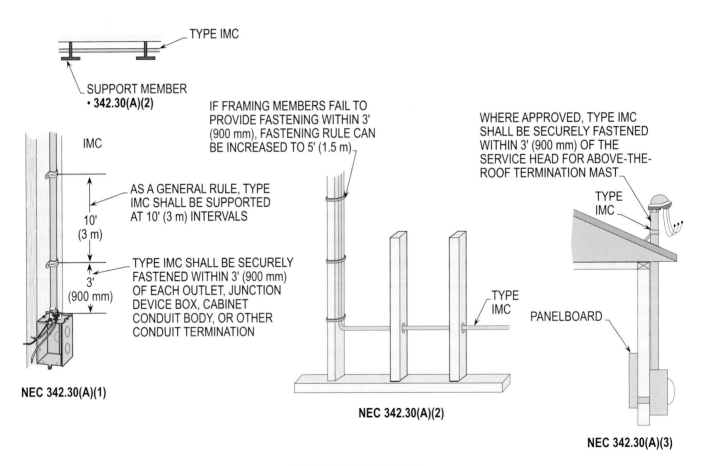

TYPE IMC

SUPPORT MEMBER
• 342.30(A)(2)

IMC

IF FRAMING MEMBERS FAIL TO PROVIDE FASTENING WITHIN 3' (900 mm), FASTENING RULE CAN BE INCREASED TO 5' (1.5 m)

WHERE APPROVED, TYPE IMC SHALL BE SECURELY FASTENED WITHIN 3' (900 mm) OF THE SERVICE HEAD FOR ABOVE-THE-ROOF TERMINATION MAST

AS A GENERAL RULE, TYPE IMC SHALL BE SUPPORTED AT 10' (3 m) INTERVALS

10' (3 m)

3' (900 mm)

TYPE IMC SHALL BE SECURELY FASTENED WITHIN 3' (900 mm) OF EACH OUTLET, JUNCTION DEVICE BOX, CABINET CONDUIT BODY, OR OTHER CONDUIT TERMINATION

TYPE IMC

TYPE IMC

PANELBOARD

NEC 342.30(A)(1)

NEC 342.30(A)(2)

NEC 342.30(A)(3)

**SECURELY FASTENED
NEC 342.30(A)(1) THRU (A)(3)**

Purpose of Change: To clarify the requirements for supporting and securely fastening Type IMC.

Type of Change	Revision			Committee Change	Accept			2008 NEC	348.60
ROP	pg. 408	# 8-51	log: 2345	ROC	pg. 216	# 8-21	log: 1233	UL	1
Submitter: David Mercier				Submitter: Vince Baclawski				OSHA	-
NFPA 70B	-			NFPA 70E	-			NFPA 79	13.5.4

348.60 Grounding and Bonding.

If used to connect equipment where flexibility is necessary to minimize the transmission of vibration from equipment or to provide flexibility for equipment that requires movement after installation, an equipment grounding conductor shall be installed.

Stallcup's Comment: A revision has been made to correlate with **250.118(5)(d)** to clarify the grounding and bonding requirements for flexible metal conduit (Type FMC) where flexibility is necessary to minimize the transmission of vibration from equipment or to provide flexibility for equipment that requires movement. A revision has also been made **350.60** with the same language for liquidtight flexible metal conduit (Type LFMC).

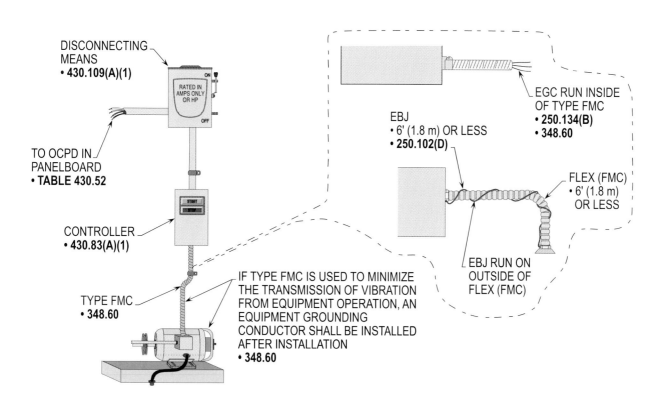

GROUNDING AND BONDING
NEC 348.60

Purpose of Change: To require an equipment grounding conductor for flexible metal conduit when connected to equipment that requires flexible metal conduit to reduce noise problems. The same revision was made to **350.60** for liquidtight flexible metal conduit (Type LFMC).

Type of Change	New Subsection			Committee Change	Accept			2008 NEC	-
ROP	pg. 410	# 8-67a	log: CP 800	ROC	pg. -	# -	log: -	UL	651
Submitter: Code-Making Panel 8				Submitter: -				OSHA	-
NFPA 70B	-			NFPA 70E	-			NFPA 79	-

352.10 Uses Permitted.

The use of PVC conduit shall be permitted in accordance with 352.10(A) through (I).

(I) Insulation Temperature Limitations. Conductors or cables rated at a temperature higher than the listed temperature rating of PVC conduit shall be permitted to be installed in PVC conduit, provided the conductors or cables are not operated at a temperature higher than the listed temperature rating of the PVC conduit.

Stallcup's Comment: A new subsection has been relocated from **352.12(E)** and revised to require that conductors marked with a rated temperature higher than that of the raceway shall be permitted to be used when the conductors are operated within the raceway temperature rating.

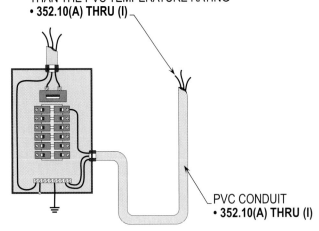

> **Stallcup's Note:** Users of PVC conduit must review **352.10(A) through (I)**.

CONDUCTORS OR CABLES WITH HIGHER TEMPERATURE RATINGS
SHALL BE PERMITTED TO BE ROUTED THROUGH PVC CONDUIT IF
THEY ARE NOT OPERATED AT A LISTED TEMPERATURE HIGHER
THAN THE PVC TEMPERATURE RATING
• **352.10(A) THRU (I)**

PVC CONDUIT
• **352.10(A) THRU (I)**

INSULATION TEMPERATURE LIMITATIONS
NEC 352.10(I)

Purpose of Change: To clarify temperature ratings and the use of conductors and PVC conduit systems.

Type of Change	New Section			Committee Change	Accept in Principle			2008 NEC	-
ROP	pg. 444	# 8-212	log: 2866	ROC	pg. -	# -	log: -	UL	83
Submitter: James M. Imlah				Submitter: -				OSHA	-
NFPA 70B	-			NFPA 70E	-			NFPA 79	-

380.23 Insulated Conductors.

For field assembled multioutlet assemblies, insulated conductors shall comply with 380.23(A) and (B).

(A) Deflected Insulated Conductors. Where insulated conductors are deflected within a multioutlet assembly, either at the ends or where conduits, fittings, or other raceways or cables enter or leave the multioutlet assembly, or where the direction of the multioutlet assembly is deflected greater than 30 degrees, dimensions corresponding to one wire per terminal in Table 312.6(A) shall apply.

(B) Multioutlet Assemblies Used as Pull Boxes. Where insulated conductors 4 AWG or larger are pulled through a multioutlet assembly, the distance between raceway and cable entries enclosing the same conductor shall not be less than that required by 314.28(A)(1) for straight pulls and 314.28(A)(2) for angle pulls. When transposing cable size into raceway size, the minimum metric designator (trade size) raceway required for the number and size of conductors in the cable shall be used.

Stallcup's Comment: A new section has been added to address the requirements for field assembled multioutlet assemblies.

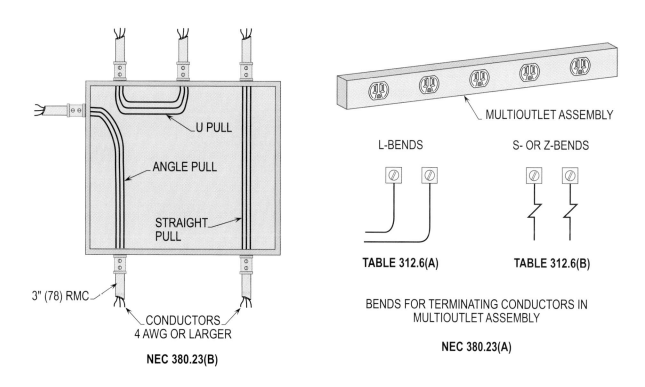

INSULATED CONDUCTORS
NEC 380.23(A) AND (B)

Purpose of Change: To provide requirements for installation of field assembled multioutlet assemblies.

Type of Change	Renumbered			Committee Change	Accept		2008 NEC	392.1 thru 392.13
ROP	pg. 458	# 8-235a	log: CP 804	ROC	pg. 231	# 8-100a	log: CC 800	UL	568
Submitter: Code-Making Panel 8				Submitter: Code-Making Panel 8			OSHA	1910.305(a)(3)
NFPA 70B	20.3			NFPA 70E	215.3		NFPA 79	13.5.10

Article 392 Cable Trays

Part I General

Part II Installation

Part III Construction Specifications

Stallcup's Comment: Article 392 has been renumbered as **392.1 through 392.100** and placed into three parts for general requirements, installation requirements, and construction specifications.

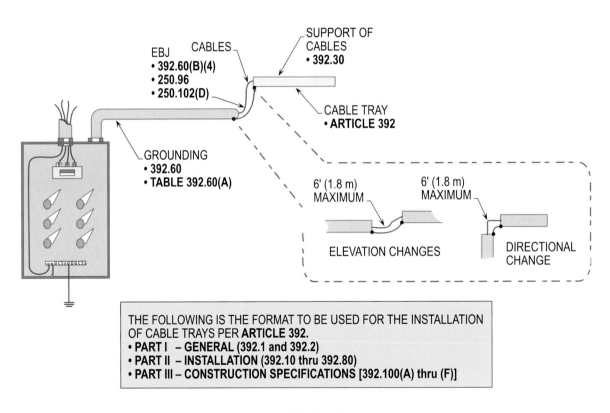

CABLE TRAYS
ARTICLE 392

Purpose of Change: To renumber sections for easy reading and better application as well as dividing into three parts as other Articles of **Chapter 3** appear.

Type of Change	New Article			Committee Change	Accept			2008 NEC	-
ROP	pg. 491	# 7-162	log: 680	ROC	pg. 257	# 7-63	log: 125	UL	-
Submitter: Technical Correlating Committee				Submitter: Technical Correlating Committee				OSHA	1910.333(c)(3)
NFPA 70B	-			NFPA 70E	130.5			NFPA 79	-

Article 399 Outdoor Overhead Conductors over 600 Volts.

399.1 Scope.

399.2 Definitions.

399.10 Uses Permitted.

399.12 Uses Not Permitted.

399.30 Support.

Stallcup's Comment: A new **Article 399** has been accepted and numbered as **399.1 through 399.30** to address outdoor overhead conductors over 600 volts.

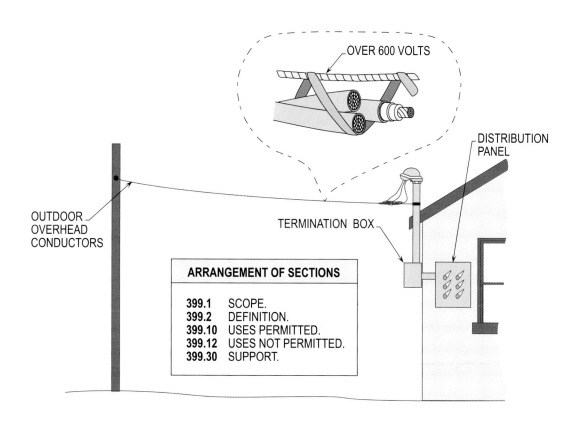

OUTDOOR OVERHEAD CONDUCTORS OVER 600 VOLTS
ARTICLE 399

Purpose of Change: To provide a new article for outdoor overhead conductors rated at over 600 volts.

Name   Date

Chapter 3
Wiring Methods and Materials

Section   Answer

1. A cable, raceway, or box, installed in exposed or concealed locations under metal-corrugated sheet roof decking, shall be installed and supported so there is not less than _____ in. measured from the lowest surface of the roof decking to the top of the cable, raceway, or box.

(A) 1   **(B)** 1-1/4
(C) 1-1/2   **(D)** 2

2. The interior of enclosures or raceways installed underground shall be considered to be a:

(A) Damp location   **(B)** Dry location
(C) Wet location   **(D)** Damp/wet location

3. Nonshielded, ozone-resistant insulated conductors with a maximum phase-to-phase voltage of _____ volts shall be permitted in Type MC clad cables in industrial establishments where the conditions of maintenance and supervision ensure that only qualified persons service the installation.

(A) 1000   **(B)** 2400
(C) 4800   **(D)** 5000

4. Where parallel equipment bonding jumpers are installed in raceways, they shall be sized and installed in accordance with _____.

(A) 250.66   **(B)** 250.102
(C) 250.106   **(D)** 250.122

5. The _____ factors shall be permitted to be applied to the ampacity for the temperature rating of the conductor, if the corrected and adjusted ampacity does not exceed the ampacity for the temperature rating of the termination in accordance with the provisions of **110.14(C)**.

(A) adjustment   **(B)** correction
(C) both (A) and (B)   **(D)** neither (A) nor (B)

6. The wiring space of conductors for switches or overcurrent devices shall be permitted to be the total of all conductors installed at any cross-section of the wiring space, provided it does not exceed _____ percent of the cross-sectional area of that space.

(A) 40   **(B)** 50
(C) 70   **(D)** 75

7. The wiring space of conductors for switches and overcurrent device shall be permitted to be the total area of all conductors, splices, and taps installed at any cross-section of the wiring space, provided it does not exceed _____ percent of the cross-sectional area of that space.

(A) 40   **(B)** 50
(C) 70   **(D)** 75

Section   Answer

8. Conduit bodies such as capped elbows and service-entrance elbows that enclose conductors _____ AWG or smaller, and are only intended to enable the installation of the raceway and the contained conductors, shall not contain splices, taps, or devices and shall be of sufficient size to provide free space for all conductors enclosed in the conduit body.

(A) 6   (B) 4
(C) 2   (D) 1

9. Boxes used at luminaire or lampholder outlets in a wall shall be marked on the interior of the box to indicate the maximum weight of the luminaire that is permitted to be supported by the box in the wall, if other than _____ lb.

(A) 15   (B) 30
(C) 50   (D) 100

10. Power distribution blocks shall be permitted in pull and junction boxes over _____ for connections of conductors where installed in boxes.

(A) 50 in.^{3} (B) 100 in.^{3}
(C) 150 in.^{3} (D) 200 in.^{3}

11. Where branch circuits and feeders are installed in thermal insulation, the ampacity shall be in accordance with the _____°C conductor temperature rating.

(A) 40   (B) 60
(C) 75   (D) 90

12. Intermediate metal conduit shall be securely fastened within _____ ft of each outlet box, junction box, device box, cabinet, conduit body, or other conduit termination.

(A) 3   (B) 5
(C) 6   (D) 10

13. Where structural members do not readily permit fastening of intermediate metal conduit within 3 ft, fastening shall be permitted to be increased to a distance of _____ ft.

(A) 3   (B) 5
(C) 6   (D) 10

14. Where insulated conductors are deflected within a multioutlet assembly, either at the ends, or where conduits, fittings, or other raceways or cables enter or leave the multioutlet assembly, or where the direction of the multioutlet assembly is deflected greater than _____ degrees, dimensions corresponding to one wire per terminal in Table 312.6(A) shall apply.

(A) 10   (B) 20
(C) 25   (D) 30

15. Where insulated conductors _____ AWG or larger are pulled through a multioutlet assembly, the distance between raceway and cable entries enclosing the same conductor shall not be less than that required for straight pulls and for angle pulls.

(A) 10   (B) 8
(C) 6   (D) 4

Equipment for General Use

Chapter 4 of the *National Electrical Code* has always been utilized by users and maintainers who have the responsibility of installing general-use electrical equipment, luminaires, motors, and similar equipment.

If an electrician or user is replacing a ballast in a fluorescent luminaire, **410.130** must be reviewed to ensure the correct ballast and installation procedure are used. Consider a U frame motor that has been replaced with a T frame motor. In order to verify that the overloads are the proper size, **430.32(A)(1)** must be used. Switches are covered in **Article 404**, receptacles in **Article 406**, and panelboards or switchboards in **Article 408**.

Electricians must not get confused when installing specialized equipment not included in **Chapter 4**. **Chapter 4** applies to general-use electrical equipment and **Chapter 6** applies to special electrical equipment.

> **For example,** where installing a crane or a hoist, electricians must reference **Article 610**, which covers special equipment. For installing a elevator, **Article 620** must be referenced, not **Chapter 4**.

Article 445 applies to the installation of generators. If generators greater than 600 volts are being installed, **Article 490** covers general equipment operating at over 600 volts.

Type of Change	New Subsection			Committee Change	Accept in Principle			2008 NEC	-
ROP	pg. 502	# 9-95	log: 1160	ROC	pg. 260	# 9-45	log: 1042	UL	20
Submitter: Vince Baclawski				Submitter: James W. Carpenter				OSHA	-
NFPA 70B	24.5			NFPA 70E	-			NFPA 79	-

404.2 Switch Connections.

(C) Switches Controlling Lighting Units. Where switches control lighting loads supplied by a grounded general purpose branch circuit, the grounded circuit conductor for the controlled lighting circuit shall be provided at the switch location.

Stallcup's Comment: A new subsection and exception has been added to address the requirements for the grounded (neutral) circuit conductor for switches controlling lighting loads.

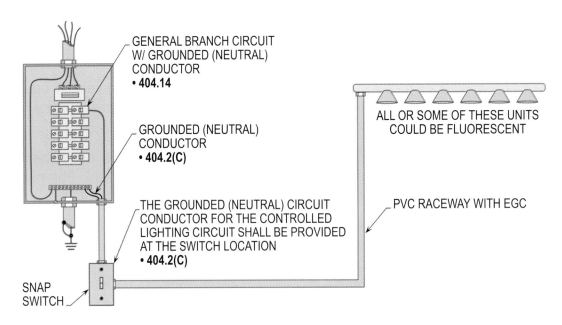

GENERAL BRANCH CIRCUIT
W/ GROUNDED (NEUTRAL)
CONDUCTOR
• 404.14

GROUNDED (NEUTRAL)
CONDUCTOR
• 404.2(C)

THE GROUNDED (NEUTRAL) CIRCUIT
CONDUCTOR FOR THE CONTROLLED
LIGHTING CIRCUIT SHALL BE PROVIDED
AT THE SWITCH LOCATION
• 404.2(C)

SNAP
SWITCH

ALL OR SOME OF THESE UNITS
COULD BE FLUORESCENT

PVC RACEWAY WITH EGC

SWITCHES CONTROLLING LIGHTING UNITS
NEC 404.2(C)

Purpose of Change: To outline requirements for the grounded circuit conductor (may be a neutral) when controlling lighting units.

Type of Change	New Subsection			Committee Change		Accept in Principle		2008 NEC	-
ROP	pg. 502	# 9-95	log: 1160	ROC	pg. 260	# 9-45	log: 1042	UL	20
Submitter: Vince Baclawski				Submitter: James W. Carpenter				OSHA	-
NFPA 70B	24.5			NFPA 70E	-			NFPA 79	-

404.2 Switch Connections.

(C) Switches Controlling Lighting Units. Where switches control lighting loads supplied by a grounded general purpose branch circuit, the grounded circuit conductor for the controlled lighting circuit shall be provided at the switch location.

Exception: The grounded circuit conductor shall be permitted to be omitted from the switch enclosure where either of the following conditions in (1) or (2) apply:

(1) Conductors for switches controlling lighting loads enter the box through a raceway. The raceway shall have sufficient cross-sectional area to accommodate the extension of the grounded circuit conductor of the lighting circuit to the switch location whether or not the conductors in the raceway are required to be increased in size to comply with 310.15(B)(3)(a).

(2) Cable assemblies for switches controlling lighting loads enter the box through a framing cavity that is open at the top or bottom on the same floor level, or through a wall, floor, or ceiling that is unfinished on one side.

Informational Note: The provision for a (future) grounded conductor is to complete a circuit path for electronic lighting control devices.

Stallcup's Comment: A new subsection and exception has been added to address the requirements for the grounded circuit conductor for switches controlling lighting loads.

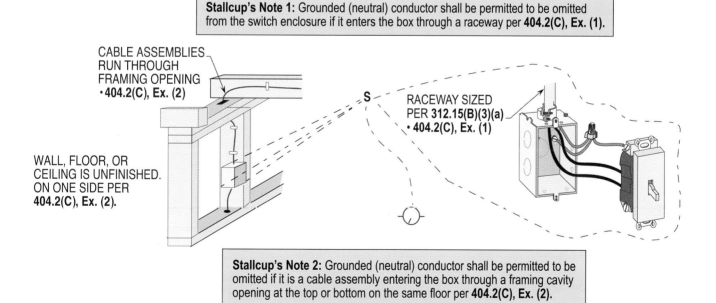

Stallcup's Note 1: Grounded (neutral) conductor shall be permitted to be omitted from the switch enclosure if it enters the box through a raceway per **404.2(C), Ex. (1).**

CABLE ASSEMBLIES
RUN THROUGH
FRAMING OPENING
• **404.2(C), Ex. (2)**

RACEWAY SIZED
PER **312.15(B)(3)(a)**
• **404.2(C), Ex. (1)**

WALL, FLOOR, OR
CEILING IS UNFINISHED.
ON ONE SIDE PER
404.2(C), Ex. (2).

Stallcup's Note 2: Grounded (neutral) conductor shall be permitted to be omitted if it is a cable assembly entering the box through a framing cavity opening at the top or bottom on the same floor per **404.2(C), Ex. (2).**

SWITCHES CONTROLLING LIGHTING UNITS
NEC 404.2(C), Ex. (1) AND (2)

Purpose of Change: To provide requirements for using the grounded (neutral) conductor in installations pertaining to switching of lighting units. *Note, cable assemblies can be routed through a wall, floor, or ceiling that is unfinished on one side.*

Type of Change	Revision			Committee Change	Accept in Principle		2008 NEC	404.4(E), Ex. (B)
ROP pg. 504		# 9-108	log: 2581	ROC pg. -	# -	log: -	UL	20
Submitter: Dan Leaf				Submitter: -			OSHA	1910.305(c)(5)
NFPA 70B 24.5				NFPA 70E -			NFPA 79	8.2.4

404.9 Provisions for General-Use Snap Switches.

(B) Grounding. Snap switches, including dimmer and similar control devices, shall be connected to an equipment grounding conductor and shall provide a means to connect metal faceplates to the equipment grounding conductor, whether or not a metal faceplate is installed. Snap switches shall be considered to be part of an effective ground-fault current path if either of the following conditions is met:

(1) The switch is mounted with metal screws to a metal box or metal cover that is connected to an equipment grounding conductor or to a nonmetallic box with integral means for connecting to an equipment grounding conductor.

(2) An equipment grounding conductor or equipment bonding jumper is connected to an equipment grounding termination of the snap switch.

Exception No. 1 to (B): Where no means exists within the snap-switch enclosure for connecting to the equipment grounding conductor, or where the wiring method does not include or provide an equipment grounding conductor, a snap switch without a connection to an equipment grounding conductor shall be permitted for replacement purposes only. A snap switch wired under the provisions of this exception and located within 2.5 m (8 ft) vertically, or 1.5 m (5 ft) horizontally, of ground or exposed grounded metal objects shall be provided with a faceplate of nonconducting noncombustible material with nonmetallic attachment screws, unless the switch mounting strap or yoke is nonmetallic or the circuit is protected by a ground-fault circuit interrupter.

Stallcup's Comment: A revision has been made to the exception to clarify the requirements for a snap switch that is wired and located within 8 ft (2.5 m) or 5 ft (1.5 m) of grounded or exposed grounded metal objects.

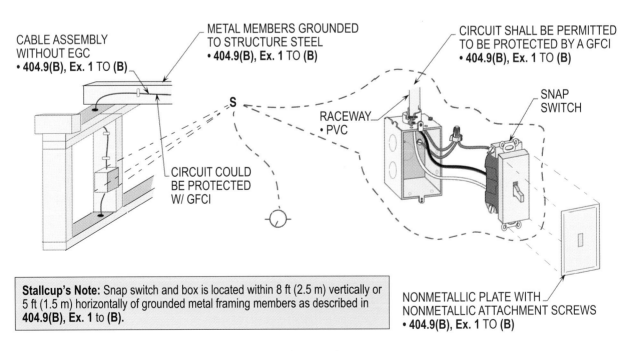

CABLE ASSEMBLY WITHOUT EGC
• **404.9(B), Ex. 1 TO (B)**

METAL MEMBERS GROUNDED TO STRUCTURE STEEL
• **404.9(B), Ex. 1 TO (B)**

CIRCUIT SHALL BE PERMITTED TO BE PROTECTED BY A GFCI
• **404.9(B), Ex. 1 TO (B)**

SNAP SWITCH

RACEWAY
• PVC

CIRCUIT COULD BE PROTECTED W/ GFCI

Stallcup's Note: Snap switch and box is located within 8 ft (2.5 m) vertically or 5 ft (1.5 m) horizontally of grounded metal framing members as described in **404.9(B), Ex. 1 to (B).**

NONMETALLIC PLATE WITH NONMETALLIC ATTACHMENT SCREWS
• **404.9(B), Ex. 1 TO (B)**

GROUNDING
NEC 404.9(B), Ex. 1 TO (B)

Purpose of Change: To provide requirements for snap switches located within specified locations of grounded objects.

Type of Change	New Exception			Committee Change	Accept			2008 NEC	-
ROP	pg. 505	# 9-110	log: 245	ROC	pg. 261	# 9-50	log: 1568	UL	20
Submitter: Douglas R. Burrell				Submitter: Robert D. Osborne				OSHA	1910.305(c)(5)
NFPA 70B	24.5			NFPA 70E	-			NFPA 79	8.2.4

404.9 Provisions for General-Use Snap Switches.

(B) Grounding. Snap switches, including dimmer and similar control devices, shall be connected to an equipment grounding conductor and shall provide a means to connect metal faceplates to the equipment grounding conductor, whether or not a metal faceplate is installed. Snap switches shall be considered to be part of an effective ground-fault current path if either of the following conditions is met:

(1) The switch is mounted with metal screws to a metal box or metal cover that is connected to an equipment grounding conductor or to a nonmetallic box with integral means for connecting to an equipment grounding conductor.

(2) An equipment grounding conductor or equipment bonding jumper is connected to an equipment grounding termination of the snap switch.

Exception No. 2 to (B): Listed kits or listed assemblies shall not be required to be connected to an equipment grounding conductor if all of the following conditions are met:

(1) The device is provided with a nonmetallic faceplate that cannot be installed on any other type of device,

(2) The device does not have mounting means to accept other configurations of faceplates,

(3) The device is equipped with a nonmetallic yoke, and

(4) All parts of the device that are accessible after installation of the faceplate are manufactured of nometallic materials.

Stallcup's Comment: A new exception had been added to clarify that listed kits or listed assemblies shall not be required to be connected to an equipment grounding conductor if certain conditions are met.

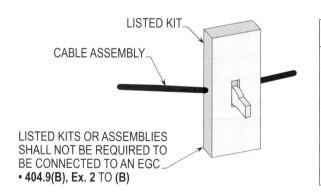

LISTED KIT

CABLE ASSEMBLY

LISTED KITS OR ASSEMBLIES
SHALL NOT BE REQUIRED TO
BE CONNECTED TO AN EGC
• 404.9(B), Ex. 2 TO (B)

REQUIREMENTS OF 404.9(B), Ex. 2 TO (B)
(1) NONMETALLIC FACEPLATE CANNOT BE CONNECTED TO ANY OTHER TYPE OF DEVICE
(2) DEVICE WILL NOT ACCEPT OTHER CONFIGURATIONS OF FACEPLATES
(3) DEVICE HAS NONMETALLIC YOKE, AND
(4) PARTS OF DEVICE ACCESSIBLE AFTER INSTALLATION OF FACEPLATE ARE OF NONMETALLIC MATERIALS

GROUNDING
NEC 404.9(B), Ex. 2 TO (B)

Purpose of Change: To clarify that listed kits or listed assemblies shall not be required to be connected to an equipment grounding conductor if certain conditions are met.

Type of Change	New Exception			Committee Change	Accept in Principle			2008 NEC	-
ROP	pg. 505	# 9-111	log: 725	ROC	pg. -	# -	log: -	UL	20
Submitter: Brian E. Rock				Submitter: -				OSHA	1910.305(c)(5)
NFPA 70B	24.5			NFPA 70E	-			NFPA 79	8.2.4

404.9 Provisions for General-Use Snap Switches.

(B) Grounding. Snap switches, including dimmer and similar control devices, shall be connected to an equipment grounding conductor and shall provide a means to connect metal faceplates to the equipment grounding conductor, whether or not a metal faceplate is installed. Snap switches shall be considered to be part of an effective ground-fault current path if either of the following conditions is met:

(1) The switch is mounted with metal screws to a metal box or metal cover that is connected to an equipment grounding conductor or to a nonmetallic box with integral means for connecting to an equipment grounding conductor.

(2) An equipment grounding conductor or equipment bonding jumper is connected to an equipment grounding termination of the snap switch.

Exception No. 3 to (B): A snap switch with integral nonmetallic enclosure complying with 300.15(E) shall be permitted without a connection to an equipment grounding conductor.

Stallcup's Comment: A new exception has been added to clarify that snap switches with integral nonmetallic enclosure shall be permitted without connection to an equipment grounding conductor.

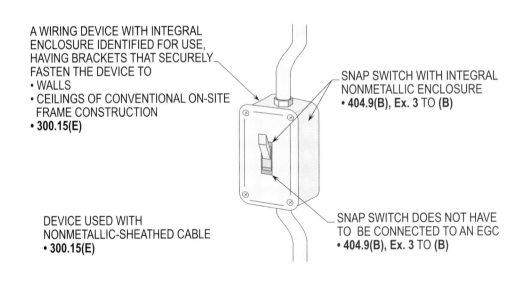

A WIRING DEVICE WITH INTEGRAL
ENCLOSURE IDENTIFIED FOR USE,
HAVING BRACKETS THAT SECURELY
FASTEN THE DEVICE TO
• WALLS
• CEILINGS OF CONVENTIONAL ON-SITE
  FRAME CONSTRUCTION
• **300.15(E)**

SNAP SWITCH WITH INTEGRAL
NONMETALLIC ENCLOSURE
• **404.9(B), Ex. 3 TO (B)**

DEVICE USED WITH
NONMETALLIC-SHEATHED CABLE
• **300.15(E)**

SNAP SWITCH DOES NOT HAVE
TO BE CONNECTED TO AN EGC
• **404.9(B), Ex. 3 TO (B)**

GROUNDING
NEC 404.9(B), Ex. 3 TO (B)

Purpose of Change: To make it clear that snap switches with integral nonmetallic enclosure shall not be required to be connected to an equipment grounding conductor.

Type of Change	New Subsection and Ex.		Committee Change	Accept			2008 NEC	-
ROP	pg. 507	# 9-118	log: 726	ROC	pg. 262	# 9-52	log: 2595	UL	20
Submitter: Brian E. Rock				Submitter: Frederic P. Hartwell			OSHA	-
NFPA 70B	24.5			NFPA 70E	-		NFPA 79	-

404.14 Rating and Use of Snap Switches.

(F) Cord-and-Plug-Connected Loads. Where a snap switch is used to control cord-and-plug-connected equipment on a general-purpose branch circuit, each snap switch controlling receptacle outlets or cord connectors that are supplied by permanently connected cord pendants shall be rated at not less than the rating of the maximum permitted ampere rating or setting of the overcurrent device protecting the receptacles or cord connectors, as provided in 210.21(B).

Informational Note: See 210.50(A) and 400.7(A)(1) for equivalency to a receptacle outlet of a cord connector that is supplied by a permanently connected cord pendant.

Exception: Where a snap switch is used to control not more than one receptacle on a branch circuit, the switch shall be permitted to be rated at not less than the rating of the receptacle.

Stallcup's Comment: A new subection has been added to address the requirements for a snap switch used to control cord-and-plug-connected equipment on a general-purpose branch circuit.

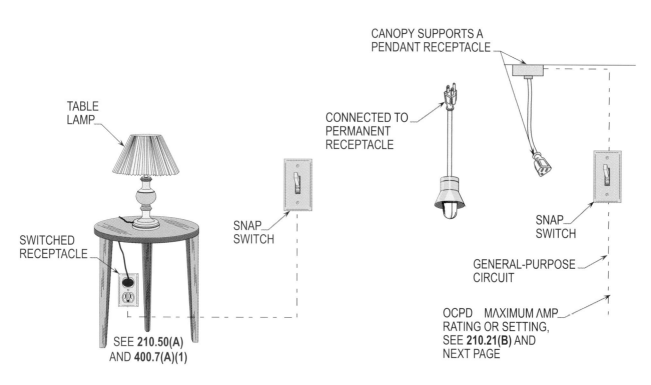

CORD-AND-PLUG-CONNECTED LOADS
NEC 404.14(F)

Purpose of Change: To provide requirements for snap switches used to control cord-and-plug-connected equipment loads.

Type of Change	New Subsection and Ex.		Committee Change	Accept		2008 NEC	-
ROP	pg. 507	# 9-118	log: 726	ROC	pg. 262	# 9-52	log: 2595	UL	20
Submitter: Brian E. Rock			Submitter: Frederic P. Hartwell			OSHA	-
NFPA 70B	24.5		NFPA 70E	-		NFPA 79	-

404.14 Rating and Use of Snap Switches.

(F) Cord-and-Plug-Connected Loads. Where a snap switch is used to control cord-and-plug-connected equipment on a general-purpose branch circuit, each snap switch controlling receptacle outlets or cord connectors that are supplied by permanently connected cord pendants shall be rated at not less than the rating of the maximum permitted ampere rating or setting of the overcurrent device protecting the receptacles or cord connectors, as provided in 210.21(B).

Informational Note: See 210.50(A) and 400.7(A)(1) for equivalency to a receptacle outlet of a cord connector that is supplied by a permanently connected cord pendant.

Exception: Where a snap switch is used to control not more than one receptacle on a branch circuit, the switch shall be permitted to be rated at not less than the rating of the receptacle.

Stallcup's Comment: A new subection has been added to address the requirements for a snap switch used to control cord-and-plug-connected equipment on a general-purpose branch circuit.

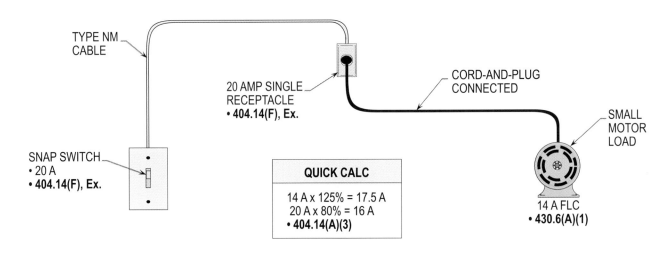

TYPE NM CABLE

20 AMP SINGLE RECEPTACLE
• 404.14(F), Ex.

CORD-AND-PLUG CONNECTED

SMALL MOTOR LOAD

SNAP SWITCH
• 20 A
• 404.14(F), Ex.

QUICK CALC

14 A x 125% = 17.5 A
20 A x 80% = 16 A
• 404.14(A)(3)

14 A FLC
• 430.6(A)(1)

CORD-AND-PLUG-CONNECTED LOADS
NEC 404.14(F), Ex.

Purpose of Change: To provide requirements for mating a snap switch to a particular size receptacle.

Type of Change	New Subdivision			Committee Change	Accept in Principle		2008 NEC	-
ROP	pg. 514	# 18-30	log: 3561	ROC	pg. 263	# 18-10 log: 488	UL	1699
Submitter: James T. Dollard, Jr.				Submitter: James T. Dollard, Jr.			OSHA	-
NFPA 70B				NFPA 70E	-		NFPA 79	-

406.4 General Installation Requirements.

(D) Replacements.

Replacement of receptacles shall comply with 406.4(D)(1) through (D)(6), as applicable.

(4) Arc-Fault Circuit-Interrupter Protection. Where a receptacle outlet is supplied by a branch circuit that requires arc-fault circuit interrupter protection as specified elsewhere in this *Code*, a replacement receptacle at this outlet shall be one of the following:

(1) A listed outlet branch circuit type arc-fault circuit interrupter receptacle

(2) A receptacle protected by a listed outlet branch circuit type arc-fault circuit interrupter type receptacle

(3) A receptacle protected by a listed combination type arc-fault circuit interrupter type circuit breaker

This requirement become effective January 1, 2014.

Stallcup's Comment: A new subdivision has been added to address the requirements for replacement of arc-fault circuit-interrupter protection.

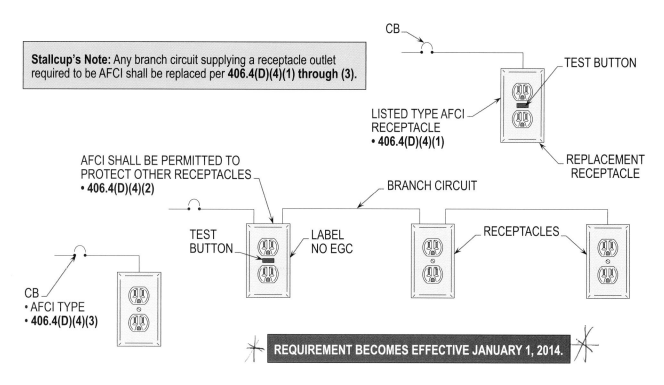

ARC-FAULT CIRCUIT-INTERRUPTER PROTECTION
NEC 406.4(D)(4)(1) THRU (D)(4)(3)

Purpose of Change: To provide requirements for replacing receptacles requiring AFCI protection.

Type of Change	New Subdivision			Committee Change		Accept in Principle		2008 NEC	-
ROP	pg. 513	# 18-24	log: 3467	ROC	pg. -	# -	log: -	UL	498
Submitter: John I. Williamson				Submitter: -				OSHA	-
NFPA 70B	-			NFPA 70E	-			NFPA 79	-

406.4 General Installation Requirements.

(D) Replacements.

Replacement of receptacles shall comply with 406.4(D)(1) through (D)(6), as applicable.

(5) Tamper-Resistant Receptacles. Listed tamper-resistant receptacles shall be provided where replacements are made at receptacle outlets that are required to be tamper-resistant elsewhere in this *Code*.

Stallcup's Comment: A new subdivision has been added to require listed tamper-resistant receptacles to be provided where replacements are made at receptacle outlets.

LOCATIONS WHERE TAMPER-RESISTANT RECEPTACLES ARE REQUIRED

- KITCHEN
- FAMILY ROOM
- LIVING ROOM
- PARLOR
- LIBRARY
- DEN
- SUNROOM
- RECREATION ROOM
- BATHROOMS
- OUTDOORS
- LAUNDRY
- BASEMENTS
- GARAGES
- HALLWAYS
- BEDROOMS

TAMPER-RESISTANT RECEPTACLES
NEC 406.4(D)(5)

Purpose of Change: To require listed tamper-resistant receptacles for replacement of existing receptacles that are required to be tamper-resistant.

Type of Change	New Subdivision			Committee Change	Accept		2008 NEC	-
ROP	pg. 515	# 18-33	log: 3847	ROC	pg. -	# -	log: -	UL	498
Submitter: Bill McGovern				Submitter: -			OSHA	-
NFPA 70B	-			NFPA 70E	-		NFPA 79	-

406.4 General Installation Requirements.

(D) Replacements.

Replacement of receptacles shall comply with 406.4(D)(1) through (D)(6), as applicable.

(6) Weather-Resistant Receptacles. Weather-resistant receptacles shall be provided where replacements are made at receptacle outlets that are required to be so protected elsewhere in this *Code*.

Stallcup's Comment: A new subdivision has been added to require weather-resistant receptacles to be provided where replacements are made at receptacle outlets.

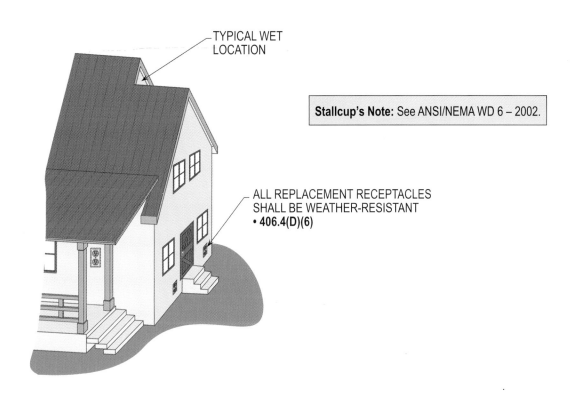

Stallcup's Note: See ANSI/NEMA WD 6 – 2002.

TYPICAL WET LOCATION

ALL REPLACEMENT RECEPTACLES SHALL BE WEATHER-RESISTANT
• 406.4(D)(6)

WEATHER-RESISTANT RECEPTACLES
NEC 406.4(D)(6)

Purpose of Change: To clarify that receptacle locations that are required to be weather-resistant shall be replaced with weather-resistant receptacles.

Type of Change	New Paragraph			Committee Change	Accept			2008 NEC	406.6
ROP	pg. 516	# 18-41	log: 1166	ROC	pg. -	# -	log: -	UL	514D
Submitter: Vince Baclawski				Submitter: -				OSHA	-
NFPA 70B	24.6			NFPA 70E	245.1(2)			NFPA 79	15.1.1(8)

406.6 Receptacle Faceplates (Cover Plates).

Receptacle faceplates shall be installed so as to completely cover the opening and seat against the mounting surface.

Receptacle faceplates mounted inside a box having a recess-mounted receptacle shall effectively close the opening and seat against the mounting surface.

Stallcup's Comment: A revision has been made to clarify the installation of receptacle faceplates mounted inside a box having a recess-mounted receptacle.

> **Stallcup's Note:** Receptacle faceplates mounted inside a box having a recess-mounted receptacle shall effectively close the opening per **406.6.**

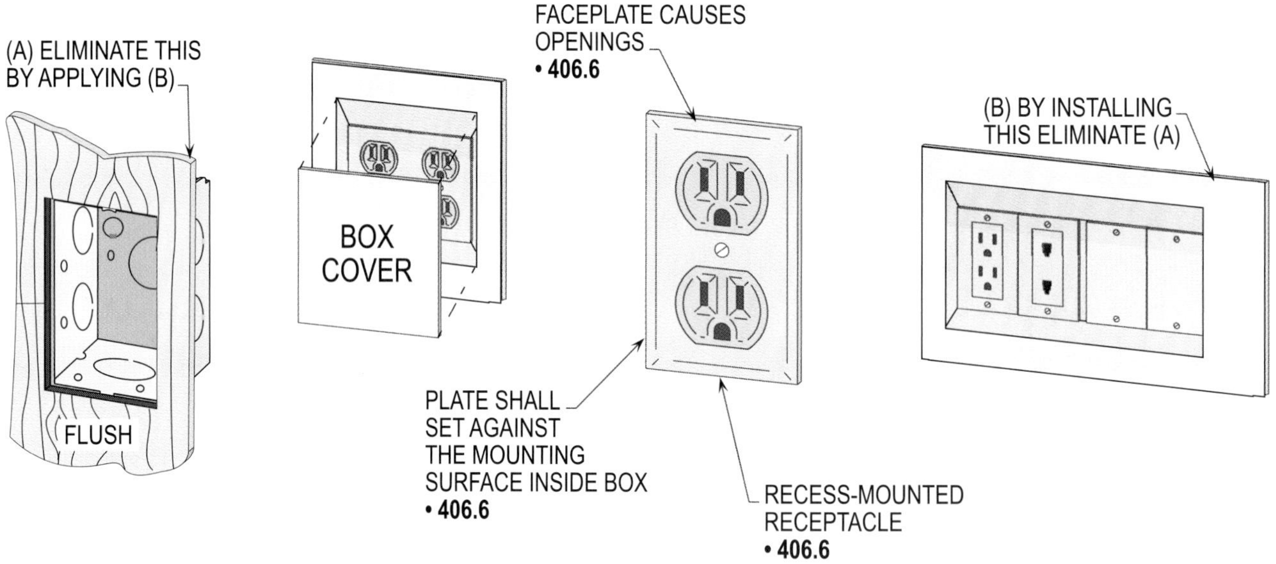

(A) ELIMINATE THIS BY APPLYING (B)

FLUSH

BOX COVER

FACEPLATE CAUSES OPENINGS
• **406.6**

PLATE SHALL SET AGAINST THE MOUNTING SURFACE INSIDE BOX
• **406.6**

RECESS-MOUNTED RECEPTACLE
• **406.6**

(B) BY INSTALLING THIS ELIMINATE (A)

RECEPTACLE FACEPLATE (COVER PLATES)
NEC 406.6

Purpose of Change: To provide requirements for a receptacle faceplate mounted inside of a box that contains a recess-mounted receptacle.

Type of Change	Revision			Committee Change	Accept			2008 NEC	406.9(B)(1)
ROP	pg. 518	# 18-54	log: 3732	ROC	pg. 267	# 18-29	log: 2597	UL	498
Submitter: Vince Baclawski				Submitter: Frederic P. Hartwell				OSHA	1910.305(j)(2)(iv) thru (vii)
NFPA 70B	-			NFPA 70E	-			NFPA 79	15.1.1(8)

406.9 Receptacles in Damp or Wet Locations.

(B) Wet Locations.

(1) 15- and 20-Ampere Receptacles in a Wet Location. 15- and 20-ampere, 125- and 250-volt receptacles installed in a wet location shall have an enclosure that is weatherproof whether or not the attachment plug cap is inserted. For other than one- or two-family dwellings, an outlet box hood installed for this purpose shall be listed, and where installed on an enclosure supported from grade as described in 314.23(B) or as described in 314.23(F) shall be identified as "extra duty." All 15- and 20-ampere, 125- and 250-volt nonlocking-type receptacles shall be listed weather-resistant type.

Stallcup's Comment: A revision has been made to clarify the installation requirements for 15- and 20-ampere receptacles in a wet location in other than one- and two-family dwellings.

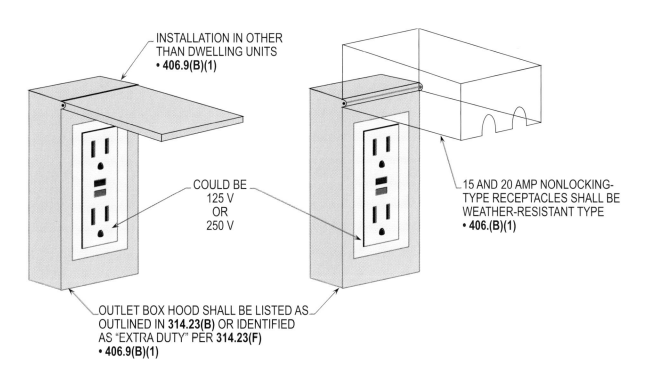

INSTALLATION IN OTHER
THAN DWELLING UNITS
• **406.9(B)(1)**

COULD BE
125 V
OR
250 V

15 AND 20 AMP NONLOCKING-
TYPE RECEPTACLES SHALL BE
WEATHER-RESISTANT TYPE
• **406.(B)(1)**

OUTLET BOX HOOD SHALL BE LISTED AS
OUTLINED IN **314.23(B)** OR IDENTIFIED
AS "EXTRA DUTY" PER **314.23(F)**
• **406.9(B)(1)**

15- AND 20-AMPERE RECEPTACLES IN A WET LOCATION
NEC 406.9(B)(1)

Purpose of Change: To clarify the requirements for replacing receptacles in other than one- or two-family dwelling units.

Type of Change	New Exception			Committee Change	Accept			2008 NEC	-
ROP	pg. 521	# 18-71	log: 3848	ROC	pg. 267	# 18-33	log: 2598	UL	498
Submitter: Bill McGovern				Submitter: Frederic P. Hartwell				OSHA	-
NFPA 70B	-			NFPA 70E	-			NFPA 79	-

406.12 Tamper-Resistant Receptacles in Dwelling Units.

In all areas specified in 210.52, all nonlocking-type 125-volt, 15- and 20-ampere receptacles shall be listed tamper-resistant receptacles.

Exception: Receptacles in the following locations shall not be required to be tamper-resistant:

(1) Receptacles located more than 1.7 m (5 1/2 ft) above the floor.

(2) Receptacles that are part of a luminaire or appliance.

Stallcup's Comment: A new exception has been added to address the locations where tamper-resistant receptacles shall not be required.

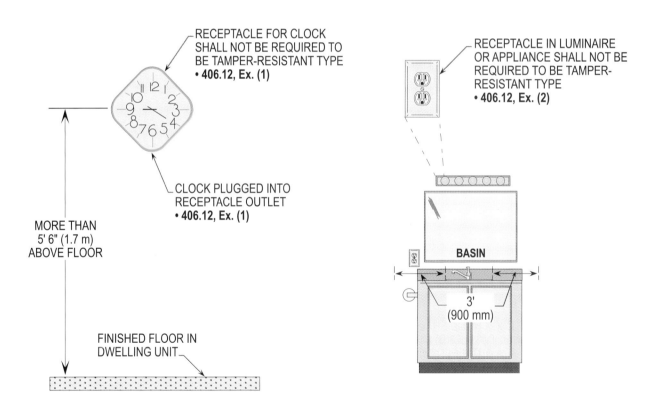

TAMPER-RESISTANT RECEPTACLES IN DWELLING UNITS
NEC 406.12, Ex. (1) AND (2)

Purpose of Change: To clarify the type of receptacles that are not required to be tamper-resistant.

Type of Change	New Exception			Committee Change	Accept		2008 NEC	-
ROP	pg. 521	# 18-71	log: 3848	ROC	pg. 267	# 18-33	log: 2598	UL	498
Submitter: Bill McGovern				Submitter: Frederic P. Hartwell			OSHA	-
NFPA 70B	-			NFPA 70E	-		NFPA 79	-

406.12 Tamper-Resistant Receptacles in Dwelling Units.

In all areas specified in 210.52, all nonlocking-type 125-volt, 15- and 20-ampere receptacles shall be listed tamper-resistant receptacles.

Exception: Receptacles in the following locations shall not be required to be tamper-resistant:

(3) A single receptacle or a duplex receptacle for two appliances located within dedicated space for each appliance that, in normal use, is not easily moved from one place to another and that is cord-and-plug connected in accordance with 400.7(A)(6), (A)(7), or (A)(8).

(4) Nongrounding receptacles used for replacements as permitted 406.4(D)(2)(a).

Stallcup's Comment: A new exception has been added to address the locations where tamper-resistant receptacles shall not be required.

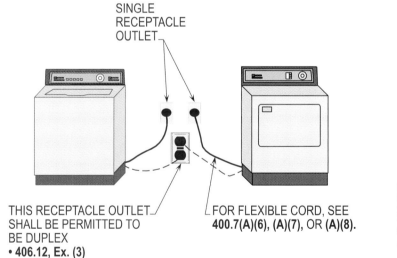

SINGLE
RECEPTACLE
OUTLET

THIS RECEPTACLE OUTLET
SHALL BE PERMITTED TO
BE DUPLEX
• **406.12, Ex. (3)**

FOR FLEXIBLE CORD, SEE
400.7(A)(6), (A)(7), OR (A)(8).

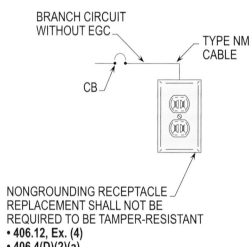

BRANCH CIRCUIT
WITHOUT EGC

TYPE NM
CABLE

CB

NONGROUNDING RECEPTACLE
REPLACEMENT SHALL NOT BE
REQUIRED TO BE TAMPER-RESISTANT
• **406.12, Ex. (4)**
• **406.4(D)(2)(a)**

Stallcup's Note: The above replacement receptacles shall not be required to be of the tamper-resistant type per **406.12, Ex. (3)** and **(4).**

TAMPER-RESISTANT RECEPTACLES IN DWELLING UNITS
NEC 406.12, Ex. (3) and (4)

Purpose of Change: To clarify the type of receptacles that are not required to be tamper-resistant.

Type of Change	New Section			Committee Change	Accept			2008 NEC	-
ROP	pg. 524	# 18-87	log: 1167	ROC	pg. 268	# 18-35	log: 2308	UL	498
Submitter: Vince Baclawski				Submitter: Mike Holt				OSHA	-
NFPA 70B	-			NFPA 70E	-			NFPA 79	-

406.13 Tamper-Resistant Receptacles in Guest Rooms and Guest Suites.

All nonlocking-type, 125-volt, 15- and 20-ampere receptacles located in guest rooms and guest suites shall be listed tamper-resistant receptacles.

Stallcup's Comment: A new section has been added to require all 15 and 20 amp receptacles installed in guest rooms and guest suites to be tamper-resistant receptacles.

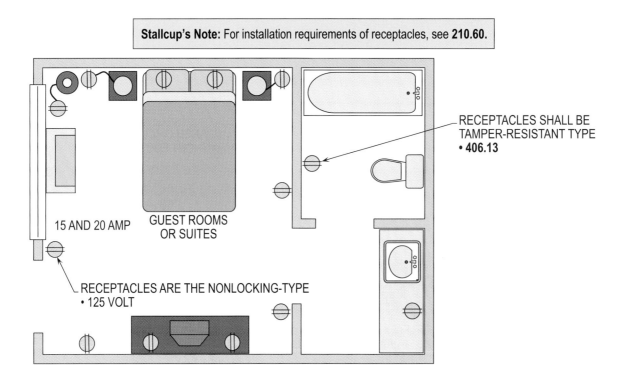

Stallcup's Note: For installation requirements of receptacles, see **210.60.**

TAMPER-RESISTANT RECEPTACLES IN GUEST ROOMS AND GUEST SUITES
NEC 406.13

Purpose of Change: To require all 125 volt, 15 and 20 amp receptacles located in guest rooms and guest suites to be listed as the tamper-resistant type.

Type of Change	New Section			Committee Change	Accept in Principle			2008 NEC	-
ROP	pg. 525	# 18-90	log: 1168	ROC	pg. -	# -	log: -	UL	498
Submitter: Vince Baclawski				Submitter: -				OSHA	-
NFPA 70B	-			NFPA 70E	-			NFPA 79	-

406.14 Tamper-Resistant Receptacles in Child Care Facilities.

In all child care facilities, all nonlocking-type, 125-volt, 15- and 20-ampere receptacles shall be listed tamper-resistant receptacles.

Stallcup's Comment: A new section has been added to require all 15- and 20-ampere receptacles installed in child care facilities to be tamper-resistant receptacles.

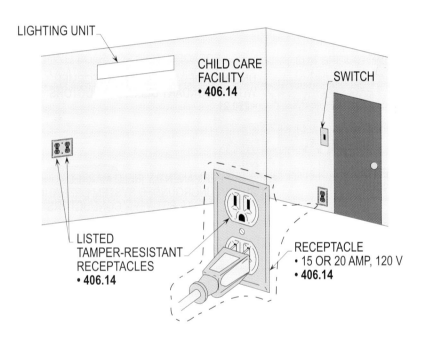

TAMPER-RESISTANT RECEPTACLES IN CHILD CARE FACILITIES
NEC 406.14

Purpose of Change: To require tamper-resistant receptacles to be installed in child care facilities.

Type of Change	New Section			Committee Change	Accept			2008 NEC	-
ROP	pg. 533	# 11-9	log: 4404	ROC	pg. 271	# 11-2	log: 1241	UL	508A
Submitter: Jay Tamblingson				Submitter: Vince Baclawski				OSHA	-
NFPA 70B	-			NFPA 70E	Article 100 (definitions)			NFPA 79	7.2.9

409.22 Short-Circuit Current Rating.

An industrial control panel shall not be installed where the available fault current exceeds its short-circuit current rating as marked in accordance with 409.110(4).

Stallcup's Comment: A new section has been added to require an industrial control panel not to be installed where the available fault current exceeds its short-circuit current rating.

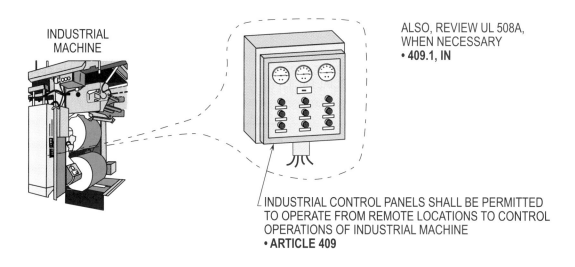

ALSO, REVIEW UL 508A, WHEN NECESSARY
• **409.1, IN**

INDUSTRIAL CONTROL PANELS SHALL BE PERMITTED TO OPERATE FROM REMOTE LOCATIONS TO CONTROL OPERATIONS OF INDUSTRIAL MACHINE
• **ARTICLE 409**

Stallcup's Note: Where available fault current exceeds industrial control panel's short-circuit rating, it shall be marked in accordance with **409.22** and **409.110(4)** to alert personnel.

SHORT-CIRCUIT CURRENT RATING
NEC 409.22

Purpose of Change: To outline short-circuit rating requirements for industrial control panels.

Type of Change	Revision			Committee Change	Accept in Principle			2008 NEC	409.110(3)
ROP	pg. 535	# 11-23	log: 4465	ROC	pg. 271	# 11-4	log: 580	UL	508A
Submitter: Robert G. Fahey				Submitter: Bob Fahey				OSHA	-
NFPA 70B	-			NFPA 70E	-			NFPA 79	Chapter 16

409.110 Marking.

An industrial control panel shall be marked with the following information that is plainly visible after installation:

(3) Industrial control panels supplied by more than one power source such that more than one disconnecting means is required to disconnect all power within the control panel shall be marked to indicate that more than one disconnecting means is required to de-energize the equipment.

Stallcup's Comment: A revision has been made to clarify that industrial control panels supplied by more than one power source shall be marked to indicate that more than one disconnecting means is required to deenergize the equipment.

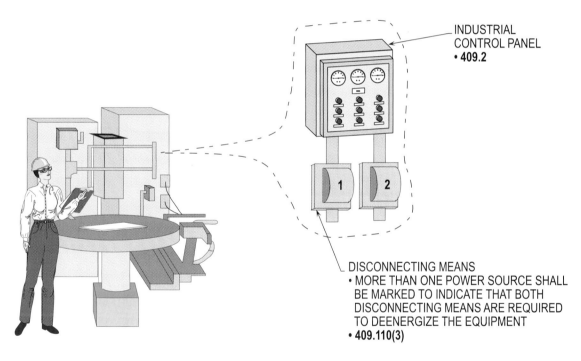

INDUSTRIAL
CONTROL PANEL
• **409.2**

DISCONNECTING MEANS
• MORE THAN ONE POWER SOURCE SHALL
BE MARKED TO INDICATE THAT BOTH
DISCONNECTING MEANS ARE REQUIRED
TO DEENERGIZE THE EQUIPMENT
• **409.110(3)**

**MARKING
NEC 409.110(3)**

Purpose of Change: To require industrial control panels that have more than one disconnecting means to disconnect power to be marked as such.

Type of Change	New Section			Committee Change	Accept			2008 NEC	250.112(J)
ROP	pg. 540	# 18-122	log: 4541	ROC	pg. 273	# 18-46	log: 1862	UL	1598
Submitter: Phil Simmons				Submitter: Phil Simmons				OSHA	-
NFPA 70B	-			NFPA 70E	-			NFPA 79	-

410.44 Methods of Grounding.

Luminaires and equipment shall be mechanically connected to an equipment grounding conductor as specified in 250.118 and sized in accordance with 250.122.

Exception No. 1: Luminaires made of insulating material that is directly wired or attached to outlets supplied by a wiring method that does not provide a ready means for grounding attachment to an equipment grounding conductor shall be made of insulating material and shall have no exposed conductive parts.

Stallcup's Comment: A new section has been added to address the methods of grounding for luminaires and equipment.

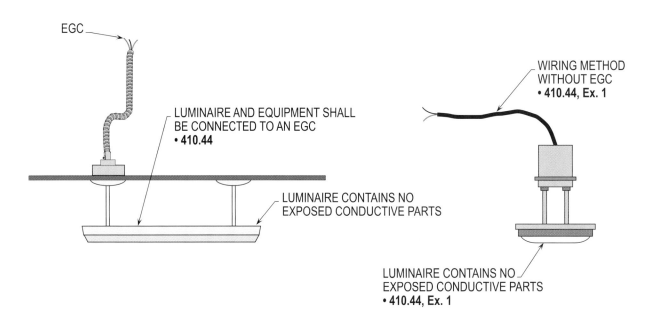

EGC

LUMINAIRE AND EQUIPMENT SHALL
BE CONNECTED TO AN EGC
• **410.44**

LUMINAIRE CONTAINS NO
EXPOSED CONDUCTIVE PARTS

WIRING METHOD
WITHOUT EGC
• **410.44, Ex. 1**

LUMINAIRE CONTAINS NO
EXPOSED CONDUCTIVE PARTS
• **410.44, Ex. 1**

**METHODS OF GROUNDING
NEC 410.44 AND Ex. 1**

Purpose of Change: To address the grounding and nongrounding procedures for metal and nonmetal luminaires.

Type of Change	New Subsections	Committee Change	Accept		2008 NEC	410.64
ROP	pg. 542	# 18-136a	log: CP 1803	ROC	pg. -	# -	log: -		UL	1598
Submitter: Code-Making Panel 18		Submitter:			OSHA	-
NFPA 70B	-	NFPA 70E	-		NFPA 79	-

410.64 Luminaires as Raceways.

Luminaires shall not be used as a raceway for circuit conductors unless they comply with 410.64(A), (B), or (C).

(A) Listed. Luminaires listed and marked for use as a raceway shall be permitted to be used as a raceway.

(B) Through-Wiring. Luminaires identified for through-wiring, as permitted by 410.21, shall be permitted to be used as a raceway.

Stallcup's Comment: A revision has been made to improve the usability of this section. Luminaires shall not be used as raceways unless they are listed, are identified for through-wiring, or are connected together.

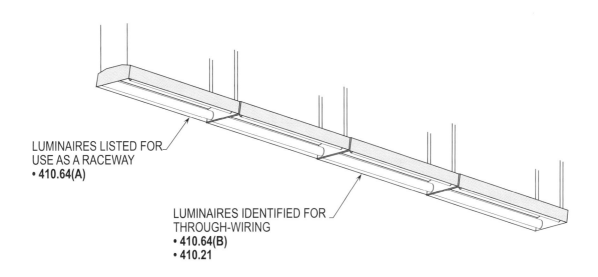

LUMINAIRES LISTED FOR
USE AS A RACEWAY
• 410.64(A)

LUMINAIRES IDENTIFIED FOR
THROUGH-WIRING
• 410.64(B)
• 410.21

**LISTED AND THROUGH-WIRING
NEC 410.64(A) AND (B)**

Purpose of Change: To provide requirements for luminaires used as a raceway for through-wiring installations.

Type of Change	New Subsections		Committee Change	Accept		2008 NEC	410.64
ROP	pg. 542	# 18-136a log: CP 1803	ROC	pg. -	# - log: -	UL	1598
Submitter: Code-Making Panel 18			Submitter:			OSHA	-
NFPA 70B	-		NFPA 70E	-		NFPA 79	-

410.64 Luminaires as Raceways.

Luminaires shall not be used as a raceway for circuit conductors unless they comply with 410.64(A), (B), or (C).

(C) Luminaires Connected Together. Luminaires designed for end-to-end connection to form a continuous assembly, or luminaires connected together by recognized wiring methods, shall be permitted to contain the conductors of a 2-wire branch circuit, or one multiwire branch circuit, supplying the connected luminaires and shall not be required to be listed as a raceway. One additional 2-wire branch circuit separately supplying one or more of the connected luminaires shall also be permitted.

Stallcup's Comment: A revision has been made to improve the usability of this section. Luminaires shall not be used as raceways unless they are listed, are identified for through-wiring, or are connected together.

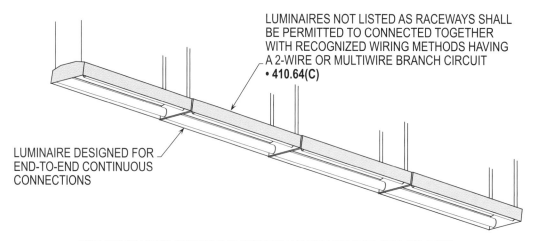

LUMINAIRES NOT LISTED AS RACEWAYS SHALL BE PERMITTED TO CONNECTED TOGETHER WITH RECOGNIZED WIRING METHODS HAVING A 2-WIRE OR MULTIWIRE BRANCH CIRCUIT
• **410.64(C)**

LUMINAIRE DESIGNED FOR END-TO-END CONTINUOUS CONNECTIONS

FOR LUMINAIRES CONTAINING MORE THAN ONE CIRCUIT, SEE **410.64(C)**.

> **Stallcup's Note:** One additional two-wire branch circuit that supplies one or more connected luminaires shall also be permitted per **410.64(C)**.

LUMINAIRES CONNECTED TOGETHER
NEC 410.64(C)

Purpose of Change: To revise requirements for using luminaires as raceways or when connected together for through-wiring.

Type of Change	New Section			Committee Change	Accept in Principle in Part		2008 NEC	-
ROP	pg. 537	# 18-101	log: 4624	ROC	pg. -	# -	log: -	UL	496
Submitter: Frederic P. Hartwell				Submitter: -			OSHA	-
NFPA 70B	-			NFPA 70E	-		NFPA 79	-

410.97 Lampholders Near Combustible Material.

Lampholders shall be constructed, installed, or equipped with shades or guards so that combustible material is not subjected to temperatures in excess of 90°C (194°F).

Stallcup's Comment: A new section has been added to address the requirements for lampholders near combustible material.

LAMPHOLDERS MOUNTED ON WALLS
OF COMBUSTIBLE MATERIAL ARE
CONSTRUCTED FOR SUCH USE
• **410.97**

COMBUSTIBLE
MATERIAL

SHADE OR GUARD
• **410.97**

> **Stallcup's Note:** Lampholders shall be equipped with shades or guards so that combustile material is not subjected to temperatures exceeding 90°C (194°F) per **410.97.**

LAMPHOLDERS NEAR COMBUSTIBLE MATERIAL
NEC 410.97

Purpose of Change: To address the requirements for lampholders near combustible material.

Type of Change	Revision			Committee Change	Accept in Principle		2008 NEC	410.116(B)
ROP	pg. 546	# 18-168	log: 2305	ROC	pg. -	# -	log: -	UL	1598
Submitter: John Marshall				Submitter: -			OSHA	-
NFPA 70B	-			NFPA 70E	-		NFPA 79	-

410.116 Clearance and Installation.

(B) Installation. Thermal insulation shall not be installed above a recessed luminaire or within 75 mm (3 in.) of the recessed luminaire's enclosure, wiring compartment, ballast, transformer, LED driver, or power supply unless the luminaire is identified as Type IC for insulation contact.

Stallcup's Comment: A revision has been made to require that thermal insulation not be installed above a recessed luminaire or within 3 in. (75 mm) of the recessed luminaire's enclosure, wiring compartment, ballast, transformer, LED driver, or power supply.

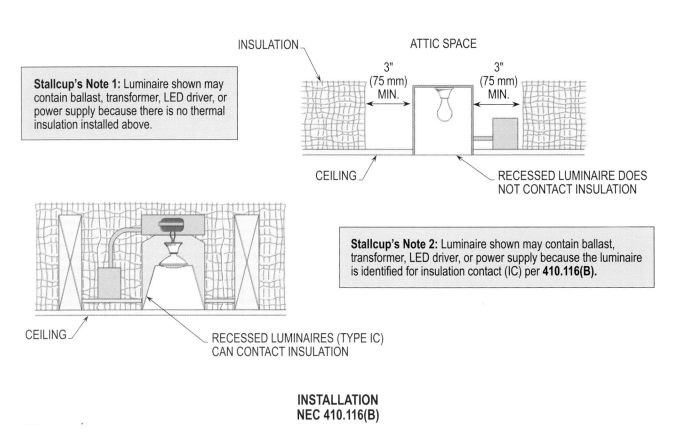

Stallcup's Note 1: Luminaire shown may contain ballast, transformer, LED driver, or power supply because there is no thermal insulation installed above.

INSULATION

ATTIC SPACE

3" (75 mm) MIN.

3" (75 mm) MIN.

CEILING

RECESSED LUMINAIRE DOES NOT CONTACT INSULATION

Stallcup's Note 2: Luminaire shown may contain ballast, transformer, LED driver, or power supply because the luminaire is identified for insulation contact (IC) per **410.116(B)**.

CEILING

RECESSED LUMINAIRES (TYPE IC) CAN CONTACT INSULATION

**INSTALLATION
NEC 410.116(B)**

Purpose of Change: To address requirements for recessed luminaires with insulation installed above such work.

Type of Change	Revision			Committee Change	Accept in Principle in Part		2008 NEC	410.130(G)(1)
ROP	pg. 547	# 18-175	log: 12	ROC	pg. -	# -	log: -	UL	1598
Submitter: Gregory J. Steinman				Submitter: -			OSHA	-
NFPA 70B	-			NFPA 70E	-		NFPA 79	-

410.130 General.

(G) Disconnecting Means.

(1) General. In indoor locations other than dwellings and associated accessory structures, fluorescent luminaires that utilize double-ended lamps and contain ballast(s) that can be serviced in place shall have a disconnecting means either internal or external to each luminaire. For existing installed luminaires without disconnecting means, at the time a ballast is replaced, a disconnecting means shall be installed. The line side terminals of the disconnecting means shall be guarded.

Stallcup's Comment: A revision has been made to require a disconnecting means to be installed at the time a ballast is being replaced in an existing luminaire.

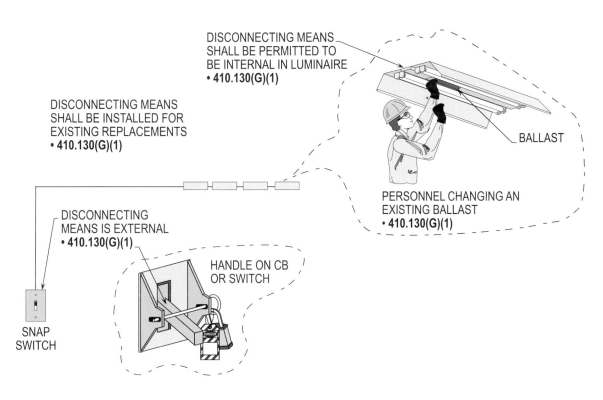

DISCONNECTING MEANS – GENERAL
NEC 410.130(G)(1)

Purpose of Change: To clarify that a disconnecting means shall be installed when changing out an existing ballast in a luminaire.

Type of Change	New Subsection			Committee Change	Accept			2008 NEC	-
ROP	pg. 555	# 17-21	log: 463	ROC	pg. 278	# 17-14	log: 2606	UL	73
Submitter: Lanny G. McMahill				Submitter: Frederic P. Hartwell				OSHA	1910.305(j)(3)(ii)
NFPA 70B	-			NFPA 70E	-			NFPA 79	5.5

422.31 Disconnection of Permanently Connected Appliances.

(C) Motor-Operated Appliances Rated over 1/8 Horsepower. For permanently connected motor-operated appliances with motors rated over 1/8 horsepower, the branch-circuit switch or circuit breaker shall be permitted to serve as the disconnecting means where the switch or circuit breaker is within sight from the appliance. The disconnecting means shall comply with 430.109 and 430.110.

Stallcup's Comment: A new subsection has been added to permit the branch-circuit switch or circuit breaker to serve as the disconnecting means where the switch or circuit breaker is within sight of the appliance for permanently connected motor-operated appliances with motors rated over 1/8 horsepower.

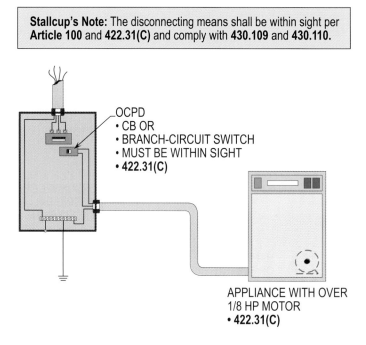

Stallcup's Note: The disconnecting means shall be within sight per **Article 100** and **422.31(C)** and comply with **430.109** and **430.110**.

OCPD
• CB OR
• BRANCH-CIRCUIT SWITCH
• MUST BE WITHIN SIGHT
• **422.31(C)**

APPLIANCE WITH OVER
1/8 HP MOTOR
• **422.31(C)**

**MOTOR-OPERATED APPLIANCES RATED OVER 1/8 HORSEPOWER
NEC 422.31(C)**

Purpose of Change: To permit a branch-circuit switch or circuit breaker within sight to serve as a disconnecting means for an appliance with a motor rated over 1/8 horsepower.

Type of Change	New Subdivisions			Committee Change	Accept			2008 NEC	424.19(A)(2)
ROP	pg. 559	# 17-44	log: 2156	ROC	pg. -	# -	log: -	UL	499
Submitter: James W. Carpenter				Submitter: -				OSHA	-
NFPA 70B	-			NFPA 70E	-			NFPA 79	-

424.19 Disconnecting Means.

(A) Heating Equipment with Supplementary Overcurrent Protection.

(2) Heater Containing a Motor(s) Rated over 1/8 Horsepower. The above disconnecting means shall be permitted to serve as the required disconnecting means for both the motor controller(s) and heater under either of the following conditions:

(1) Where the disconnecting means is in sight from the motor controller(s) and the heater and complies with Part IX of Article 430.

Stallcup's Comment: A revision has been made to clarify that the disconnecting means shall be permitted to serve as the required disconnecting means for both the motor conroller(s) and heater under certain conditions.

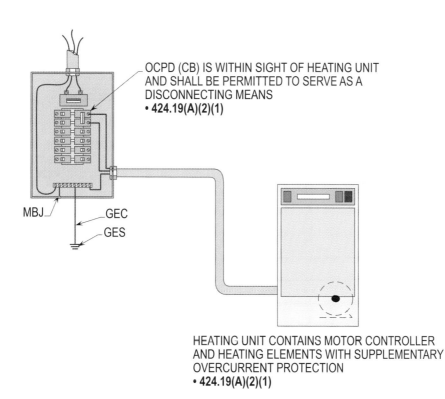

OCPD (CB) IS WITHIN SIGHT OF HEATING UNIT
AND SHALL BE PERMITTED TO SERVE AS A
DISCONNECTING MEANS
• **424.19(A)(2)(1)**

MBJ — GEC — GES

HEATING UNIT CONTAINS MOTOR CONTROLLER
AND HEATING ELEMENTS WITH SUPPLEMENTARY
OVERCURRENT PROTECTION
• **424.19(A)(2)(1)**

HEATER CONTAINING A MOTOR(S) RATED OVER 1/8 HORSEPOWER
NEC 424.19(A)(2)(1)

Purpose of Change: To clarify the location of a disconnecting means that shall be permitted to be used to deenergize the unit and components.

Type of Change	New Subdivisions			Committee Change	Accept			2008 NEC	424.19(A)(2)
ROP	pg. 559	# 17-44	log: 2156	ROC	pg. -	# -	log: -	UL	499
Submitter: James W. Carpenter				Submitter: -				OSHA	-
NFPA 70B	-			NFPA 70E	-			NFPA 79	-

424.19 Disconnecting Means.

(A) Heating Equipment with Supplementary Overcurrent Protection.

(2) Heater Containing a Motor(s) Rated over 1/8 Horsepower. The above disconnecting means shall be permitted to serve as the required disconnecting means for both the motor controller(s) and heater under either of the following conditions:

(2) Where a motor(s) of more than 1/8 hp and the heater are provided with a single unit switch that complies with 422.34(A), (B), (C), or (D), the disconnecting means shall be permitted to be out of sight from the motor controller.

Stallcup's Comment: A revision has been made to clarify that the disconnecting means shall be permitted to serve as the required disconnecting means for both the motor conroller(s) and heater under certain conditions.

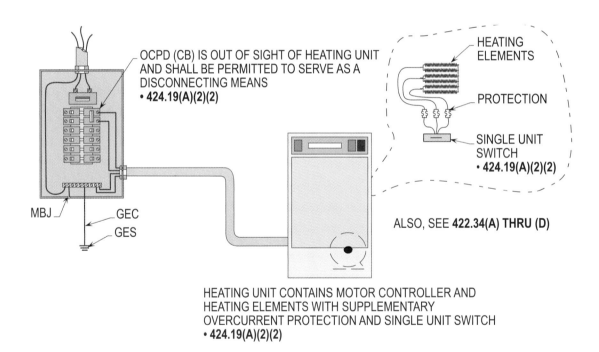

OCPD (CB) IS OUT OF SIGHT OF HEATING UNIT
AND SHALL BE PERMITTED TO SERVE AS A
DISCONNECTING MEANS
• **424.19(A)(2)(2)**

HEATING
ELEMENTS

PROTECTION

SINGLE UNIT
SWITCH
• **424.19(A)(2)(2)**

MBJ

GEC

GES

ALSO, SEE **422.34(A) THRU (D)**

HEATING UNIT CONTAINS MOTOR CONTROLLER AND
HEATING ELEMENTS WITH SUPPLEMENTARY
OVERCURRENT PROTECTION AND SINGLE UNIT SWITCH
• **424.19(A)(2)(2)**

**HEATER CONTAINING A MOTOR(S) RATED OVER 1/8 HORSEPOWER
NEC 424.19(A)(2)(2)**

Purpose of Change: To clarify the location of a disconnecting means that shall be permitted to be used to deenergize the unit and components.

Type of Change	New Subsection			Committee Change	Accept in Principle			2008 NEC	-
ROP	pg. 572	# 11-26	log: 2216	ROC	pg. -	# -	log: -	UL	1004
Submitter: Paul Guidry				Submitter: -				OSHA	-
NFPA 70B	-			NFPA 70E	-			NFPA 79	14.5

430.6 Ampacity and Motor Rating Determination.

(D) Valve Actuator Motor Assemblies. For valve actuator motor assemblies (VAMs), the rated current shall be the nameplate full-load current, and this current shall be used to determine the maximum rating or setting of the motor branch-circuit short-circuit and ground-fault protective device and the ampacity of the conductors.

Stallcup's Comment: A new subsection has been added to clarify that the nameplate value shall be used for valve actuator motor assemblies to determine the maximum rating or setting of the motor branch-circuit short-circuit and ground-fault protective device and the ampacity of the conductors.

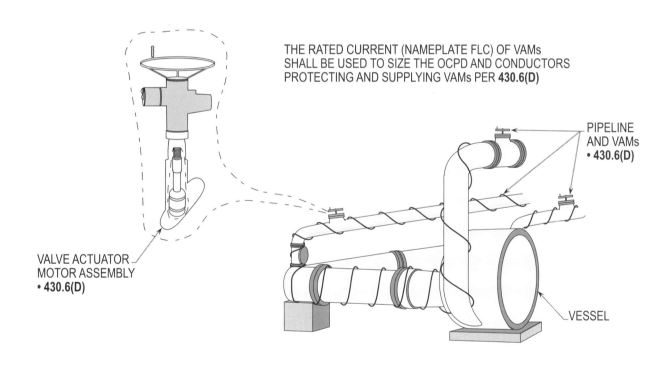

THE RATED CURRENT (NAMEPLATE FLC) OF VAMs
SHALL BE USED TO SIZE THE OCPD AND CONDUCTORS
PROTECTING AND SUPPLYING VAMs PER **430.6(D)**

PIPELINE
AND VAMs
• **430.6(D)**

VALVE ACTUATOR
MOTOR ASSEMBLY
• **430.6(D)**

VESSEL

VALVE ACTUATOR MOTOR ASSEMBLIES
NEC 430.6(D)

Purpose of Change: To clarify that the nameplate current in full-load amps of the valve actuator motor assemblies shall be used to determine the size of the overcurrent protection device and conductors protecting and supplying the system.

Type of Change	Relocated and Revised			Committee Change	Accept			**2008 NEC**	430.22(A), Ex. (a) and (b)
ROP pg. 575	# 11-48a	log: CP 1100		**ROC** pg. -	# -	log: -		**UL**	1004
Submitter: Code-Making Panel 11				Submitter: -				**OSHA**	-
NFPA 70B -				**NFPA 70E** -				**NFPA 79**	14.5

430.22 Single Motor.

(A) Direct-Current Motor-Rectifier Supplied. For dc motors operating from a rectified power supply, the conductor ampacity on the input of the rectifier shall not be less than 125 percent of the rated input current to the rectifier. For dc motors operating from a rectified single-phase power supply, the conductors between the field wiring output terminals of the rectifier and the motor shall have an ampacity of not less than the following percentages of the motor full-load current rating:

(1) Where a rectifier bridge of the single-phase, half-wave type is used, 190 percent.

(2) Where a rectifier bridge of the single-phase, full-wave type is used, 150 percent.

Stallcup's Comment: A relocation of material and revision has been made to address the requirements for direct-current motors operating from a rectified power supply.

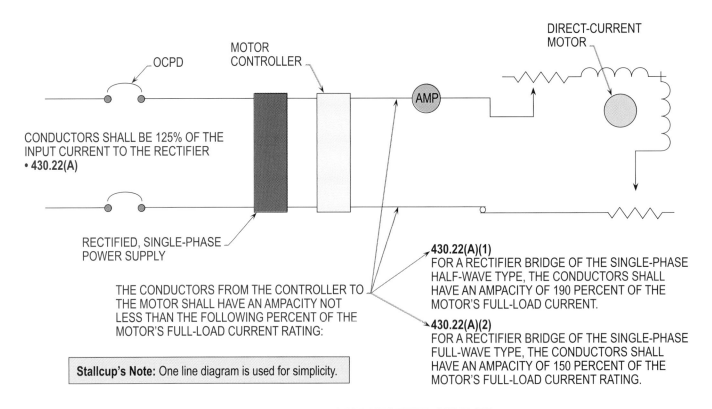

CONDUCTORS SHALL BE 125% OF THE
INPUT CURRENT TO THE RECTIFIER
• 430.22(A)

RECTIFIED, SINGLE-PHASE
POWER SUPPLY

THE CONDUCTORS FROM THE CONTROLLER TO
THE MOTOR SHALL HAVE AN AMPACITY NOT
LESS THAN THE FOLLOWING PERCENT OF THE
MOTOR'S FULL-LOAD CURRENT RATING:

430.22(A)(1)
FOR A RECTIFIER BRIDGE OF THE SINGLE-PHASE
HALF-WAVE TYPE, THE CONDUCTORS SHALL
HAVE AN AMPACITY OF 190 PERCENT OF THE
MOTOR'S FULL-LOAD CURRENT.

430.22(A)(2)
FOR A RECTIFIER BRIDGE OF THE SINGLE-PHASE
FULL-WAVE TYPE, THE CONDUCTORS SHALL
HAVE AN AMPACITY OF 150 PERCENT OF THE
MOTOR'S FULL-LOAD CURRENT RATING.

Stallcup's Note: One line diagram is used for simplicity.

**DIRECT-CURRENT MOTOR-RECTIFIER SUPPLIED
NEC 430.22(A)(1) AND (A)(2)**

Purpose of Change: To relocate material for direct-current motor-rectifier design in an effort to make this section easy to read and understand.

Type of Change	Revision			Committee Change	Accept			2008 NEC	430.22(C)
ROP	pg. 575	# 11-48a	log: CP 1100	ROC	pg. -	# -	log: -	UL	1004
Submitter: Code-Making Panel 11				Submitter: -				OSHA	-
NFPA 70B	-			NFPA 70E	-			NFPA 79	14.5

430.22 Single Motor.

(C) Wye-Start, Delta-Run Motor. For a wye-start, delta-run connected motor, the ampacity of the branch-circuit conductors on the line side of the controller shall not be less than 125 percent of the motor full-load current as determined by 430.6(A)(1). The ampacity of the conductors between the controller and the motor shall not be less than 72 percent of the motor full-load current rating as determined by 430.6(A)(1).

Informational Note: The individual motor circuit conductors of a wye-start, delta-run connected motor carry 58 percent of the rated load current. The multiplier of 72 percent is obtained by multiplying 58 percent by 1.25.

Stallcup's Comment: A revision has been made to clarify the requirements for a wye-start, delta-run connected motor.

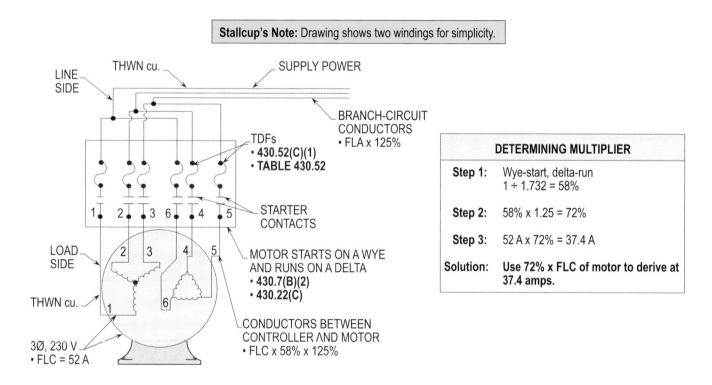

WYE-START, DELTA-RUN MOTOR
NEC 430.22(C) AND IN

Purpose of Change: To clarify how to obtain and apply the 72 percent requirement to derive full-load current in amps to size conductors.

Type of Change	Revision			Committee Change	Accept			2008 NEC	430.22(D)
ROP	pg. 575	# 11-48a	log: CP 1100	ROC	pg. -	# -	log: -	UL	1004
Submitter: Code-Making Panel 11				Submitter: -				OSHA	-
NFPA 70B	-			NFPA 70E	-			NFPA 79	14.5

430.22 Single Motor.

(D) Part-Winding Motor. For a part-winding connected motor, the ampacity of the branch-circuit conductors on the line side of the controller shall not be less than 125 percent of the motor full-load current as determined by 430.6(A)(1). The ampacity of the conductors between the controller and the motor shall not be less than 62.5 percent of the motor full-load current rating as determined by 430.6(A)(1).

Informational Note: The multiplier of 62.5 percent is obtained by multiplying 50 percent by 1.25.

Stallcup's Comment: A revision has been made to clarify the requirements for a part-winding connected motor.

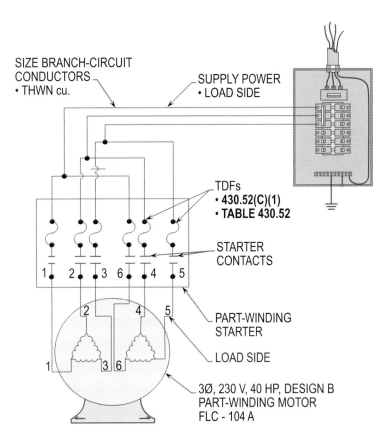

Stallcup's Note: For simplicity, illustration shows two windings in motor.

SIZE BRANCH-CIRCUIT
CONDUCTORS
• THWN cu.

SUPPLY POWER
• LOAD SIDE

TDFs
• 430.52(C)(1)
• TABLE 430.52

STARTER
CONTACTS

PART-WINDING
STARTER

LOAD SIDE

3Ø, 230 V, 40 HP, DESIGN B
PART-WINDING MOTOR
FLC - 104 A

DETERMINING MULTIPLIER	
Step 1:	Part Winding 100% ÷ 2 = 50%
Step 2:	50% x 1.25 = 62.5%
Step 3:	104 A x 62.5% = 65 A
Solution:	**Use 62.5% x FLC of motor to derive at 65 amps.**

**PART-WINDING MOTOR
NEC 430.22(D) AND IN**

Purpose of Change: To clarify how to obtain and apply the 62.5 percent requirement to derive full-load current in amps to size conductors.

Type of Change	New Subsection			Committee Change	Accept			2008 NEC	-
ROP	pg. 576	# 11-50	log: 4794	ROC	pg. 284	# 11-17a	log: CC 1101	UL	1004
Submitter: David Drennan				Submitter: Code-Making Panel 11				OSHA	-
NFPA 70B	-			NFPA 70E	-			NFPA 79	14.5

430.22 Single Motor.

(G) Conductors for Small Motors. Conductors for small motors shall not be smaller than 14 AWG unless otherwise permitted in 430.22(G)(1) or (G)(2).

(1) 18 AWG Copper. Where installed in a cabinet or enclosure, 18 AWG individual copper conductors, copper conductors that are part of a jacketed multiconductor cable assembly, or copper conductors in a flexible cord shall be permitted, under either of the following sets of conditions:

(1) Motor circuits with a full-load ampacity greater than 3.5 amperes or less than or equal to 5 amperes if all the following conditions are met:

Stallcup's Comment: A new subsection has been added to address the requirements for 18 AWG copper conductors used for small motors.

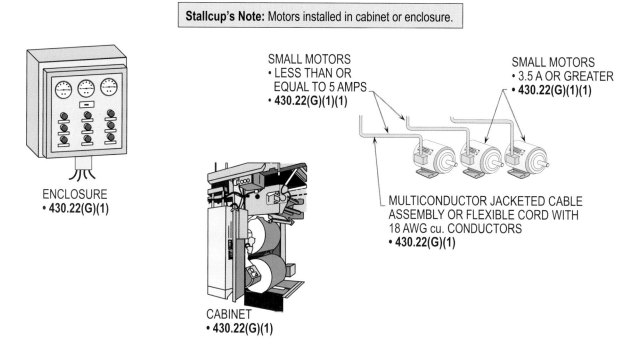

Stallcup's Note: Motors installed in cabinet or enclosure.

ENCLOSURE
• 430.22(G)(1)

CABINET
• 430.22(G)(1)

SMALL MOTORS
• LESS THAN OR EQUAL TO 5 AMPS
• 430.22(G)(1)(1)

SMALL MOTORS
• 3.5 A OR GREATER
• 430.22(G)(1)(1)

MULTICONDUCTOR JACKETED CABLE ASSEMBLY OR FLEXIBLE CORD WITH 18 AWG cu. CONDUCTORS
• 430.22(G)(1)

CONDUCTORS FOR SMALL MOTORS – 18 AWG COPPER
NEC 430.22(G)(1)(1)

Purpose of Change: To clarify the requirements for small motor conductors with 18 AWG copper.

Type of Change	New Subsection			Committee Change	Accept		2008 NEC	-
ROP	pg. 576	# 11-50	log: 4794	ROC	pg. 284	# 11-17a	log: CC 1101	UL	1004
Submitter: David Drennan				Submitter: Code-Making Panel 11			OSHA	-
NFPA 70B	-			NFPA 70E	-		NFPA 79	14.5

430.22 Single Motor.

(G) Conductors for Small Motors.

(1) 18 AWG Copper.

(1) Motor circuits with a full-load ampacity greater than 3.5 amperes or less than or equal to 5 amperes if all the following conditions are met:

a. The circuit is protected in accordance with 430.52.

b. The circuit is provided with maximum Class 10 overload protection in accordance with 430.32.

c. Overcurrent protection is provided in accordance with 240.4(D)(1)(2).

(2) Motor circuits with a full-load ampacity of 3.5 amperes or less if all the following conditions are met:

a. The circuit is protected in accordance with 430.52.

b. The circuit is provided with maximum Class 20 overload protection in accordance with 430.32.

c. Overcurrent protection is provided in accordance with 240.4(D)(1)(2).

Stallcup's Comment: A new subsection has been added to address the requirements for 18 AWG copper conductors used for small motors.

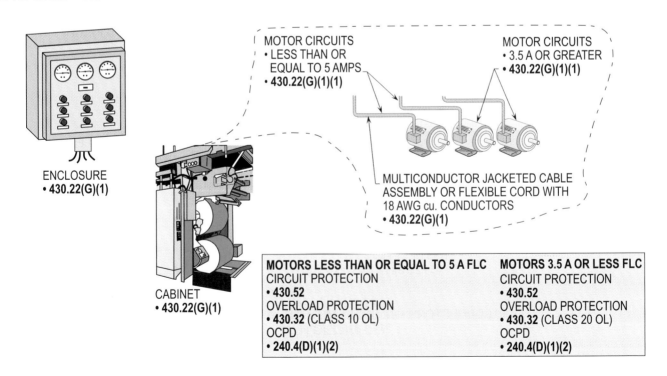

ENCLOSURE
• 430.22(G)(1)

CABINET
• 430.22(G)(1)

MOTOR CIRCUITS
• LESS THAN OR EQUAL TO 5 AMPS
• 430.22(G)(1)(1)

MOTOR CIRCUITS
• 3.5 A OR GREATER
• 430.22(G)(1)(1)

MULTICONDUCTOR JACKETED CABLE ASSEMBLY OR FLEXIBLE CORD WITH 18 AWG cu. CONDUCTORS
• 430.22(G)(1)

MOTORS LESS THAN OR EQUAL TO 5 A FLC	MOTORS 3.5 A OR LESS FLC
CIRCUIT PROTECTION	CIRCUIT PROTECTION
• 430.52	• 430.52
OVERLOAD PROTECTION	OVERLOAD PROTECTION
• 430.32 (CLASS 10 OL)	• 430.32 (CLASS 20 OL)
OCPD	OCPD
• 240.4(D)(1)(2)	• 240.4(D)(1)(2)

**CONDUCTORS FOR SMALL MOTORS – 18 AWG COPPER
NEC 430.22(G)(1)(1) AND (G)(1)(2)**

Purpose of Change: To clarify the requirements for small motor conductors with 18 AWG copper.

Type of Change	New Subsection			Committee Change	Accept		2008 NEC	-
ROP	pg. 576	# 11-50	log: 4794	ROC	pg. 284	# 11-17a	log: CC 1101	UL	1004
Submitter: David Drennan				Submitter: Code-Making Panel 11			OSHA	-
NFPA 70B	-			NFPA 70E	-		NFPA 79	14.5

430.22 Single Motor.

(G) Conductors for Small Motors. Conductors for small motors shall not be smaller than 14 AWG unless otherwise permitted in 430.22(G)(1) or (G)(2).

(2) 16 AWG Copper. Where installed in a cabinet or enclosure, 16 AWG individual copper conductors, copper conductors that are part of a jacketed multiconductor cable assembly, or copper conductors in a flexible cord shall be permitted under either of the following sets of conditions:

(1) Motor circuits with a full-load ampacity greater than 5.5 amperes or less than or equal to 8 amperes if all the following conditions are met:

Stallcup's Comment: A new subsection has been added to address the requirements for 16 AWG copper conductors used for small motors.

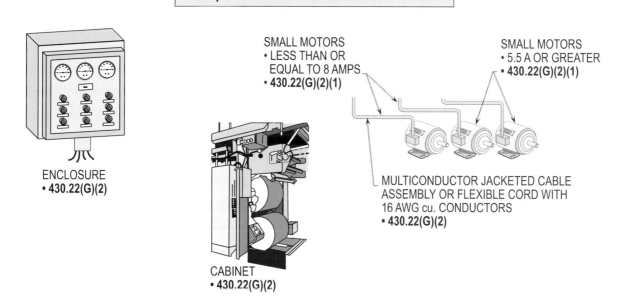

CONDUCTORS FOR SMALL MOTORS – 16 AWG COPPER
NEC 430.22(G)(2)(1)

Purpose of Change: To clarify the requirements for small motor conductors with 16 AWG copper.

Type of Change	New Subsection		Committee Change	Accept		2008 NEC	-
ROP	pg. 576	# 11-50	log: 4794	ROC	pg. 284	# 11-17a	log: CC 1101	UL	1004
Submitter: David Drennan			Submitter: Code-Making Panel 11			OSHA	-
NFPA 70B	-		NFPA 70E	-			NFPA 79	14.5

430.22 Single Motor.

(G) Conductors for Small Motors.

(2) 16 AWG Copper.

(1) Motor circuits with a full-load ampacity greater than 5.5 amperes or less than or equal to 8 amperes if all the following conditions are met:

a. The circuit is protected in accordance with 430.52.

b. The circuit is provided with maximum Class 10 overload protection in accordance with 430.32.

c. Overcurrent protection is provided in accordance with 240.4(D)(2)(2).

(2) Motor circuits with a full-load ampacity of 5.5 amperes or less if all the following conditions are met:

a. The circuit is protected in accordance with 430.52.

b. The circuit is provided with maximum Class 20 overload protection in accordance with 430.32.

c. Overcurrent protection is provided in accordance with 240.4(D)(2)(2).

Stallcup's Comment: A new subsection has been added to address the requirements for 16 AWG copper conductors used for small motors.

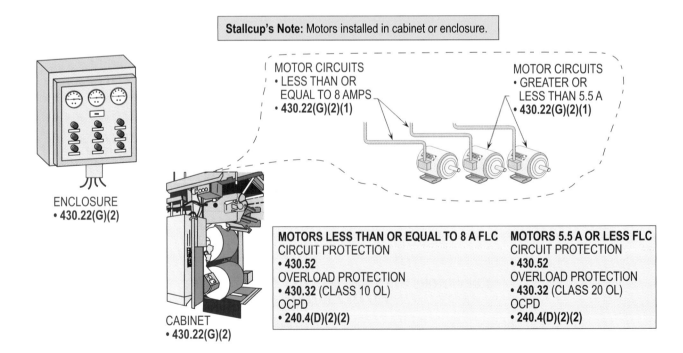

Stallcup's Note: Motors installed in cabinet or enclosure.

MOTOR CIRCUITS
• LESS THAN OR EQUAL TO 8 AMPS
• 430.22(G)(2)(1)

MOTOR CIRCUITS
• GREATER OR LESS THAN 5.5 A
• 430.22(G)(2)(1)

ENCLOSURE
• 430.22(G)(2)

CABINET
• 430.22(G)(2)

MOTORS LESS THAN OR EQUAL TO 8 A FLC	MOTORS 5.5 A OR LESS FLC
CIRCUIT PROTECTION	CIRCUIT PROTECTION
• 430.52	• 430.52
OVERLOAD PROTECTION	OVERLOAD PROTECTION
• 430.32 (CLASS 10 OL)	• 430.32 (CLASS 20 OL)
OCPD	OCPD
• 240.4(D)(2)(2)	• 240.4(D)(2)(2)

CONDUCTORS FOR SMALL MOTORS – 16 AWG COPPER
NEC 430.22(G)(2)(1) AND (G)(2)(2)

Purpose of Change: To clarify the requirements for small motor conductors with 16 AWG copper.

Type of Change	Revision			Committee Change	Accept			2008 NEC	430.24
ROP	pg. 576	# 11-50a	log: CP 1101	ROC	pg. -	# -	log: -	UL	1004
Submitter: Code-Making Panel 11				Submitter:				OSHA	-
NFPA 70B	-			NFPA 70E	-			NFPA 79	14.5

430.24 Several Motors or a Motor(s) and Other Load(s).

Conductors supplying several motors, or a motor(s) and other load(s), shall have an ampacity not less than the sum of each of the following:

(1) 125 percent of the full-load current rating of the highest rated motor, as determined by 430.6(A)

(2) Sum of the full-load current ratings of all the other motors in the group, as determined by 430.6(A)

(3) 100 percent of the noncontinuous non-motor load

(4) 125 percent of the continuous non-motor load

Stallcup's Comment: A revision has been made to clarify the calculation ampacity method for conductors supplying several motors.

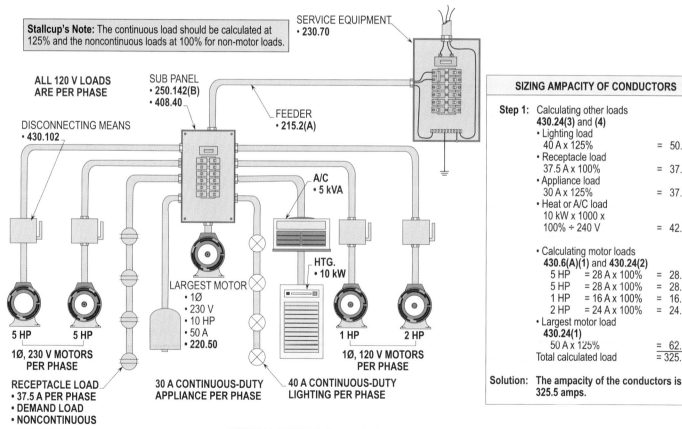

SEVERAL MOTORS OR A MOTOR(S) AND OTHER LOAD(S)
NEC 430.24(1) THRU (4)

Purpose of Change: To clarify the requirements for determining the ampacity of motor(s) and other loads.

Type of Change	Revision			Committee Change	Accept			2008 NEC	430.53(C)(1)
ROP	pg. 580	# 11-69	log: 3715	ROC	pg. -	# -	log: -	UL	1004
Submitter: Vince Baclawski				Submitter:				OSHA	-
NFPA 70B	16.6			NFPA 70E	-			NFPA 79	14.5

430.53 Several Motors or Loads on One Branch Circuit.

(C) Other Group Installations. Two or more motors of any rating or one or more motors and other load(s), with each motor having individual overload protection, shall be permitted to be connected to one branch circuit where the motor controller(s) and overload device(s) are (1) installed as a listed factory assembly and the motor branch-circuit short-circuit and ground-fault protective device either is provided as part of the assembly or is specified by a marking on the assembly, or (2) the motor branch-circuit short-circuit and ground-fault protective device, the motor controller(s), and overload device(s) are field-installed as separate assemblies listed for such use and provided with manufacturers' instructions for use with each other, and (3) all of the following conditions are complied with:

(1) Each motor overload device is either (a) listed for group installation with a specified maximum rating of fuse, inverse time circuit breaker, or both, or (b) selected such that the ampere rating of the motor-branch short-circuit and ground-fault protective device does not exceed that permitted by 430.52 for that individual motor overload device and corresponding motor load.

Stallcup's Comment: A revision has been made to clarify the requirements for each motor overload device used in other group installations.

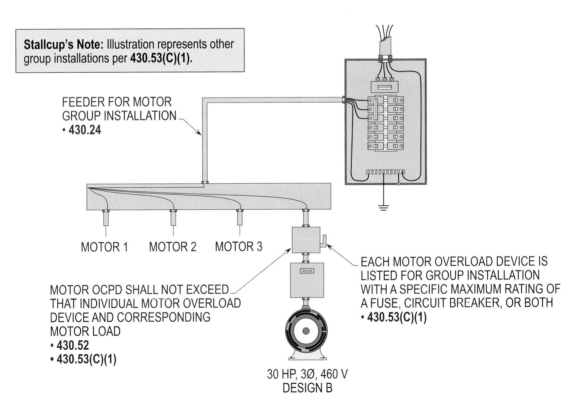

Stallcup's Note: Illustration represents other group installations per **430.53(C)(1).**

FEEDER FOR MOTOR GROUP INSTALLATION
• **430.24**

MOTOR 1   MOTOR 2   MOTOR 3

MOTOR OCPD SHALL NOT EXCEED THAT INDIVIDUAL MOTOR OVERLOAD DEVICE AND CORRESPONDING MOTOR LOAD
• **430.52**
• **430.53(C)(1)**

EACH MOTOR OVERLOAD DEVICE IS LISTED FOR GROUP INSTALLATION WITH A SPECIFIC MAXIMUM RATING OF A FUSE, CIRCUIT BREAKER, OR BOTH
• **430.53(C)(1)**

30 HP, 3Ø, 460 V
DESIGN B

OTHER GROUP INSTALLATIONS
NEC 430.53(C)(1)

Purpose of Change: To clarify the requirements for each motor overload device when used in other group installations.

Type of Change	Revision			Committee Change	Accept			2008 NEC	430.53(C)(2)
ROP	pg. 580	# 11-69	log: 3715	ROC	pg. -	# -	log: -	UL	1004
Submitter: Vince Baclawski				Submitter:				OSHA	-
NFPA 70B	16.6			NFPA 70E	-			NFPA 79	14.5

430.53 Several Motors or Loads on One Branch Circuit.

(C) Other Group Installations. Two or more motors of any rating or one or more motors and other load(s), with each motor having individual overload protection, shall be permitted to be connected to one branch circuit where the motor controller(s) and overload device(s) are (1) installed as a listed factory assembly and the motor branch-circuit short-circuit and ground-fault protective device either is provided as part of the assembly or is specified by a marking on the assembly, or (2) the motor branch-circuit short-circuit and ground-fault protective device, the motor controller(s), and overload device(s) are field-installed as separate assemblies listed for such use and provided with manufacturers' instructions for use with each other, and (3) all of the following conditions are complied with:

(2) Each motor controller is either (a) listed for group installation with a specified maximum rating of fuse, circuit breaker, or both, or (b) selected such that the ampere rating of the motor-branch short-circuit and ground-fault protective device does not exceed that permitted by 430.52 for that individual controller and corresponding motor load.

Stallcup's Comment: A revision has been made to clarify the requirements for each motor overload device used in other group installations.

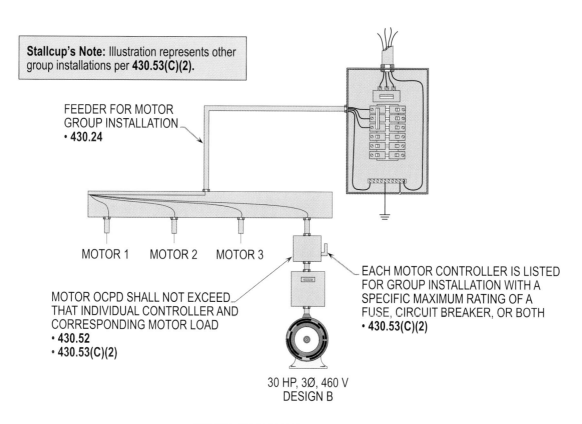

Stallcup's Note: Illustration represents other group installations per **430.53(C)(2).**

FEEDER FOR MOTOR
GROUP INSTALLATION
• **430.24**

MOTOR 1   MOTOR 2   MOTOR 3

MOTOR OCPD SHALL NOT EXCEED
THAT INDIVIDUAL CONTROLLER AND
CORRESPONDING MOTOR LOAD
• **430.52**
• **430.53(C)(2)**

EACH MOTOR CONTROLLER IS LISTED
FOR GROUP INSTALLATION WITH A
SPECIFIC MAXIMUM RATING OF A
FUSE, CIRCUIT BREAKER, OR BOTH
• **430.53(C)(2)**

30 HP, 3Ø, 460 V
DESIGN B

OTHER GROUP INSTALLATIONS
NEC 430.53(C)(2)

Purpose of Change: To clarify the requirements for each motor overload device when used in other group installations.

Type of Change	New Section			Committee Change	Accept in Principle			2008 NEC	-
ROP	pg. 599	# 5-176	log: 3821	ROC	pg. -	# -	log: -	UL	1561
Submitter: James J. Rogers				Submitter: -				OSHA	-
NFPA 70B	-			NFPA 70E	-			NFPA 79	7.2.7.1

450.14 Disconnecting Means.

Transformers, other than Class 2 or Class 3 transformers, shall have a disconnecting means located either in sight of the transformer or in a remote location. Where located in a remote location, the disconnecting means shall be lockable, and the location shall be field marked on the transformer.

Stallcup's Comment: A new section has been added to require a disconnecting means to be located either in sight of the transformer or in a remote location for other than Class 2 or Class 3 transformers. Transformers located in a remote location shall be of the lockable type and the location shall be field marked on the transformer.

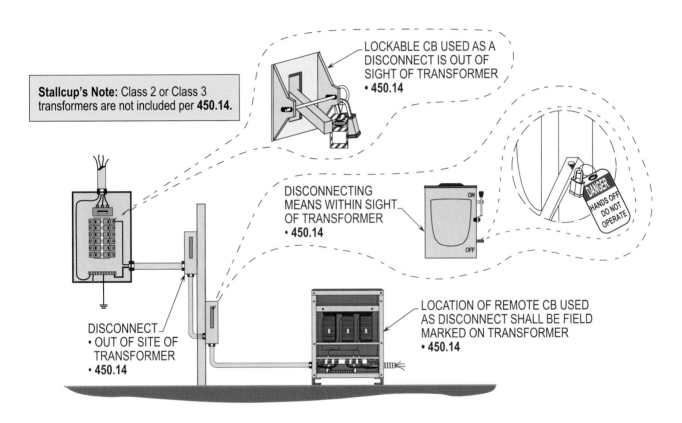

Stallcup's Note: Class 2 or Class 3 transformers are not included per **450.14**.

LOCKABLE CB USED AS A DISCONNECT IS OUT OF SIGHT OF TRANSFORMER
• 450.14

DISCONNECTING MEANS WITHIN SIGHT OF TRANSFORMER
• 450.14

DISCONNECT
• OUT OF SITE OF TRANSFORMER
• 450.14

LOCATION OF REMOTE CB USED AS DISCONNECT SHALL BE FIELD MARKED ON TRANSFORMER
• 450.14

DISCONNECTING MEANS
NEC 450.14

Purpose of Change: To require a disconnecting means for transformers other than Class 2 and Class 3.

Name   Date

Chapter 4
Equipment for General Use

Section   Answer

1. Where switches control lighting loads supplied by a grounded general-purpose branch circuit, the _____ circuit conductor for the controlled lighting circuit shall be provided at the switch location.

(A) bonding   **(B)** equipment
(C) grounded   **(D)** grounding

2. Where no means exist within the snap-switch enclosure for connecting to the equipment grounding conductor, a snap-switch wired and located within _____ ft vertically, or 5 ft horizontally, of ground or exposed grounded metal objects shall be provided with a faceplate of nonconducting noncombustible material with nonmetallic attachment screws, unless the switch mounting strap or yoke is nonmetallic or the circuit is protected by a ground-fault circuit interrupter.

(A) 5   **(B)** 8
(C) 10   **(D)** 12

3. Where a receptacle outlet is supplied by a branch circuit that requires arc-fault circuit interrupter protection as specified elsewhere in the *National Electrical Code*, a replacement receptacle at this outlet shall become effective _____.

(A) January 1, 2011   **(B)** January 1, 2012
(C) January 1, 2013   **(D)** January 1, 2014

4. For other than one- and two-family dwellings, an outlet box hood installed for this purpose shall be listed, and where installed on an enclosure supported from grade, it shall be identified as _____.

(A) extra duty   **(B)** extra hard
(C) extra usage   **(D)** all of the above

5. Receptacles located more than _____ ft above the floor shall not be required to be tamper-resistant.

(A) 5   **(B)** 5-1/2
(C) 6   **(D)** 6-1/2

6. In which of the following locations shall tamper-resistant receptacle be provided:

(A) Dwelling units   **(B)** Guest rooms
(C) Child care facilities   **(D)** All of the above

7. Luminaires made of insulating material that is directly wired or attached to outlets supplied by a wiring method that does not provide a ready means for grounding attachment to an _____ conductor shall be made of insulating material and shall have no exposed conductive parts.

(A) bonding   **(B)** equipment grounding
(C) grounded   **(D)** ungrounded

Section   Answer

8. Luminaires _____ and marked for use as a raceway shall be permitted to be used as a raceway.

(A) approved   (B) identified
(C) listed   (D) labeled

9. Lampholders shall be constructed, installed, or equipped with shades or guards so that combustible material is not subjected to temperatures in excess of _____ °C.

(A) 60   (B) 65
(C) 75   (D) 90

10. Thermal insulation shall not be installed above a recessed luminaire or within _____ in. of the recessed luminaire's enclosure, wiring compartment, ballast, transformer, LED driver, or power supply unless the luminaire is identified as Type IC for insulation contact.

(A) 3   (B) 4
(C) 6   (D) 12

11. For permanently connected motor-operated appliances with motors rated over _____ horsepower, the branch-circuit switch or circuit breaker shall be permitted to serve as the disconnecting means where the switch or circuit breaker is within sight from the appliance.

(A) 1/16   (B) 1/8
(C) 1/4   (D) 1/2

12. For dc motors operating from a rectified power supply, the conductor ampacity on the input of the rectifier shall not be less than _____ percent of the rated input current to the rectifier.

(A) 100   (B) 125
(C) 135   (D) 150

13. For a wye-start, delta-run motor, the ampacity of the conductors between the controller and the motor shall not be less than _____ percent of the motor full-load current rating.

(A) 33   (B) 47
(C) 62.5   (D) 72

14. For a part-winding connected motor, the ampacity of the conductors between the controller and the motor shall not be less than _____ percent of the motor full-load current rating.

(A) 33   (B) 47
(C) 62.5   (D) 72

15. Conductors for small motors shall not be smaller than _____ AWG unless other conditions of use are permitted.

(A) 14   (B) 12
(C) 10   (D) 8

5

Special Occupancies

Chapter 5 of the *National Electrical Code* has always been known as the special chapter, covering specific subjects that have particular functions or purposes. **Chapter 5** deals with places or locations where people work, and such occupancies can have built-in electrical hazards.

The main function of this chapter is to protect personnel and equipment from electrical hazards that could occur in a particular work area, based on the type of occupancy.

For example, designers, installers, and inspectors use the requirements of **Chapter 5** to design, install, and inspect the electrical wiring methods and equipment located in hazardous areas.

Chapter 5 is special because it is based on occupancies that prohibit a large portion of the design and installation techniques from falling under the requirements of **Chapters 1 through 4**.

Type of Change	Revision and Exception			Committee Change	Accept in Principle			2008 NEC	500.8(E)(1)
ROP	pg. 621	# 14-33	log: 2826	**ROC**	pg. 323	# 14-20	log: 2619	**UL**	1604
Submitter: Donald W. Ankele				Submitter: Frederic P. Hartwell				**OSHA**	1910.307(d)
NFPA 496	-			**NFPA 497**	-			**NFPA 499**	-

500.8 Equipment.

(E) Threading.

(1) Equipment Provided with Threaded Entries for NPT Threaded Conduit or Fittings. For equipment provided with threaded entries for NPT threaded conduit or fittings, listed conduit, conduit fittings, or cable fittings shall be used. All NPT threaded conduit and fittings shall be threaded with a National (American) Standard Pipe Taper (NPT) thread.

NPT threaded entries into explosionproof equipment shall be made up with at least five threads fully engaged.

Exception: For listed explosionproof equipment, joints with factory threaded NPT entries shall be made up with at least 4 1/2 threads fully engaged.

Stallcup's Comment: A revision has been made to clarify the threading requirements for equipment provided with threaded entries for NPT threaded conduit or fittings.

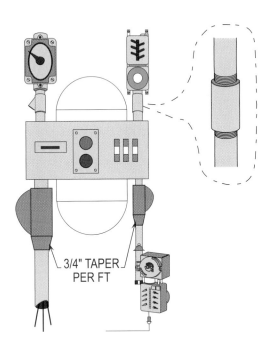

3/4" TAPER PER FT

Stallcup's Note 1: All NPT threaded conduit and fittings shall be threaded with National (American) Standard Pipe Taper (NPT) thread per **500.8(E)(1).**

Stallcup's Note 2: Threaded entries into explosionproof equipment shall be made with at least five threads fully engaged per **500.8(E)(1).**

Stallcup's Note 3: Listed factory explosionproof equipment shall be permitted to have 4-1/2 threads fully engaged per **500.8(E)(1), Ex.**

EQUIPMENT PROVIDED WITH THREADED ENTRIES FOR NPT THREADED CONDUIT OR FITTINGS
NEC 500.8(E)(1) AND Ex.

Purpose of Change: Revised to add requirements for threading that pertains to factory or other threaded NPT entries.

Type of Change	Revision			Committee Change	Accept in Principle			2008 NEC	500.8(E)(2)
ROP	pg. 621	# 14-33	log: 2826	ROC	pg. 323	# 14-21	log: 2157	UL	1604
Submitter: Donsald W. Ankele				Submitter: Jeremy Neagle				OSHA	1910.307(d)
NFPA 496	-			NFPA 497	-			NFPA 499	-

500.8 Equipment.

(E) Threading.

(2) Equipment Provided with Threaded Entries for Metric Threaded Conduit or Fittings. For equipment with metric threaded entries, listed conduit fittings or listed cable fittings shall be used. Such entries shall be identified as being metric, or listed adapters to permit connection to conduit or NPT threaded fittings shall be provided with the equipment and shall be used for connection to conduit or NPT threaded fittings.

Metric threaded entries into explosionproof equipment shall have a class of fit of at least 6g/6H and shall be made up with at least five threads fully engaged for Group C and Group D, and at least eight threads fully engaged for Group A and Group B.

Stallcup's Comment: A revision has been made to clarify the threading requirements for equipment provided with threaded entries for metric conduit or fittings.

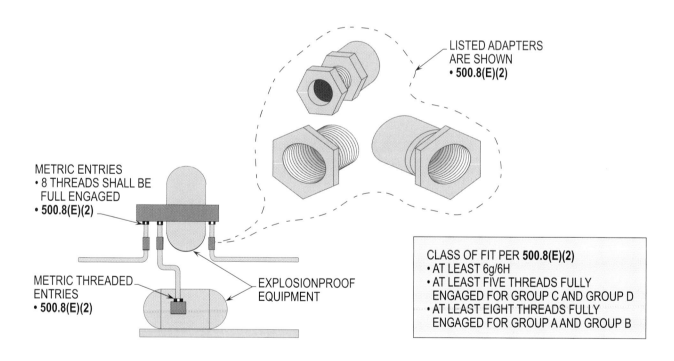

LISTED ADAPTERS
ARE SHOWN
• **500.8(E)(2)**

METRIC ENTRIES
• 8 THREADS SHALL BE
  FULL ENGAGED
• **500.8(E)(2)**

METRIC THREADED
ENTRIES
• **500.8(E)(2)**

EXPLOSIONPROOF
EQUIPMENT

CLASS OF FIT PER **500.8(E)(2)**
• AT LEAST 6g/6H
• AT LEAST FIVE THREADS FULLY
  ENGAGED FOR GROUP C AND GROUP D
• AT LEAST EIGHT THREADS FULLY
  ENGAGED FOR GROUP A AND GROUP B

**EQUIPMENT PROVIDED WITH THREADED ENTRIES FOR METRIC THREADED CONDUIT OR FITTINGS
NEC 500.8(E)(2)**

Purpose of Change: Revised to add requirements for threading that pertains to factory or other threaded NPT entries

Type of Change	New Subdivision		Committee Change	Accept in Principle		2008 NEC	-
ROP pg. 621	# 14-33	log: 2826	**ROC** pg. -	# -	log: -	**UL**	1604
Submitter: Donald W. Ankele			Submitter: -			**OSHA**	-
NFPA 496 -			**NFPA 497** -			**NFPA 499**	-

500.8 Equipment.

(E) Threading.

(3) Unused Openings. All unused openings shall be closed with listed metal close-up plugs. The plug engagement shall comply with 500.8(E)(1) or 500.8(E)(2).

Stallcup's Comment: A new subdivision has been added to address the requirements for unused openings in hazardous (classified) locations.

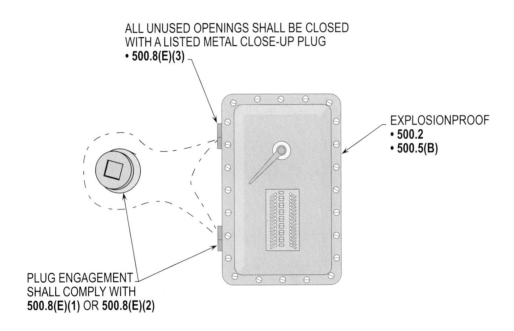

ALL UNUSED OPENINGS SHALL BE CLOSED
WITH A LISTED METAL CLOSE-UP PLUG
• **500.8(E)(3)**

EXPLOSIONPROOF
• **500.2**
• **500.5(B)**

PLUG ENGAGEMENT
SHALL COMPLY WITH
500.8(E)(1) OR **500.8(E)(2)**

UNUSED OPENINGS
NEC 500.8(E)(3)

Purpose of Change: To add requirements for unused openings for equipment located in hazardous (classified) locations.

Type of Change	New Section			Committee Change	Accept in Principle			2008 NEC	-
ROP	pg. 626	# 14-60	log: 4441	ROC	pg. 326	# 14-38	log: 6	UL	1604
Submitter: Eliana Beattie				Submitter: Ted H. Schnaare				OSHA	-
NFPA 496	-			NFPA 497	-			NFPA 499	-

501.17 Process Sealing.

This section shall apply to process-connected equipment, which includes, but is not limited to, canned pumps, submersible pumps, flow, pressure, temperature, or analysis measurement instruments. A process seal is a device to prevent the migration of process fluids from the designed containment into the external electrical system. Process connected electrical equipment that incorporates a single process seal, such as a single compression seal, diaphragm, or tube to prevent flammable or combustible fluids from entering a conduit or cable system capable of transmitting fluids, shall be provided with an additional means to mitigate a single process seal failure. The additional means may include, but is not limited to the following:

Stallcup's Comment: A new section has been added to address the requirements for process sealing in Class I locations.

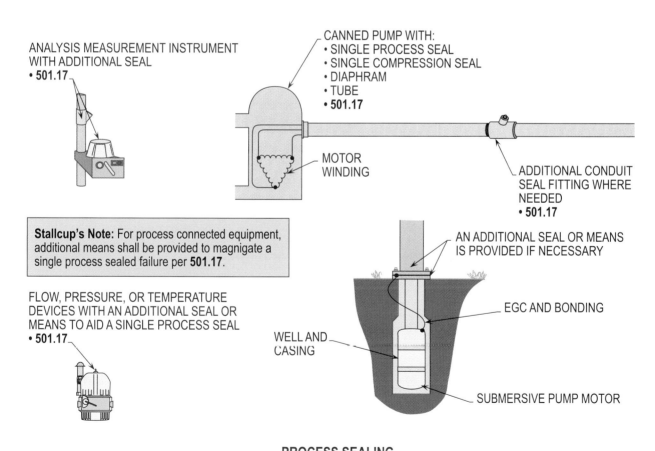

ANALYSIS MEASUREMENT INSTRUMENT
WITH ADDITIONAL SEAL
• **501.17**

CANNED PUMP WITH:
• SINGLE PROCESS SEAL
• SINGLE COMPRESSION SEAL
• DIAPHRAM
• TUBE
• **501.17**

MOTOR
WINDING

ADDITIONAL CONDUIT
SEAL FITTING WHERE
NEEDED
• **501.17**

Stallcup's Note: For process connected equipment, additional means shall be provided to magnigate a single process sealed failure per **501.17**.

FLOW, PRESSURE, OR TEMPERATURE
DEVICES WITH AN ADDITIONAL SEAL OR
MEANS TO AID A SINGLE PROCESS SEAL
• **501.17**

AN ADDITIONAL SEAL OR MEANS
IS PROVIDED IF NECESSARY

EGC AND BONDING

WELL AND
CASING

SUBMERSIVE PUMP MOTOR

**PROCESS SEALING
NEC 501.17**

Purpose of Change: To clarify the requirements for process sealing in Class 1 locations.

Type of Change	New Section			Committee Change	Accept in Principle			2008 NEC	-
ROP	pg. 626	# 14-60	log: 4441	ROC	pg. 326	# 14-38	log: 6	UL	1604
Submitter: Eliana Beattie				Submitter: Ted H. Schnaar				OSHA	-
NFPA 70B	-			NFPA 70E	-			NFPA 79	-

501.17 Process Sealing.

This section shall apply to process-connected equipment, which includes, but is not limited to, canned pumps, submersible pumps, flow, pressure, temperature, or analysis measurement instruments. A process seal is a device to prevent the migration of process fluids from the designed containment into the external electrical system. Process connected electrical equipment that incorporates a single process seal, such as a single compression seal, diaphragm, or tube to prevent flammable or combustible fluids from entering a conduit or cable system capable of transmitting fluids, shall be provided with an additional means to mitigate a single process seal failure. The additional means may include, but is not limited to the following:

(1) A suitable barrier meeting the process temperature and pressure conditions that the barrier will be subjected to upon failure of the single process seal. There shall be a vent or drain between the single process seal and the suitable barrier. Indication of the single process seal failure shall be provided by visible leakage, an audible whistle, or other means of monitoring.

(2) A listed Type MI cable assembly, rated at not less than 125 percent of the process pressure and not less than 125 percent of the maximum process temperature (in degrees Celsius), installed between the cable or conduit and the single process seal.

Stallcup's Comment: A new section has been added to address the requirements for process sealing in Class I locations.

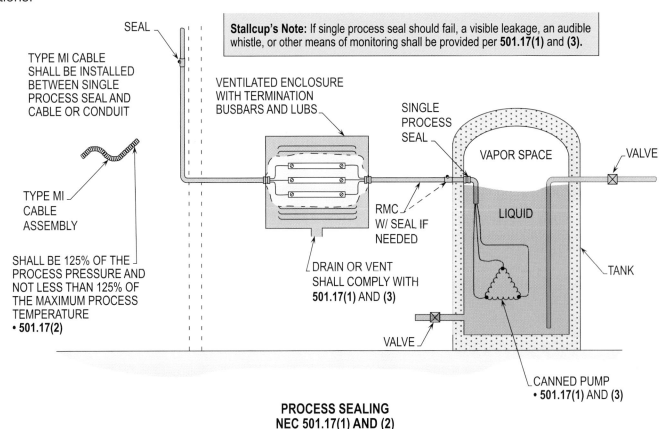

PROCESS SEALING
NEC 501.17(1) AND (2)

Purpose of Change: To clarify the requirements for process sealing in Class 1 locations.

Type of Change	New Section			Committee Change	Accept in Principle			2008 NEC	-
ROP	pg. 626	# 14-60	log: 4441	ROC	pg. 326	# 14-38	log: 6	UL	1604
Submitter: Eliana Beattie				Submitter: Ted H. Schnaare				OSHA	-
NFPA 70B	-			NFPA 70E	-			NFPA 79	-

501.17 Process Sealing.

This section shall apply to process-connected equipment, which includes, but is not limited to, canned pumps, submersible pumps, flow, pressure, temperature, or analysis measurement instruments. A process seal is a device to prevent the migration of process fluids from the designed containment into the external electrical system. Process connected electrical equipment that incorporates a single process seal, such as a single compression seal, diaphragm, or tube to prevent flammable or combustible fluids from entering a conduit or cable system capable of transmitting fluids, shall be provided with an additional means to mitigate a single process seal failure. The additional means may include, but is not limited to the following:

(3) A drain or vent located between the single process seal and a conduit or cable seal. The drain or vent shall be sufficiently sized to prevent overpressuring the conduit or cable seal above 6 in. water column (1493 Pa). Identification of the single process seal failure shall be provided by visible leakage, an audible whistle, or other means of monitoring.

Process-connected electrical equipment that does not rely on a single process seal or is listed and marked "single seal" or "dual seal" shall not be required to be provided with an additional means of sealing.

Stallcup's Comment: A new section has been added to address the requirements for process sealing in Class I locations.

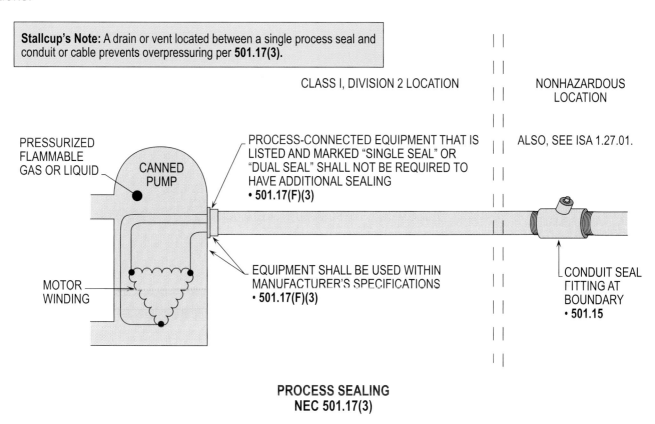

Stallcup's Note: A drain or vent located between a single process seal and conduit or cable prevents overpressuring per **501.17(3)**.

CLASS I, DIVISION 2 LOCATION

NONHAZARDOUS LOCATION

PRESSURIZED FLAMMABLE GAS OR LIQUID

CANNED PUMP

PROCESS-CONNECTED EQUIPMENT THAT IS LISTED AND MARKED "SINGLE SEAL" OR "DUAL SEAL" SHALL NOT BE REQUIRED TO HAVE ADDITIONAL SEALING
• **501.17(F)(3)**

ALSO, SEE ISA 1.27.01.

MOTOR WINDING

EQUIPMENT SHALL BE USED WITHIN MANUFACTURER'S SPECIFICATIONS
• **501.17(F)(3)**

CONDUIT SEAL FITTING AT BOUNDARY
• **501.15**

**PROCESS SEALING
NEC 501.17(3)**

Purpose of Change: To recognize a listed and marked "single seal" or "dual seal" that eliminates the requirement for an additional means of sealing.

Type of Change	New Subdivision			Committee Change	Accept			2008 NEC	-
ROP	pg. 631	# 14-85	log: 3192	ROC	pg. -	# -	log: -	UL	1604
Submitter: A. W. Ballard				Submitter:				OSHA	1910.307(c)
NFPA 496	-			NFPA 497	-			NFPA 499	-

501.140 Flexible Cords, Class I, Divisions 1 and 2.

(A) Permitted Uses. Flexible cord shall be permitted:

(5) For temporary portable assemblies consisting of receptacles, switches, and other devices that are not considered portable utilization equipment but are individually listed for the location.

Stallcup's Comment: A new subdivision has been added to address the permitted use of flexible cord for temporary portable assemblies.

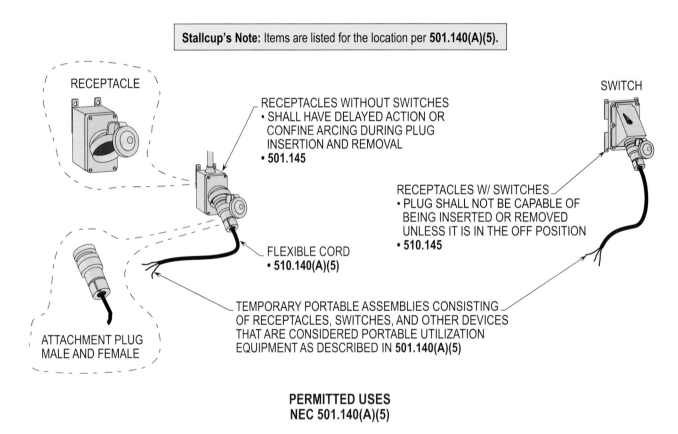

Stallcup's Note: Items are listed for the location per **501.140(A)(5)**.

RECEPTACLE

RECEPTACLES WITHOUT SWITCHES
• SHALL HAVE DELAYED ACTION OR CONFINE ARCING DURING PLUG INSERTION AND REMOVAL
• **501.145**

SWITCH

RECEPTACLES W/ SWITCHES
• PLUG SHALL NOT BE CAPABLE OF BEING INSERTED OR REMOVED UNLESS IT IS IN THE OFF POSITION
• **510.145**

FLEXIBLE CORD
• **510.140(A)(5)**

ATTACHMENT PLUG MALE AND FEMALE

TEMPORARY PORTABLE ASSEMBLIES CONSISTING OF RECEPTACLES, SWITCHES, AND OTHER DEVICES THAT ARE CONSIDERED PORTABLE UTILIZATION EQUIPMENT AS DESCRIBED IN **501.140(A)(5)**

PERMITTED USES
NEC 501.140(A)(5)

Purpose of Change: To add requirements for the permitted use of flexible cord for temporary portable assemblies under certain conditions of use.

Type of Change	New Subdivision			Committee Change	Accept			2008 NEC	-
ROP	pg. 631	# 14-88	log: 2830	ROC	pg. 329	# 14-53a	log: CC 1400	UL	1604
Submitter: Donald W. Ankele				Submitter: Code-Making Panel 14				OSHA	1910.307(c)
NFPA 496	-			NFPA 497	-			NFPA 499	-

501.140 Flexible Cords, Class I, Divisions 1 and 2.

(B) Installation. Where flexible cords are used, the cords shall comply with all of the following:

(4) In Division 1 locations or in Division 2 locations where the boxes, fittings, or enclosures are required to be explosionproof, the cord shall be terminated with a cord connector or attachment plug listed for the location or a cord connector installed with a seal listed for the location. In Division 2 locations where explosionproof equipment is not required, the cord shall be terminated with a listed cord connector or listed attachment plug.

Stallcup's Comment: A new subdivision has been added to address the installation requirements for flexible cords in Division 1 or Division 2 locations.

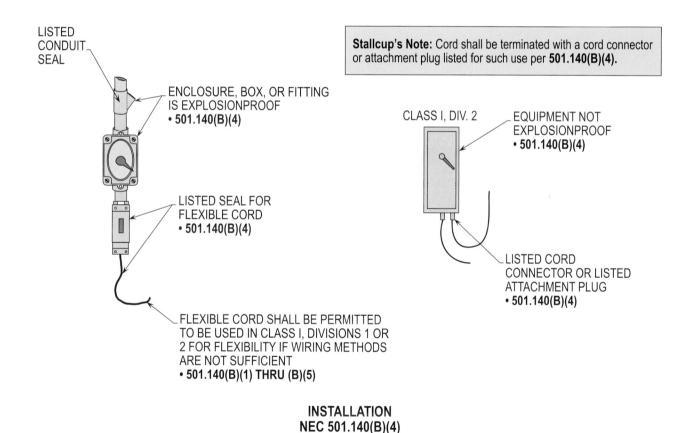

Stallcup's Note: Cord shall be terminated with a cord connector or attachment plug listed for such use per **501.140(B)(4).**

LISTED CONDUIT SEAL

ENCLOSURE, BOX, OR FITTING IS EXPLOSIONPROOF
• **501.140(B)(4)**

LISTED SEAL FOR FLEXIBLE CORD
• **501.140(B)(4)**

FLEXIBLE CORD SHALL BE PERMITTED TO BE USED IN CLASS I, DIVISIONS 1 OR 2 FOR FLEXIBILITY IF WIRING METHODS ARE NOT SUFFICIENT
• **501.140(B)(1) THRU (B)(5)**

CLASS I, DIV. 2

EQUIPMENT NOT EXPLOSIONPROOF
• **501.140(B)(4)**

LISTED CORD CONNECTOR OR LISTED ATTACHMENT PLUG
• **501.140(B)(4)**

**INSTALLATION
NEC 501.140(B)(4)**

Purpose of Change: To recognize new material pertaining to the installation requirements for flexible cords and accessories in Class I, Divisions 1 and 2 locations.

Type of Change	New Subdivision			Committee Change	Accept in Principle			2008 NEC	-
ROP	pg. 631	# 14-89	log: 3193	ROC	pg. 329	# 14-54	log: 1648	UL	1604
Submitter: A. W. Ballard				Submitter: A. W. Ballard				OSHA	1910.307(c)
NFPA 496	-			NFPA 497	-			NFPA 499	-

501.140 Flexible Cords, Class I, Divisions 1 and 2.

(B) Installation. Where flexible cords are used, the cords shall comply with all of the following:

(5) Be of continuous length. Where 501.140(A)(5) is applied, cords shall be of continuous length from the power source to the temporary portable assembly and from the temporary portable assembly to the utilization equipment.

Stallcup's Comment: A new subdivision has been added to address the installation of flexible cords in Class I, Division 1 and 2 locations.

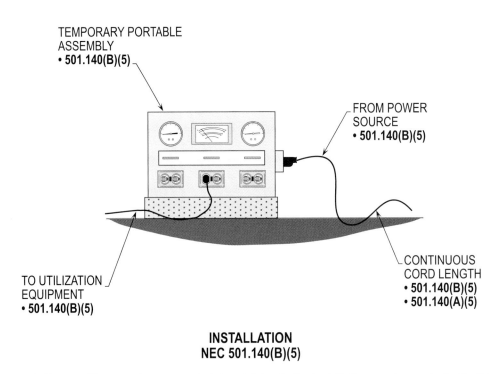

INSTALLATION
NEC 501.140(B)(5)

Purpose of Change: To recognize new material pertaining to the installation requirements for flexible cords and accessories in Class I, Divisions 1 and 2 locations.

Type of Change	New Section			Committee Change	Accept			2008 NEC	-
ROP	pg. 634	# 14-93	log: 4442	ROC	pg. -	# -	log: -	UL	1604
Submitter: Eliana Beattie				Submitter: -				OSHA	-
NFPA 496	-			NFPA 497	-			NFPA 499	-

502.6 Zone Equipment.

Equipment listed and marked in accordance with 506.9(C)(2) for Zone 20 locations shall be permitted in Class II, Division 1 locations for the same dust atmosphere; and with a suitable temperature class.

Equipment listed and marked in accordance with 506.9(C)(2) for Zone 20, 21, or 22 locations shall be permited in Class II, Division 2 locations for the same dust atmosphere and with a suitable temperature class.

Stallcup's Comment: A new section has been added to address the requirements for zone equipment in Class II, Division 1 and 2 locations.

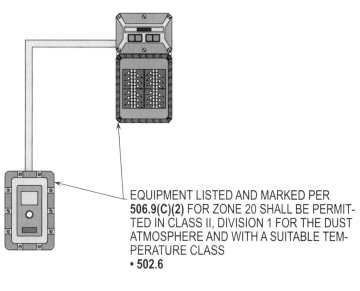

Stallcup's Note: The same requirement applies for Zone 20, 21, or 22 where installed in a Class II Division 1 location per **502.6.**

EQUIPMENT LISTED AND MARKED PER **506.9(C)(2)** FOR ZONE 20 SHALL BE PERMITTED IN CLASS II, DIVISION 1 FOR THE DUST ATMOSPHERE AND WITH A SUITABLE TEMPERATURE CLASS
• **502.6**

ZONE EQUIPMENT
NEC 502.6

Purpose of Change: To add requirements for zone equipment permitted to be installed in Class II, Divisions 1 and 2 locations.

Type of Change	New Section			Committee Change	Accept			2008 NEC	-
ROP	pg. 639	# 14-130	log: 4444	ROC	pg. -	# -	log: -	UL	1604
Submitter: Eliana Beattie				Submitter: -				OSHA	-
NFPA 496	-			NFPA 497	-			NFPA 499	-

503.6 Zone Equipment.

Equipment listed and marked in accordance with 506.9(C)(2) for Zone 20 locations and with a temperature class of not greater than T120°C (for equipment that may be overloaded) or not greater than T165°C (for equipment not subject to overloading) shall be permitted in Class III, Division 1 locations.

Equipment listed and marked in accordance with 506.9(C)(2) for Zone 20, 21, or 22 locations and with a temperature class of not greater than T120°C (for equipment that may be overloaded) or not greater than T165°C (for equipment not subject to overloading) shall be permitted in Class III, Division 2 locations.

Stallcup's Comment: A new section has been added to address the requirements for zone equipment in Class III, Division 1 and 2 locations.

Stallcup's Note 1: The same requirement applies for Zone 20, 21, or 22 and with a temperature class not greater than T165°C per **503.6.**

Stallcup's Note 2: Equipment is not subjected to overloading conditions per **503.6.**

EQUIPMENT LISTED AND MARKED PER **506.9(C)(2)** FOR ZONE 20 SHALL BE PERMITTED IN CLASS III, DIVISION 1 WITH A TEMPERATURE CLASS NOT GREATER THAN T120°C OR NOT GREATER THAN THAN T165°C
• **503.6.**
• **506.9(C)(2)**

ZONE EQUIPMENT
NEC 503.6

Purpose of Change: To add requirements for zone equipment permitted to be installed in Class III, Divisions 1 and 2 locations.

Type of Change	New Subdivision			Committee Change	Accept in Part			2008 NEC	-
ROP	pg. 640	# 14-131	log: 2837	ROC	pg. 334	# 14-69	log: 594	UL	1604
Submitter: Donald W. Ankele				Submitter: John L. Simmons				OSHA	1910.307(c)
NFPA 496	-			NFPA 497	-			NFPA 400	-

503.10 Wiring Methods.

(A) Class III, Division 1.

(1) General. In Class III, Division 1 locations, the wiring method shall be in accordance with (1) through (4):

(1) Rigid metal conduit, Type PVC conduit, Type RTRC conduit, intermediate metal conduit, electrical metallic tubing, dusttight wireways, or Type MC or MI cable with listed termination fittings.

(2) Type PLTC and PLTC-ER cable in accordance with the provisions of Article 725 including installation in cable tray systems. The cable shall be terminated with listed fittings.

(3) Type ITC and Type ITC-ER cable as permitted in 727.4 and terminated with listed fittings.

(4) Type MC, MI, or TC cable installed in ladder, ventilated trough, or ventilated channel cable trays in a single layer, with a space not less than the larger cable diameter between the two adjacent cables, shall be the wiring method employed.

Exception to (4): Type MC cable listed for use in Class II, Division 1 locations shall be permitted to be installed without the spacings required by 503.10(A)(1)(4).

Stallcup's Comment: A new section has been added to address the requirements for wiring methods in Class III, Division 1 locations.

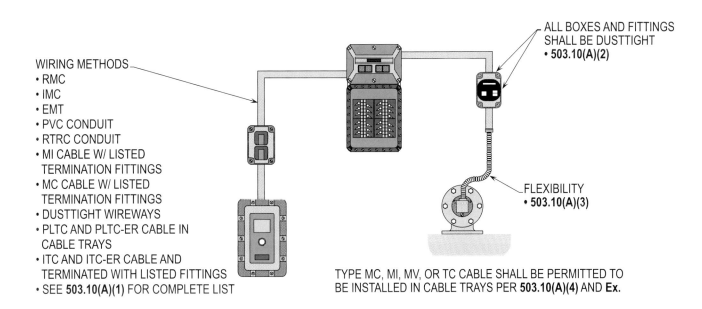

WIRING METHODS
• RMC
• IMC
• EMT
• PVC CONDUIT
• RTRC CONDUIT
• MI CABLE W/ LISTED
  TERMINATION FITTINGS
• MC CABLE W/ LISTED
  TERMINATION FITTINGS
• DUSTTIGHT WIREWAYS
• PLTC AND PLTC-ER CABLE IN
  CABLE TRAYS
• ITC AND ITC-ER CABLE AND
  TERMINATED WITH LISTED FITTINGS
• SEE **503.10(A)(1)** FOR COMPLETE LIST

ALL BOXES AND FITTINGS
SHALL BE DUSTTIGHT
• **503.10(A)(2)**

FLEXIBILITY
• **503.10(A)(3)**

TYPE MC, MI, MV, OR TC CABLE SHALL BE PERMITTED TO
BE INSTALLED IN CABLE TRAYS PER **503.10(A)(4)** AND **Ex.**

**CLASS III, DIVISION 1 – GENERAL
NEC 503.10(A)(1)(1) THRU (A)(1)(4)**

Purpose of Change: To provide a list of wiring methods for Class III, Division 1 locations.

Type of Change	Revision			Committee Change	Accept			2008 NEC	-
ROP	pg. 644	# 14-160	log: 2840	ROC	pg. -	# -	log: -	UL	913
Submitter: Donald W. Ankele				Submitter: -				OSHA	1910.307(c)(1)
NFPA 496	-			NFPA 497	4.1.5.1			NFPA 499	4.1.5.1

504.30 Separation of Intrinsically Safe Conductors.

(B) From Different Intrinsically Safe Circuit Conductors. The clearance between two terminals for connection of field wiring of different intrinsically safe circuits shall be at least 6 mm (0.25 in.), unless this clearance is permitted to be reduced by the control drawing. Different intrinsically safe circuits shall be separated from each other by one of the following means:

(1) The conductors of each circuit are within a grounded metal shield.

(2) The conductors of each circuit have insulation with a minimum thickness of 0.25 mm (0.01 in.).

Stallcup's Comment: A revision has been made to clarify the clearance requirements between two terminals for connection of field wiring of different intrinsically safe circuits.

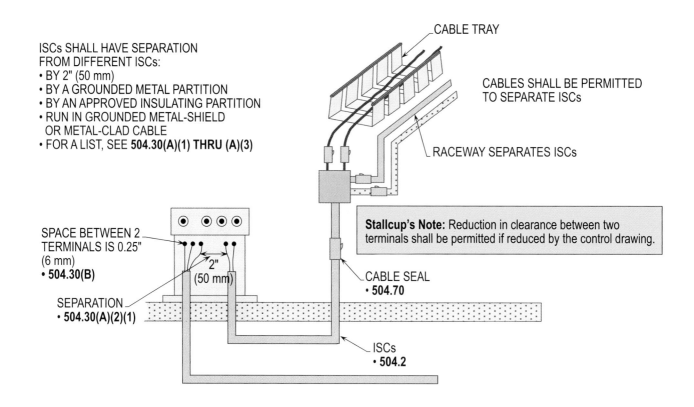

ISCs SHALL HAVE SEPARATION
FROM DIFFERENT ISCs:
• BY 2" (50 mm)
• BY A GROUNDED METAL PARTITION
• BY AN APPROVED INSULATING PARTITION
• RUN IN GROUNDED METAL-SHIELD
 OR METAL-CLAD CABLE
• FOR A LIST, SEE **504.30(A)(1) THRU (A)(3)**

CABLE TRAY

CABLES SHALL BE PERMITTED
TO SEPARATE ISCs

RACEWAY SEPARATES ISCs

SPACE BETWEEN 2
TERMINALS IS 0.25"
(6 mm)
• **504.30(B)**

2"
(50 mm)

SEPARATION
• **504.30(A)(2)(1)**

Stallcup's Note: Reduction in clearance between two terminals shall be permitted if reduced by the control drawing.

CABLE SEAL
• **504.70**

ISCs
• **504.2**

**FROM DIFFERENT INTRINSICALLY SAFE CIRCUIT CONDUCTORS
NEC 504.30(B)(1) AND (B)(2)**

Purpose of Change: To clarify separation requirements between intrinsically safe conductors and different intrinsically safe circuit conductors as well as terminations.

Type of Change	New Subsection			Committee Change	Accept in Principle			2008 NEC	-
ROP	pg. 647	# 14-175	log: 3201	ROC	pg. -	# -	log: -	UL	-
Submitter: A. W. Ballard				Submitter: -				OSHA	1910.307(c)(4)
NFPA 496	-			NFPA 497	552			NFPA 499	414

505.7 Special Precaution.

(E) Simultaneous Presence of Flammable Gases and Combustible Dusts or Fibers/Flyings. Where flammable gases, combustible dusts, or fibers/flyings are or may be present at the same time, the simultaneous presence shall be considered during the selection and installation of the electrical equipment and the wiring methods, including the determination of the safe operating temperature of the electrical equipment.

Stallcup's Comment: A new subsection has been added to address the requirements for simultaneous presence of flammable gases, combustible dusts, or fibers/flyings.

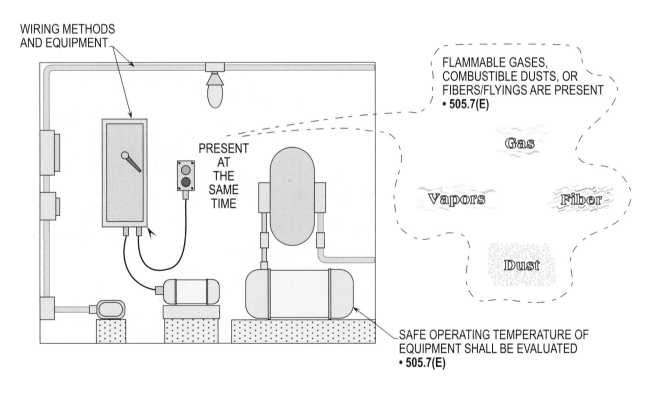

SIMULTANEOUS PRESENCE OF FLAMMABLE GASES AND COMBUSTIBLE DUSTS OR FIBERS/FLYINGS
NEC 505.7(E)

Purpose of Change: To clarify the requirements for selection of electrical equipment and wiring methods when gases, dust, or fiber/flyings are present at the same time.

Type of Change	New Paragraphs and Ex.		Committee Change	Accept			2008 NEC	505.9(E)(1)
ROP	pg. 649	# 14-188	log: 2842	ROC	pg. -	# -	log: -	UL	-
Submitter: Donald W. Ankele			Submitter: -				OSHA	1910.305(g)(2)(iv)
NFPA 496	-		NFPA 497	-			NFPA 499	-

505.9 Equipment.

(E) Threading. The supply connection entry thread form shall be NPT or metric. Conduit and fittings shall be made wrenchtight to prevent sparking when fault current flows through the conduit system, and to ensure the explosionproof or flameproof integrity of the conduit system where applicable. Equipment provided with threaded entries for field wiring connections shall be installed in accordance with 505.9(E)(1) or (E)(2) and with (E)(3).

(1) Equipment Provided with Threaded Entries for NPT Threaded Conduit or Fittings. For equipment provided with threaded entries for NPT threaded conduit or fittings, listed conduit, conduit fittings, or cable fittings shall be used.

All NPT threaded conduit and fittings referred to herein shall be threaded with a National (American) Standard Pipe Taper (NPT) thread.

NPT threaded entries into explosionproof or flameproof equipment shall be made up with at least five threads fully engaged.

Exception: For listed explosionproof or flameproof equipment, factory threaded NPT entries shall be made up with at least 4 1/2 threads fully engaged.

Stallcup's Comment: A revision has been made to clarify the threading requirements for equipment provided with threaded entries for NPT threaded conduit or fittings.

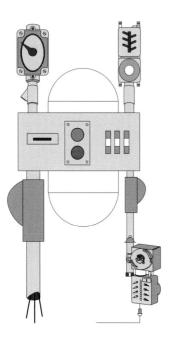

Stallcup's Note 1: All NPT threaded conduit and fittings shall be threaded with National (American) Standard Pipe Taper (NPT) thread per **505.9(E)(1).**

Stallcup's Note 2: Threaded entries into explosionproof equipment shall be made with at least five threads fully engaged per **505.9(E)(1).**

Stallcup's Note: Listed factory explosionproof equipment shall be permitted to have 4-1/2 threads fully engaged per **505.9(E)(1), Ex.**

**EQUIPMENT PROVIDED WITH THREADED ENTRIES FOR NPT THREADED CONDUIT OR FITTINGS
NEC 505.9(E)(1) AND Ex.**

Purpose of Change: To add requirements for threading pertaining to factory or other threaded NPT entries.

Type of Change	New Paragraph			Committee Change	Accept			2008 NEC	505.9(E)(2)
ROP	pg. 649	# 14-188	log: 2842	ROC	pg. -	# -	log: -	UL	-
Submitter: Donald W. Ankele				Submitter: -				OSHA	1910.305(g)(2)(iv)
NFPA 496	-			NFPA 497	-			NFPA 499	-

505.9 Equipment.

(E) Threading. The supply connection entry thread form shall be NPT or metric. Conduit and fittings shall be made wrenchtight to prevent sparking when fault current flows through the conduit system, and to ensure the explosionproof or flameproof integrity of the conduit system where applicable. Equipment provided with threaded entries for field wiring connections shall be installed in accordance with 505.9(E)(1) or (E)(2) and with (E)(3).

(2) Equipment Provided with Threaded Entries for Metric Threaded Conduit or Fittings. For equipment with metric threaded entries, listed conduit fittings or listed cable fittings shall be used. Such entries shall be identified as being metric, or listed adapters to permit connection to conduit or NPT threaded fittings shall be provided with the equipment and shall be used for connection to conduit or NPT threaded fittings.

Metric threaded entries into explosionproof or flameproof equipment shall have a class of fit of at least 6g/6H and be made up with at least five threads fully engaged for Groups C, D, IIB, or IIA and not less than eight threads fully engaged for Groups A, B, IIC, or IIB + H_2.

Stallcup's Comment: A revision has been made to clarify the threading requirements for equipment provided with threaded entries for metric threaded conduit or fittings.

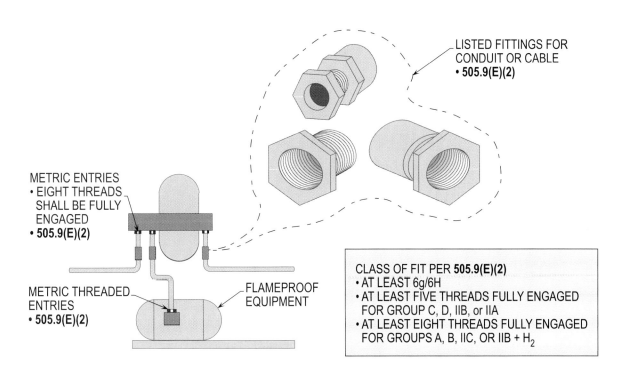

METRIC ENTRIES
• EIGHT THREADS SHALL BE FULLY ENGAGED
• 505.9(E)(2)

LISTED FITTINGS FOR CONDUIT OR CABLE
• 505.9(E)(2)

METRIC THREADED ENTRIES
• 505.9(E)(2)

FLAMEPROOF EQUIPMENT

CLASS OF FIT PER **505.9(E)(2)**
• AT LEAST 6g/6H
• AT LEAST FIVE THREADS FULLY ENGAGED FOR GROUP C, D, IIB, or IIA
• AT LEAST EIGHT THREADS FULLY ENGAGED FOR GROUPS A, B, IIC, OR IIB + H_2

**EQUIPMENT PROVIDED WITH THREADED ENTRIES FOR
METRIC THREADED CONDUIT OR FITTINGS
NEC 505.9(E)(2)**

Purpose of Change: To add requirements for threading pertaining to equipment with threaded entries for metric conduit or fittings.

Type of Change	New Subdivision			Committee Change	Accept			2008 NEC	-
ROP	pg. 649	# 14-188	log: 2842	ROC	pg. -	# -	log: -	UL	-
Submitter: Donald W. Ankele				Submitter: -				OSHA	-
NFPA 496	-			NFPA 497	-			NFPA 499	-

505.9 Equipment.

(E) Threading. The supply connection entry thread form shall be NPT or metric. Conduit and fittings shall be made wrenchtight to prevent sparking when fault current flows through the conduit system, and to ensure the explosionproof or flameproof integrity of the conduit system where applicable. Equipment provided with threaded entries for field wiring connections shall be installed in accordance with 505.9(E)(1) or (E)(2) and with (E)(3).

(3) Unused Openings. All unused openings shall be closed with close-up plugs listed for the location and shall maintain the type of protection. The plug engagement shall comply with 505.9(E)(1) or 505.9(E)(2).

Stallcup's Comment: A new subdivision has been added to address the requirements for unused openings in Class I, Zone 0, 1, and 2 locations.

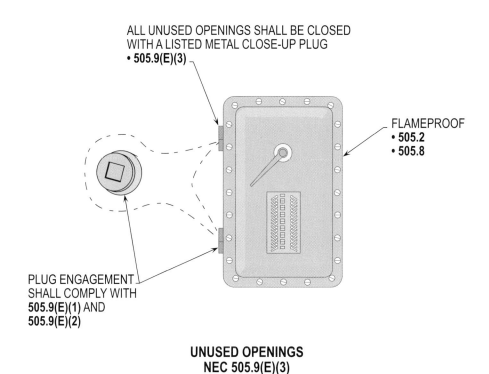

ALL UNUSED OPENINGS SHALL BE CLOSED
WITH A LISTED METAL CLOSE-UP PLUG
• **505.9(E)(3)**

FLAMEPROOF
• **505.2**
• **505.8**

PLUG ENGAGEMENT
SHALL COMPLY WITH
505.9(E)(1) AND
505.9(E)(2)

UNUSED OPENINGS
NEC 505.9(E)(3)

Purpose of Change: To add requirements for unused openings for equipment located in Class I, Zone 0, 1, and 2 locations.

Type of Change	Revision			Committee Change		Accept in Principle		2008 NEC	505.16(C)(1)(b)
ROP	pg. 653	# 14-202	log: 3206	ROC	pg. -	# -	log: -	UL	-
Submitter: A. W. Ballard				Submitter: -				OSHA	-
NFPA 496	-			NFPA 497	-			NFPA 499	

505.16 Sealing and Drainage.

(C) Zone 2. In Class I, Zone 2 locations, seals shall be located in accordance with 505.16(C)(1) and (C)(2).

(1) Conduit Seals. Conduit seals shall be located in accordance with (C)(1)(a) and (C)(1)(b).

(b) In each conduit run passing from a Class I, Zone 2 location into an unclassified location. The sealing fitting shall be permitted on either side of the boundary of such location within 3.05 m (10 ft) of the boundary and shall be designed and installed so as to minimize the amount of gas or vapor within the Zone 2 portion of the conduit from being communicated to the conduit beyond the seal. Rigid metal conduit or threaded steel intermediate metal conduit shall be used between the sealing fitting and the point at which the conduit leaves the Zone 2 location, and a threaded connection shall be used at the sealing fitting. Except for listed explosionproof reducers at the conduit seal, there shall be no union, coupling, box, or fitting between the conduit seal and the point at which the conduit leaves the Zone 2 location. Conduits shall be sealed to minimize the amount of gas or vapor within the Class I, Zone 2 portion of the conduit from being communicated to the conduit beyond the seal. Such seals shall not be required to be flameproof or explosionproof but shall be identified for the purpose of minimizing passage of gases under normal operating conditions and shall be accessible.

Stallcup's Comment: A revision has been made to clarify that conduits shall be sealed to minimize the amount of gas or vapor within the Class I, Zone 2 portion of the conduit from being communicated to the conduit beyond the seal.

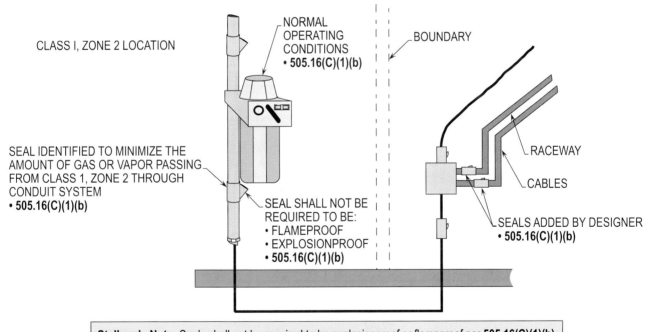

CONDUIT SEALS
NEC 505.16(C)(1)(b)

Purpose of Change: To require sealing to minimize the movement of gas or vapor through cables and raceways from the Class I, Zone 2 location.

Type of Change	New Section			Committee Change	Accept in Principle		2008 NEC	-
ROP	pg. 653	# 14-204	log: 4452	ROC	pg. 341	# 14-103	log: CC 1409	UL	-
Submitter: Eliana Beattie				Submitter: Code-Making Panel 14			OSHA	-
NFPA 70B	-			NFPA 70E	-		NFPA 79	-

505.26 Process Sealing.

This section shall apply to process-connected equipment, which includes, but is not limited to, canned pumps, submersible pumps, flow, pressure, temperature, or analysis measurement instruments. A process seal is a device to prevent the migration of process fluids from the designed containment into the external electrical system. Process connected electrical equipment that incorporates a single process seal, such as a single compression seal, diaphragm, or tube to prevent flammable or combustible fluids from entering a conduit or cable system capable of transmitting fluids, shall be provided with an additional means to mitigate a single process seal failure. The additional means may include, but is not limited to the following:

(3) A drain or vent located between the single process seal and a conduit or cable seal. The drain or vent shall be sufficiently sized to prevent overpressuring the conduit or cable seal above 6 in. water column (1493 Pa). Indication of the single process seal failure shall be provided by visible leakage, an audible whistle, or other means of monitoring.

Process-connected electrical equipment that does not rely on a single process seal or is listed and marked "single seal" or "dual seal" shall not be required to be provided with an additional means of sealing.

Informational Note: For construction and testing requirements for process sealing for listed and marked "single seal" or "dual seal" requirements, refer to ANSI/ISA-12.27.01-2003, *Requirements for Process Sealing Between Electrical Systems and Potentially Flammable or Combustible Process Fluids.*

Stallcup's Comment: A new section has been added to address the requirements for process sealing in Class I, Zone 0, 1, and 2 locations.

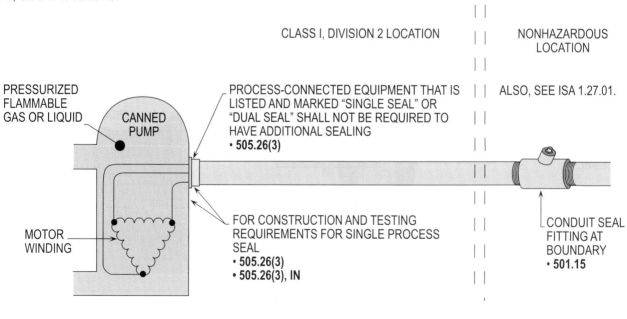

PROCESS SEALING
NEC 505.26(3) AND IN

Purpose of Change: To recognize a listed and marked "single seal" or "dual seal" that eliminates the requirement for an additional means of sealing.

Type of Change	Revision			Committee Change	Accept in Principle			2008 NEC	506.9(E)(1)
ROP	pg. 660	# 14-232	log: 379	ROC	pg. -	# -	log: -	UL	-
Submitter: James M. Daly				Submitter: -				OSHA	1910.305(g)(2)(iv)
NFPA 496	-			NFPA 497	-			NFPA 499	-

506.9 Equipment Requirements.

(E) Threading. The supply connection entry thread form shall be NPT or metric. Conduit and fittings shall be made wrenchtight to prevent sparking when the fault current flows through the conduit system and to ensure the integrity of the conduit system. Equipment provided with threaded entries for field wiring connections shall be installed in accordance with 506.9(E)(1) or (E)(2) and with (E)(3).

(1) Equipment Provided with Threaded Entries for NPT Threaded Conduit or Fittings. For equipment provided with threaded entries for NPT threaded conduit or fittings, listed conduit fittings, or cable fittings shall be used. All NPT threaded conduit and fittings referred to herein shall be threaded with a National (American) Standard Pipe Taper (NPT) thread.

Informational Note: Thread specifications for NPT threads are located in ANSI/ASME B1.20.1-1983, *Pipe Threads, General Purpose (Inch)*.

Stallcup's Comment: A revision has been made to clarify the threading requirements for equipment provided with threaded entries for NPT threaded conduit or fittings.

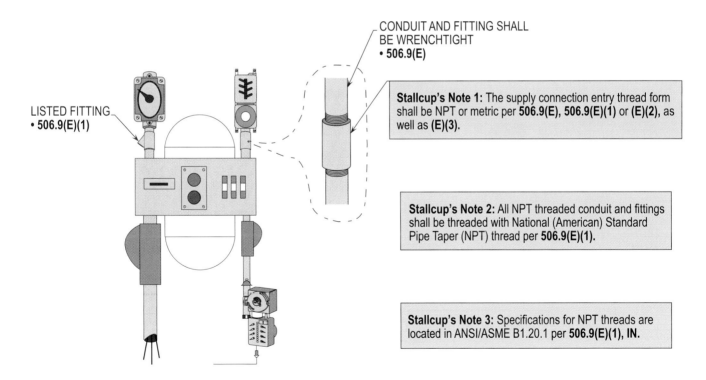

CONDUIT AND FITTING SHALL BE WRENCHTIGHT
• **506.9(E)**

LISTED FITTING
• **506.9(E)(1)**

Stallcup's Note 1: The supply connection entry thread form shall be NPT or metric per **506.9(E)**, **506.9(E)(1)** or **(E)(2)**, as well as **(E)(3)**.

Stallcup's Note 2: All NPT threaded conduit and fittings shall be threaded with National (American) Standard Pipe Taper (NPT) thread per **506.9(E)(1)**.

Stallcup's Note 3: Specifications for NPT threads are located in ANSI/ASME B1.20.1 per **506.9(E)(1), IN**.

**EQUIPMENT PROVIDED WITH THREADED ENTRIES FOR NPT THREADED CONDUIT OR FITTINGS
NEC 506.9(E)(1)**

Purpose of Change: To add threading requirements for equipment provided with threaded entries for NPT threaded conduit or fittings used in zone equipment and wiring methods.

Type of Change	Revision			Committee Change	Accept in Principle			**2008 NEC**	505.9(E)(2)
ROP	pg. 660	# 14-232	log: 379	**ROC**	pg. -	# -	log: -	**UL**	-
Submitter: James M. Daly				Submitter: -				**OSHA**	1910.305(g)(2)(iv)
NFPA 496	-			**NFPA 497**	-			**NFPA 499**	-

506.9 Equipment Requirements.

(E) Threading. The supply connection entry thread form shall be NPT or metric. Conduit and fittings shall be made wrenchtight to prevent sparking when the fault current flows through the conduit system and to ensure the integrity of the conduit system. Equipment provided with threaded entries for field wiring connections shall be installed in accordance with 506.9(E)(1) or (E)(2) and with (E)(3).

(2) Equipment Provided with Threaded Entries for Metric Threaded Conduit or Fittings. For equipment with metric threaded entries, listed conduit fittings or listed cable fittings shall be used. Such entries shall be identified as being metric, or listed adapters to permit connection to conduit or NPT threaded fittings shall be provided with the equipment and shall be used for connection to conduit or NPT threaded fittings. Metric threaded entries shall be made up with at least five threads fully engaged.

Stallcup's Comment: A revision has been made to clarify the threading requirements for equipment provided with threaded entries for metric threaded conduit or fittings.

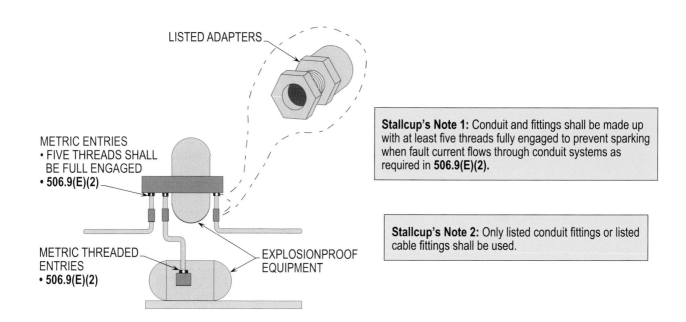

EQUIPMENT PROVIDED WITH THREADED ENTRIES FOR METRIC THREADED CONDUIT OR FITTINGS
NEC 506.9(E)(2)

Purpose of Change: To add requirements for equipment with metric threaded entries for conduit or fittings.

Type of Change	New Subdivision		Committee Change	Accept in Principle		2008 NEC	-
ROP pg. 660	# 14-232	log: 379	ROC pg. -	# -	log: -	UL	-
Submitter: James M. Daly			Submitter: -			OSHA	-
NFPA 496 -			NFPA 497 -			NFPA 499	-

506.9 Equipment Requirements.

(E) Threading. The supply connection entry thread form shall be NPT or metric. Conduit and fittings shall be made wrenchtight to prevent sparking when the fault current flows through the conduit system and to ensure the integrity of the conduit system. Equipment provided with threaded entries for field wiring connections shall be installed in accordance with 506.9(E)(1) or (E)(2) and with (E)(3).

(3) Unused Openings. All unused openings shall be closed with listed metal close-up plugs. The plug engagement shall comply with 506.9(E)(1) or (E)(2).

Stallcup's Comment: A revision has been made to clarify the requirements for unused openings in Zone 20, 21, and 22 locations.

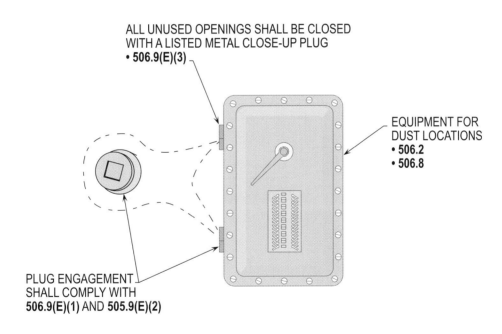

ALL UNUSED OPENINGS SHALL BE CLOSED
WITH A LISTED METAL CLOSE-UP PLUG
• **506.9(E)(3)**

EQUIPMENT FOR
DUST LOCATIONS
• **506.2**
• **506.8**

PLUG ENGAGEMENT
SHALL COMPLY WITH
506.9(E)(1) AND **505.9(E)(2)**

UNUSED OPENINGS
NEC 506.9(E)(3)

Purpose of Change: To add requirements for unused openings for equipment located in Zone 20, 21, and 22 locations.

Type of Change	New Subdivision		Committee Change	Accept in Principle		2008 NEC	-
ROP	pg. 660	# 14-234	log: 2846	ROC	pg. -	# -	log: -	UL	-
Submitter: Donald W. Ankele			Submitter: -			OSHA	-
NFPA 496	-		NFPA 497	-		NFPA 499	-

506.15 Wiring Methods.

(A) Zone 20. In Zone 20 locations, the following wiring methods shall be permitted.

(4) In industrial establishments with restricted public access, where the conditions of maintenance and supervision ensure that only qualified persons service the installation, and where the cable is not subject to physical damage, Type ITC-HL cable listed for use in Zone 1 or Class I, Division 1 locations, with a gas/vaportight continuous corrugated metallic sheath and an overall jacket of suitable polymeric material, and terminated with fittings listed for the application. Type ITC-HL cable shall be installed in accordance with the provisions of Article 727.

Stallcup's Comment: A new subdivision has been added to address the requirements for wiring methods in Zone 20 locations for industrial establishments with restricted public access.

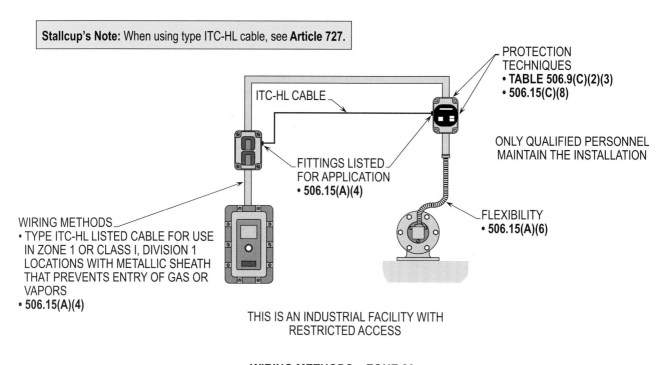

Stallcup's Note: When using type ITC-HL cable, see **Article 727**.

ITC-HL CABLE

PROTECTION TECHNIQUES
• **TABLE 506.9(C)(2)(3)**
• **506.15(C)(8)**

ONLY QUALIFIED PERSONNEL MAINTAIN THE INSTALLATION

FITTINGS LISTED FOR APPLICATION
• **506.15(A)(4)**

FLEXIBILITY
• **506.15(A)(6)**

WIRING METHODS
• TYPE ITC-HL LISTED CABLE FOR USE IN ZONE 1 OR CLASS I, DIVISION 1 LOCATIONS WITH METALLIC SHEATH THAT PREVENTS ENTRY OF GAS OR VAPORS
• **506.15(A)(4)**

THIS IS AN INDUSTRIAL FACILITY WITH RESTRICTED ACCESS

WIRING METHODS – ZONE 20
NEC 506.15(A)(4)

Purpose of Change: To add requirements for wiring methods installed in Zone 20 locations.

Type of Change	New Subdivision			Committee Change	Accept			**2008 NEC**	517.13(B)
ROP	pg. 678	# 15-25	log: 3755	**ROC**	pg. 360	# 15-35	log: 2445	**UL**	-
Submitter: Jim Pauley				Submitter: Phil Simmons				**OSHA**	-
NFPA 406	-			**NFPA 497**	-			**NFPA 499**	-

517.13 Grounding of Receptacles and Fixed Electrical Equipment in Patient Care Areas.

(B) Insulated Equipment Grounding Conductor.

(1) General. The following shall be directly connected to an insulated copper equipment grounding conductor that is installed with the branch circuit conductors in the wiring methods as provided in 517.13(A).

(1) The grounding terminals of all receptacles.

(2) Metal boxes and enclosures containing receptacles.

(3) All non-current-carrying conductive surfaces of fixed electrical equipment likely to become energized that are subject to personal contact, operating at over 100 volts.

Exception: An insulated equipment bonding jumper that directly connects to the equipment grounding conductor is permitted to connect the box and receptacle(s) to the equipment grounding conductor.

Stallcup's Comment: A new subdivision has been added to address the requirements for an insulated copper equipment grounding conductor in health care facilities.

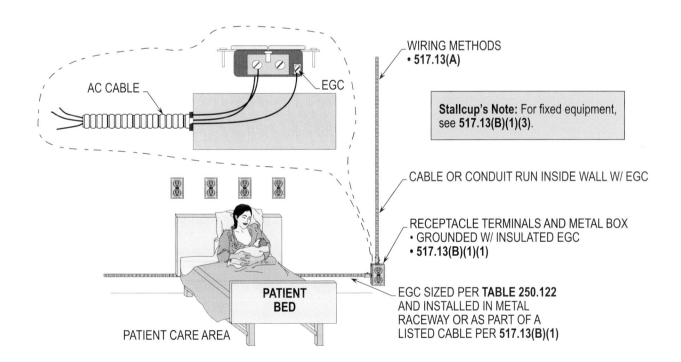

INSULATED EQUIPMENT GROUNDING CONDUCTOR – GENERAL
NEC 517.13(B)(1)(1) THRU (B)(1)(3)

Purpose of Change: To more appropriately address the grounding requirements in patient care areas.

Type of Change	Revision			Committee Change	Accept			2008 NEC	517.20(A)
ROP	pg. 688	# 15-64	log: 2926	ROC	pg. 365	# 15-69	log: 189	UL	-
Submitter: Burton R. Klein				Submitter: Technical Correlating Committee				OSHA	-
NFPA 496	-			NFPA 497	-			NFPA 499	-

517.20 Wet Procedure Locations.

(A) Receptacles and Fixed Equipment. Wet procedure location patient care areas shall be provided with special protection against electric shock by one of the following means:

(1) Power distribution system that inherently limits the possible ground-fault current due to a first fault to a low value, without interrupting the power supply

(2) Power distribution system in which the power supply is interrupted if the ground-fault current does, in fact, exceed a value of 6 mA

Stallcup's Comment: A revision has been made to clarify the requirements for receptacles and fixed equipment in wet procedure location patient care areas.

GROUND-FAULT PROTECTION IN WET PROCEDURE LOCATION PATIENT
CARE AREAS SHALL BE APPLIED AS FOLLOWS:

(1) POWER DISTRIBUTION SYSTEMS THAT LIMIT GROUND-FAULT
CURRENT OF A FIRST FAULT TO A LOW VALUE WITHOUT
INTERRUPTING THE POWER SUPPLY PER **517.20(A)(1)**

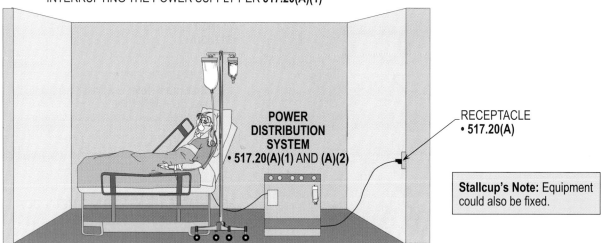

POWER
DISTRIBUTION
SYSTEM
• **517.20(A)(1)** AND **(A)(2)**

RECEPTACLE
• **517.20(A)**

Stallcup's Note: Equipment could also be fixed.

(2) POWER DISTRIBUTION SYSTEM INTERRUPTS THE POWER SUPPLY
IF 6 mA OF CURRENT IS EXCEEDED PER **517.20(A)(2)**

RECEPTACLES AND FIXED EQUIPMENT
NEC 517.20(A)(1) AND (A)(2)

Purpose of Change: To include requirements for reducing shock hazards pertaining to receptacles and fixed equipment located in wet procedure location patient care areas.

Type of Change	New Paragraphs			Committee Change	Accept			**2008 NEC**	517.30(C)(1)
ROP	pg. 690	# 15-74	log: 2798	**ROC**	pg. 369	# 15-90	log: 672	**UL**	-
Submitter: James W. Carpenter				Submitter: Thomas Guida				**OSHA**	-
NFPA 490				**NFPA 407**	-			**NFPA 499**	-

517.30 Essential Electrical Systems for Hospitals.

(C) Wiring Requirements.

(1) Separation from Other Circuits. The life safety branch and critical branch of the emergency system shall be kept entirely independent of all other wiring and equipment and shall not enter the same raceways, boxes, or cabinets with each other or other wiring.

Where general care locations are served from two separate transfer switches on the emergency system in accordance with 517.18(A), Exception No. 3, the general care circuits from the two separate systems shall be kept independent of each other.

Where critical care locations are served from two separate transfer switches on the emergency system in accordance with 517.19(A), Exception No. 2, the critical care circuits from the two separate systems shall be kept independent of each other.

Stallcup's Comment: A new paragraph has been added to address the requirements for separation from other circuits in general care locations and critical care locations.

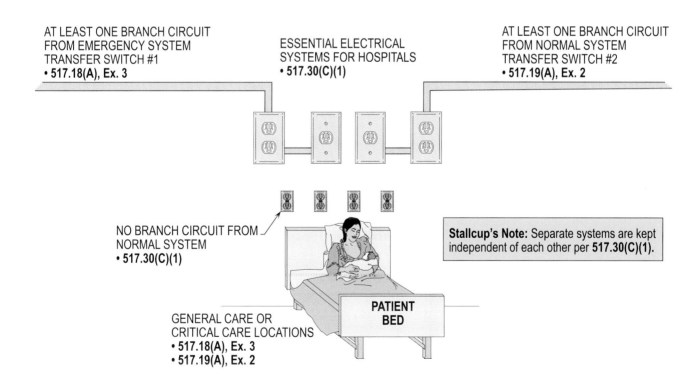

SEPARATION FROM OTHER CIRCUITS
NEC 517.30(C)(1)

Purpose of Change: To permit the same requirements for general care and critical care locations.

Type of Change	Revision			Committee Change	Accept in Principle		2008 NEC	517.63(A)
ROP	pg. 698	# 15-109	log: 215	ROC	pg. -	# -	log: -	UL	-
Submitter: Chris Pogorzelski				Submitter: -			OSHA	-
NFPA 496	-			NFPA 497	-		NFPA 499	-

517.63 Grounded Power Systems in Anesthetizing Locations.

(A) Battery-Powered Lighting Units. One or more battery-powered lighting units shall be provided and shall be permitted to be wired to the critical lighting circuit in the area and connected ahead of any local switches.

Stallcup's Comment: A revision has been made to clarify the requirements for battery-powered lighting units.

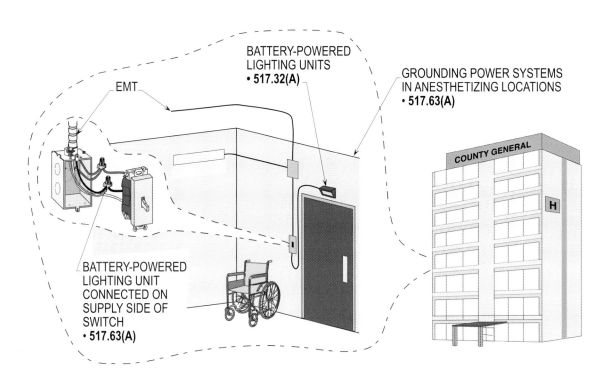

BATTERY-POWERED LIGHTING UNITS
NEC 517.63(A)

Purpose of Change: To add requirements for powering battery-powered lighting units.

Type of Change	Revision			Committee Change	Accept in Principle			2008 NEC	517.80
ROP	pg. 700	# 15-120	log: 67	ROC	pg. 375	# 15-132	log: 1070	UL	-
Submitter: Hugh O. Nash, Jr.				Submitter: Noel Williams				OSHA	-
NFPA 496	-			NFPA 497	-			NFPA 499	-

517.80 Patient Care Areas.

Equivalent insulation and isolation to that required for the electrical distribution systems in patient care areas shall be provided for communications, signaling systems, data system circuits, fire alarm systems, and systems less than 120 volts, nominal.

Class 2 and Class 3 signaling and communications systems and power-limited fire alarm systems shall not be required to comply with the grounding requirements of 517.13, to comply with the mechanical protection requirements of 517.30(C)(3)(5), or to be enclosed in raceways, unless otherwise specified by Chapter 7 or 8.

Secondary circuits of transformer-powered communications or signaling systems shall not be required to be enclosed in raceways unless otherwise specified by Chapter 7 or 8.

Stallcup's Comment: A revision has been made to clarify the requirements for Class 2 and Class 3 signaling and communications systems and power-limited fire alarm systems in patient care areas.

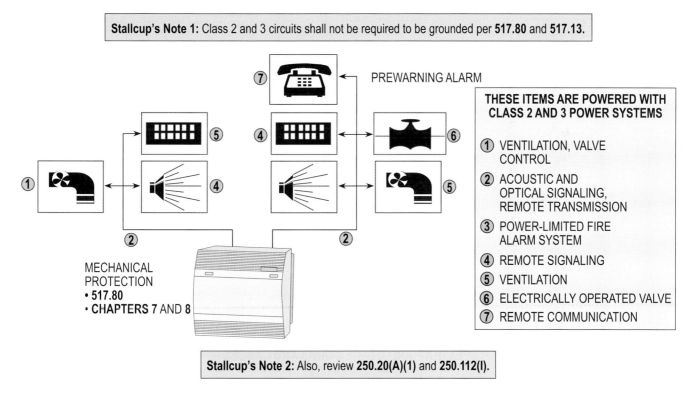

Stallcup's Note 1: Class 2 and 3 circuits shall not be required to be grounded per **517.80** and **517.13**.

PREWARNING ALARM

THESE ITEMS ARE POWERED WITH CLASS 2 AND 3 POWER SYSTEMS

① VENTILATION, VALVE CONTROL
② ACOUSTIC AND OPTICAL SIGNALING, REMOTE TRANSMISSION
③ POWER-LIMITED FIRE ALARM SYSTEM
④ REMOTE SIGNALING
⑤ VENTILATION
⑥ ELECTRICALLY OPERATED VALVE
⑦ REMOTE COMMUNICATION

MECHANICAL PROTECTION
• **517.80**
• **CHAPTERS 7** AND **8**

Stallcup's Note 2: Also, review **250.20(A)(1)** and **250.112(I)**.

PATIENT CARE AREAS
NEC 517.80

Purpose of Change: To provide requirements that do not require Class 2 and Class 3 circuits to be grounded when complying with **517.13**.

Type of Change	New Subdivision			Committee Change	Accept			2008 NEC	-
ROP	pg. 705	# 15-151	log: 3937	ROC	pg. -	# -	log: -	UL	-
Submitter: Mitchell K. Hefter				Submitter: -				OSHA	-
NFPA 496	-			NFPA 497	-			NFPA 499	-

520.44 Borders, Proscenium Sidelights, Drop Boxes, and Connector Strips.

(C) Cords and Cables for Border Lights, Drop Boxes, and Connector Strips.

(3) Identification of Conductors in Multiconductor Extra-hard Usage Cords and Cables. Grounded (neutral) conductors shall be white without stripe or shall be identified by a distinctive white marking at their terminations. Grounding conductors shall be green with or without yellow stripe or shall be identified by a distinctive green marking at their terminations.

Stallcup's Comment: A new subdivision has been added to address the requirements for identification of conductors in multiconductor extra-hard usage cords and cables.

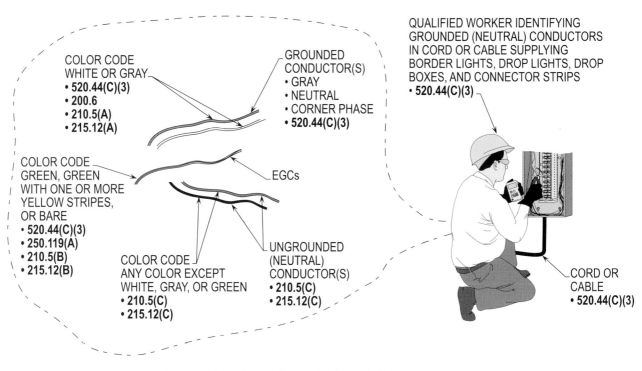

**IDENTIFICATION OF CONDUCTORS IN MULTICONDUCTOR
EXTRA-HARD USAGE CORDS AND CABLES
NEC 520.44(C)(3)**

Purpose of Change: To clarify the requirements permitted to be used to identify the grounded (neutral) conductor(s) and equipment grounding conductor(s).

Type of Change	New Paragraph			Committee Change	Accept			**2008 NEC**	525.23(A)
ROP pg. 714	# 15-213		log: 2188	**ROC** pg. -		# -	log: -	**UL**	943
Submitter: James W. Carpenter				Submitter: -				**OSHA**	-
NFPA 496 -				**NFPA 497**				**NFPA 499**	-

525.23 Ground-Fault Circuit-Interrupter (GFCI) Protection.

(A) Where GFCI Protection Is Required. GFCI protection for personnel shall be provided for the following:

(1) All 125-volt, single-phase, 15- and 20-ampere nonlocking-type receptacles used for disassembly and reassembly or readily accessible to the general public

(2) Equipment that is readily accessible to the general public and supplied from a 125-volt, single-phase, 15- or 20-ampere branch circuit

The ground-fault circuit-interrupter shall be permitted to be an integral part of the attachment plug or located in the power-supply cord within 300 mm (12 in.) of the attachment plug. Listed cord sets incorporating ground-fault circuit-interrupter for personnel shall be permitted.

Stallcup's Comment: A new paragraph has been added to address the requirements for ground-fault circuit-interrupter (GFCI) protection for carnivals, circuses, fairs, and similar events.

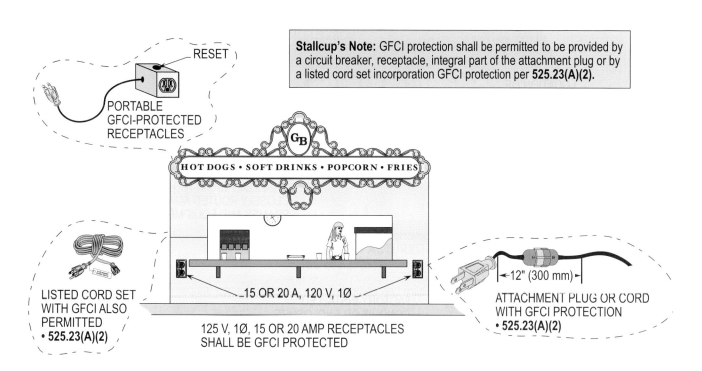

Stallcup's Note: GFCI protection shall be permitted to be provided by a circuit breaker, receptacle, integral part of the attachment plug or by a listed cord set incorporation GFCI protection per **525.23(A)(2)**.

RESET

PORTABLE
GFCI-PROTECTED
RECEPTACLES

HOT DOGS • SOFT DRINKS • POPCORN • FRIES

LISTED CORD SET
WITH GFCI ALSO
PERMITTED
• **525.23(A)(2)**

15 OR 20 A, 120 V, 1Ø

125 V, 1Ø, 15 OR 20 AMP RECEPTACLES
SHALL BE GFCI PROTECTED

←12" (300 mm)→

ATTACHMENT PLUG OR CORD
WITH GFCI PROTECTION
• **525.23(A)(2)**

**WHERE GFCI PROTECTION IS REQUIRED
NEC 525.23(A)(2)**

Purpose of Change: To clarify the types of GFCI protection allowed to be used for protecting personnel.

Type of Change	Revision			Committee Change	Accept			2008 NEC	553.4
ROP	pg. 753	# 19-241	log: 4765	ROC	pg. 410	# 19-170	log: 1613	UL	869A
Submitter: Joseph P. Fello				Submitter: Joseph P. Fello				OSHA	-
NFPA 496	-			NFPA 497	-			NFPA 499	-

553.4 Location of Service Equipment.

The service equipment for a floating building shall be located adjacent to, but not in or on, the building or any floating structure. The main overcurrent protective device that feeds the floating structure shall have ground fault protection not exceeding 100 mA. Ground fault protection of each individual branch or feeder circuit shall be permitted as a suitable alternative.

Stallcup's Comment: A revision has been made to clarify the requirements for the main overcurrent protective device and ground fault protection of each individual branch or feeder circuit.

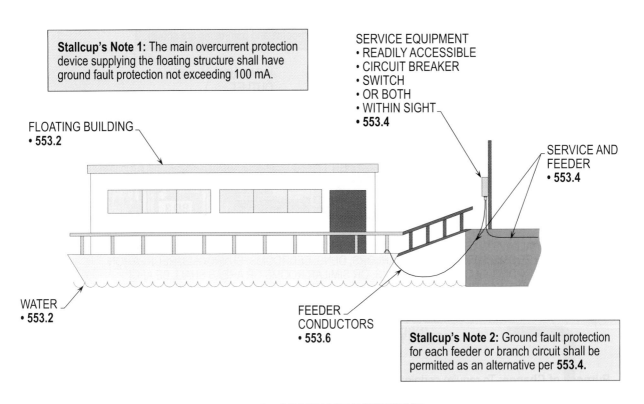

LOCATION OF SERVICE EQUIPMENT
NEC 553.4

Purpose of Change: To add requirements for locating service equipment to supply a floating building.

Type of Change	New Section			Committee Change	Accept			2008 NEC	-
ROP	pg. 755	# 19-252	log: 4766	ROC	pg. 412	# 19-189	log: 1612	UL	1053
Submitter: Joseph P. Fello				Submitter: Joseph P. Fello				OSHA	-
NFPA 496	-			NFPA 497	-			NFPA 499	-

555.3 Ground-Fault Protection.

The main overcurrent protective device which feeds the marina shall have ground fault protection not exceeding 100 mA. Ground-fault protection of each individual branch or feeder circuit shall be permitted as a suitable alternative.

Stallcup's Comment: A new section has been added to address the requirements for ground fault protection in marinas and boatyards.

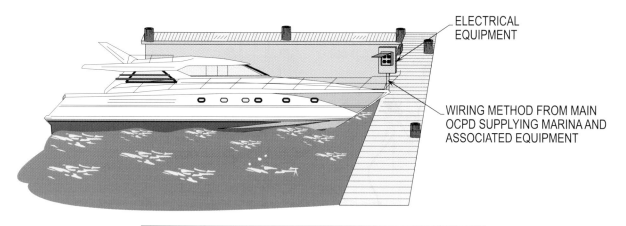

Stallcup's Note 1: The main overcurrent protection device that supplies power to the marina shall have ground fault protection that does not exceed 100 mA per **555.3**.

ELECTRICAL
EQUIPMENT

WIRING METHOD FROM MAIN
OCPD SUPPLYING MARINA AND
ASSOCIATED EQUIPMENT

Stallcup's Note 2: Ground-fault protection for each individual branch or feeder circuit shall be permitted as an alternative method per **555.3**.

GROUND-FAULT PROTECTION
NEC 555.3

Purpose of Change: To address the need for safety, ground fault protection has been added, based on the supplying method

Type of Change	New Subsections		Committee Change	Accept		2008 NEC	-
ROP	pg. 770	# 18-206a	log: CP 1800	ROC	pg. -	# -	log: -	UL	48
Submitter: Code-Making Panel 18			Submitter: -			OSHA	-
NFPA 70B	-		NFPA 70E	-		NFPA 79	-

600.4 Markings.

(C) Visibility. The markings required in 600.4(A) and listing labels shall not be required to be visible after installation but shall be permanently applied in a location visible during servicing.

(D) Durability. Marking labels shall be permanent, durable and, when in wet locations, shall be weatherproof.

Stallcup's Comment: A new subdivision has been added to address the marking requirements for electric signs and outline lighting.

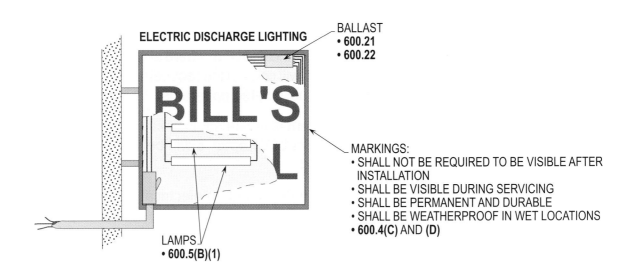

VISIBILITY AND DURABILITY
NEC 600.4(C) AND (D)

Purpose of Change: Two new subsections have been added to require markings to be visible after installation, visible during servicing, permanent and durable, and weatherproof in wet locations.

Type of Change	Revision			Committee Change	Accept			2008 NEC	600.6
ROP	pg. 770	# 18-210	log: 1013	ROC	pg. 426	# 18-72	log: 1934	UL	48
Submitter: Dan Leaf				Submitter: David Servine				OSHA	1910.306(a)(1)
NFPA 70B	-			NFPA 70E	-			NFPA 79	-

600.6 Disconnects.

Each sign and outline lighting system, feeder circuit or branch circuit supplying a sign, outline lighting system, or skeleton tubing shall be controlled by an externally operable switch or circuit breaker that opens all ungrounded conductors and controls no other load. The switch or circuit breaker shall open all ungrounded conductors simultaneously on multi-wire branch circuits in accordance with 210.4(B). Signs and outline lighting systems located within fountains shall have the disconnect located in accordance with 680.12.

Stallcup's Comment: A revision has been made to clarify that the switch or circuit breaker shall open all ungrounded conductors simultaneously on multiwire branch circuits.

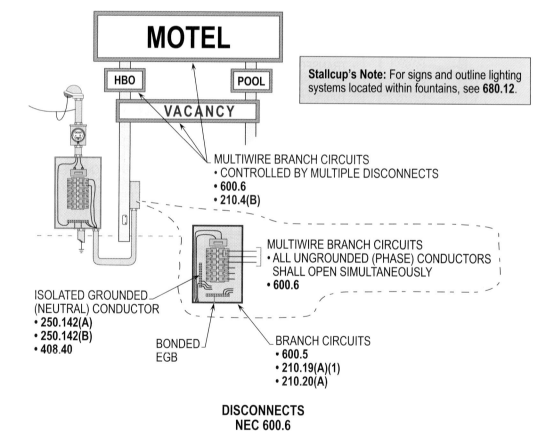

Stallcup's Note: For signs and outline lighting systems located within fountains, see **680.12**.

MULTIWIRE BRANCH CIRCUITS
• CONTROLLED BY MULTIPLE DISCONNECTS
• 600.6
• 210.4(B)

MULTIWIRE BRANCH CIRCUITS
• ALL UNGROUNDED (PHASE) CONDUCTORS SHALL OPEN SIMULTANEOUSLY
• 600.6

ISOLATED GROUNDED (NEUTRAL) CONDUCTOR
• 250.142(A)
• 250.142(B)
• 408.40

BONDED EGB

BRANCH CIRCUITS
• 600.5
• 210.19(A)(1)
• 210.20(A)

DISCONNECTS
NEC 600.6

Purpose of Change: To clarify that a switch or circuit breaker shall open all ungrounded (phase) conductors in multiwire branch circuits.

Type of Change	New Subdivisions		Committee Change	Accept in Principle		2008 NEC	690.4(B)
ROP pg. 863	# 4-184	log: 2480	ROC pg. 476	# 4-66	log: 1948	UL	1703
Submitter: John Wiles			Submitter: D. Jerry Flaherty			OSHA	-
NFPA 70B -			NFPA 70E -			NFPA 79	-

690.4 Installation.

(B) Identification and Grouping. Photovoltaic source circuits and PV output circuits shall not be contained in the same raceway, cable tray, cable, outlet box, junction box, or similar fitting as conductors, feeders, or branch circuits of other non-PV systems, unless the conductors of the different systems are separated by a partition. Photovoltaic system conductors shall be identified and grouped as required by 690.4(B)(1) through (4). The means of identification shall be permitted by separate color coding, marking tape, tagging, or other approved means.

(4) Grouping. Where the conductors of more than one PV system occupy the same junction box or raceway with a removable cover(s), the ac and dc conductors of each system shall be grouped separately by wire ties or similar means at least once, and then shall be grouped at intervals not to exceed 1.8 m (6 ft).

Exception: The requirement for grouping shall not apply if the circuit enters from a cable or raceway unique to the circuit that makes the grouping obvious.

Stallcup's Comment: A new subdivision has been added to address the requirements for identification and grouping of conductors of multiple systems.

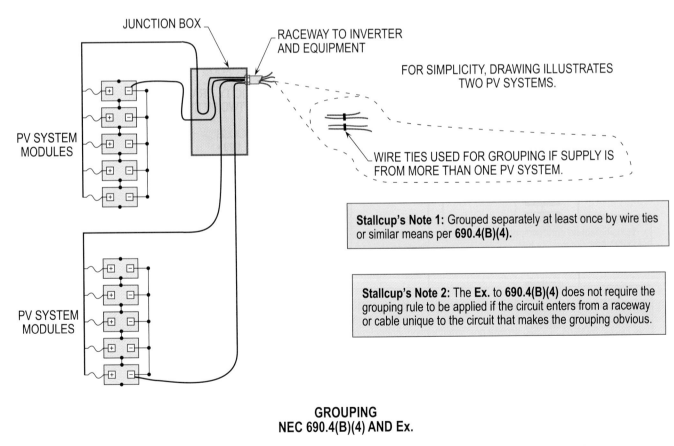

GROUPING
NEC 690.4(B)(4) AND Ex.

Purpose of Change: To address requirements for grouping PV circuits under certain conditions of use.

Type of Change	New Subsections		Committee Change	Accept		2008 NEC	-
ROP pg. 865	# 4-187	log: 2482	**ROC** pg. 477	# 4-69	log: 1783	**UL**	1703
Submitter: John Wiles			Submitter: Michael J. Johnston			**OSHA**	1910.308(f)(1)
NFPA 70B -			**NFPA 70E** -			**NFPA 79**	-

690.4 Installation.

(E) Wiring and Connections. The equipment and systems in 690.4(A) through (D) and all associated wiring and interconnections shall be installed only by qualified persons.

Informational Note: See Article 100 for the definition of *qualified person*.

(F) Circuit Routing. Photovoltaic source and PV output conductors, in and out of conduit, and inside of a building or structure, shall be routed along building structural members such as beams, rafters, trusses, and columns where the location of those structural members can be determined by observation. Where circuits are imbedded in built-up, laminate, or membrane roofing materials in roof areas not covered by PV modules and associated equipment, the location of circuits shall be clearly marked.

Stallcup's Comment: A new subsection has been added to address the requirements for wiring and connections and circuit routing.

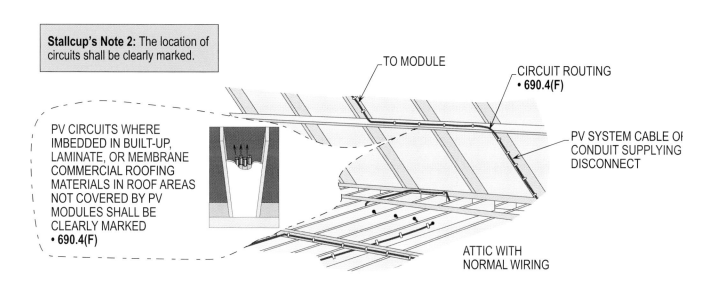

Stallcup's Note 2: The location of circuits shall be clearly marked.

TO MODULE

CIRCUIT ROUTING
• 690.4(F)

PV SYSTEM CABLE OI
CONDUIT SUPPLYING
DISCONNECT

PV CIRCUITS WHERE
IMBEDDED IN BUILT-UP,
LAMINATE, OR MEMBRANE
COMMERCIAL ROOFING
MATERIALS IN ROOF AREAS
NOT COVERED BY PV
MODULES SHALL BE
CLEARLY MARKED
• 690.4(F)

ATTIC WITH
NORMAL WIRING

Stallcup's Note 1: All PV system wiring and associated components shall be installed by a qualified person per **690.4(E)** and **IN.**

WIRING AND CONNECTIONS AND CIRCUIT ROUTING
NEC 690.4(E) AND (F)

Purpose of Change: To provide requirements for circuit routing and wiring connections of PV systems.

Type of Change	New Subsection			Committee Change	Accept in Principle		2008 NEC	-
ROP	pg. 865	# 4-188	log: 2483	ROC	pg. 478	# 4-71	log: 2671	UL	1703
Submitter: John Wiles				Submitter: Frederic P. Hartwell			OSHA	1910.308(f)(1)
NFPA 70B	-			NFPA 70E	-		NFPA 79	-

690.4 Installation.

(G) Bipolar Photovoltaic Systems. Where the sum, without consideration of polarity, of the PV system voltages of the two monopole subarrays exceeds the rating of the conductors and connected equipment, monopole subarrays in a bipolar PV system shall be physically separated, and the electrical output circuits from each monopole subarray shall be installed in separate raceways until connected to the inverter. The disconnecting means and overcurrent protective devices for each monopole subarray output shall be in separate enclosures. All conductors from each separate monopole subarray shall be routed in the same raceway.

Exception: Listed switchgear rated for the maximum voltage between circuits and containing a physical barrier separating the disconnecting means for each monopole subarray shall be permitted to be used instead of disconnecting means in separate enclosures.

Stallcup's Comment: A new subsection has been added to address the requirements for bipolar photovoltaic systems.

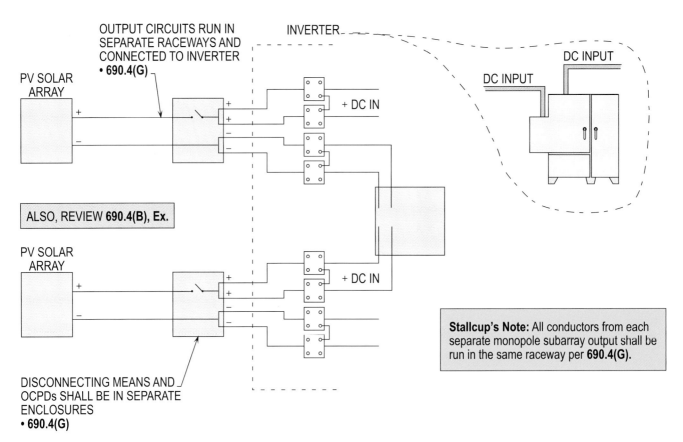

BIPOLAR PHOTOVOLTAIC SYSTEMS
NEC 690.4(G) AND Ex.

Purpose of Change: To provide requirements for installing bipolar photovoltaic systems.

Type of Change	New Subsection			Committee Change	Accept			2008 NEC	-
ROP	pg. 865	# 4-189	log: 2484	ROC	pg. -	# -	log: -	UL	1703
Submitter: John Wiles				Submitter: -				OSHA	1910.308(f)(1)
NFPA 70B	-			NFPA 70E	-			NFPA 79	-

690.4 Installation.

(H) Multiple Inverters. A PV system shall be permitted to have multiple utility-interactive inverters installed in or on a single building or structure. Where the inverters are remotely located from each other, a directory in accordance with 705.10 shall be installed at each dc PV system disconnecting means, at each ac disconnecting means, and at the main service disconnecting means showing the location of all ac and dc PV system disconnecting means in the building.

Stallcup's Comment: A new subsection has been added to address the requirements for multiple inverters.

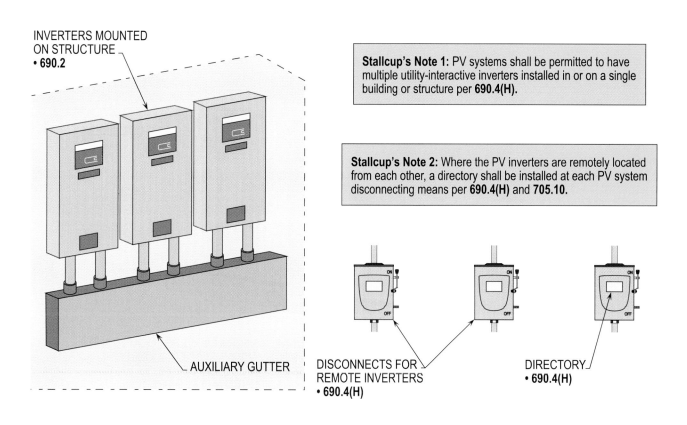

INVERTERS MOUNTED ON STRUCTURE
• 690.2

Stallcup's Note 1: PV systems shall be permitted to have multiple utility-interactive inverters installed in or on a single building or structure per **690.4(H)**.

Stallcup's Note 2: Where the PV inverters are remotely located from each other, a directory shall be installed at each PV system disconnecting means per **690.4(H)** and **705.10**.

AUXILIARY GUTTER

DISCONNECTS FOR REMOTE INVERTERS
• 690.4(H)

DIRECTORY
• 690.4(H)

MULTIPLE INVERTERS
NEC 690.4(H)

Purpose of Change: To add requirements for grouped or ungrouped multiple PV system inverters.

Type of Change	New Subsection			Committee Change	Accept			2008 NEC	-
ROP	pg. 865	# 4-189	log: 2484	ROC	pg. -	# -	log: -	UL	1703
Submitter: John Wiles				Submitter: -				OSHA	-
NFPA 70B	-			NFPA 70E	-			NFPA 79	-

690.4 Installation.

(H) Multiple Inverters. A PV system shall be permitted to have multiple utility-interactive inverters installed in or on a single building or structure. Where the inverters are remotely located from each other, a directory in accordance with 705.10 shall be installed at each dc PV system disconnecting means, at each ac disconnecting means, and at the main service disconnecting means showing the location of all ac and dc PV system disconnecting means in the building.

Exception: A directory shall not be required where all inverters and PV dc disconnecting means are grouped at the main service disconnecting means.

Stallcup's Comment: A new subsection has been added to address the requirements for multiple inverters.

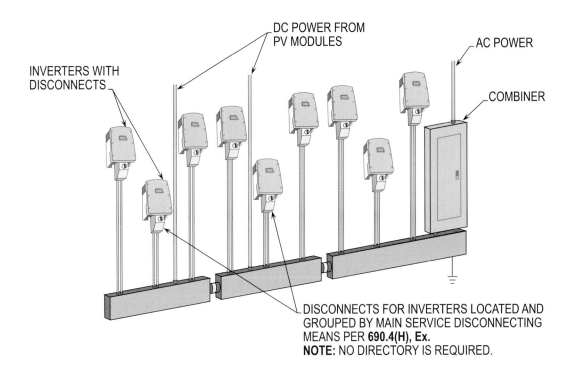

DC POWER FROM PV MODULES

AC POWER

INVERTERS WITH DISCONNECTS

COMBINER

DISCONNECTS FOR INVERTERS LOCATED AND GROUPED BY MAIN SERVICE DISCONNECTING MEANS PER **690.4(H), Ex.**
NOTE: NO DIRECTORY IS REQUIRED.

MULTIPLE INVERTERS
NEC 690.4(H), Ex.

Purpose of Change: To clarify the requirements pertaining to multiple inverters with grouped disconnects.

Type of Change	Revision			Committee Change	Accept in Principle			2008 NEC	690.8(B)(1)
ROP	pg. 867	# 4-195	log: 2488	ROC	pg. -	# -	log: -	UL	1703
Submitter: John Wiles				Submitter: -				OSHA	-
NFPA 70B	-			NFPA 70E	-			NFPA 79	-

690.8 Circuit Sizing and Current.

(B) Ampacity and Overcurrent Device Ratings.

(1) Overcurrent Devices. Overcurrent devices, where required, shall be rated as required by 690.8(B)(1)(a) through (1)(d).

(a) To carry not less than 125 percent of the maximum currents calculated in 690.8(A).

Exception: Circuits containing an assembly, together with its overcurrent device(s), that is listed for continuous operation at 100 percent of its rating shall be permitted to be used at 100 percent of its rating.

(b) Terminal temperature limits shall be in accordance with 110.3(B) and 110.14(C).

(c) Where operated at temperatures greater than 40°C (104°F), the manufacturer's temperature correction factors shall apply.

(d) The rating or setting of overcurrent devices shall be permitted in accordance with 240.4(B), (C), and (D).

Stallcup's Comment: A revision has been made to clarify the requirements of overcurrent devices for solar photovoltaic systems.

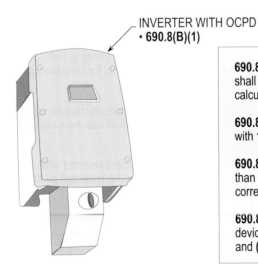

INVERTER WITH OCPD
• 690.8(B)(1)

690.8(B)(1)(a): The overcurrent protection device rating shall be at least 125 percent of the maximum current calculated per **690.8(A)**.

690.8(B)(1)(b): Terminal temperature limits shall comply with **110.3(B)** and **110.14(C)**.

690.8(B)(1)(c): Where operating temperature is greater than 40ºC (104ºF), apply manufacturer's temperature correction factor per **110.3(B)**.

690.8(B)(1)(d): The size of the overcurrent protection device shall be permitted to be sized per **240.4(B)**, **(C)**, and **(D)**.

OVERCURRENT DEVICES
NEC 690.8(B)(1)(a) THRU (B)(1)(d)

Purpose of Change: To address requirements for sizing and selecting the PV system's overcurrent device.

Type of Change	New Subdivision		Committee Change	Accept in Principle		2008 NEC	-
ROP	pg. 878	# 4-235	log: 2508	ROC	pg. -	# -	log: -	UL	1703
Submitter: John Wiles			Submitter: -			OSHA	-
NFPA 70B	-		NFPA 70E	-		NFPA 79	-

690.47 Grounding Electrode System.

(C) Systems with Alternating-Current and Direct-Current Grounding Requirements. Photovoltaic systems having dc circuits and ac circuits with no direct connection between the dc grounded conductor and ac grounded conductor shall have a dc grounding system. The dc grounding system shall be bonded to the ac grounding system by one of the methods in (1), (2), or (3).

This section shall not apply to ac PV modules.

(2) Common Direct-Current and Alternating-Current Grounding Electrode. A dc grounding electrode conductor of the size specified by 250.166 shall be run from the marked dc grounding electrode connection point to the ac grounding electrode. Where an ac grounding electrode is not accessible, the dc grounding electrode conductor shall be connected to the ac grounding electrode conductor in accordance with 250.64(C)(1). This dc grounding electrode conductor shall not be used as a substitute for any required ac equipment grounding conductors.

Stallcup's Comment: A new subdivision has been added to address the requirements of common direct-current and alternating-current grounding electrodes for solar photovoltaic systems.

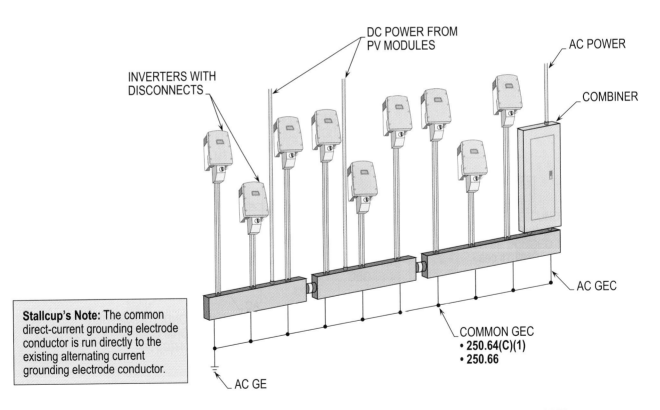

Stallcup's Note: The common direct-current grounding electrode conductor is run directly to the existing alternating current grounding electrode conductor.

COMMON DIRECT-CURRENT AND ALTERNATING-CURRENT GROUNDING ELECTRODE
NEC 690.47(C)(2)

Purpose of Change: To add requirements for the application and use of a common direct-current and grounding electrode means for PV systems.

Type of Change	New Subdivision			Committee Change	Accept in Principle			2008 NEC	-
ROP	pg. 878	# 4-235	log: 2508	ROC	pg. -	# -	log: -	UL	1703
Submitter: John Wiles				Submitter: -				OSHA	-
NFPA 70B	-			NFPA 70E	-			NFPA 79	-

690.47 Grounding Electrode System.

(C) Systems with Alternating-Current and Direct-Current Grounding Requirements. Photovoltaic systems having dc circuits and ac circuits with no direct connection between the dc grounded conductor and ac grounded conductor shall have a dc grounding system. The dc grounding system shall be bonded to the ac grounding system by one of the methods in (1), (2), or (3).

This section shall not apply to ac PV modules.

(3) Combined Direct-Current Grounding Electrode Conductor and Alternating-Current Equipment Grounding Conductor. An unspliced, or irreversibly spliced, combined grounding conductor shall be run from the marked dc grounding electrode conductor connection point along with the ac circuit conductors to the grounding busbar in the associated ac equipment. This combined grounding conductor shall be the larger of the sizes specified by 250.122 or 250.166 and shall be installed in accordance with 250.64(E).

Stallcup's Comment: A new subdivision has been added to address the requirement of a combined direct-current grounding electrode conductor and alternating-current equipment grounding conductor for solar photovoltaic systems.

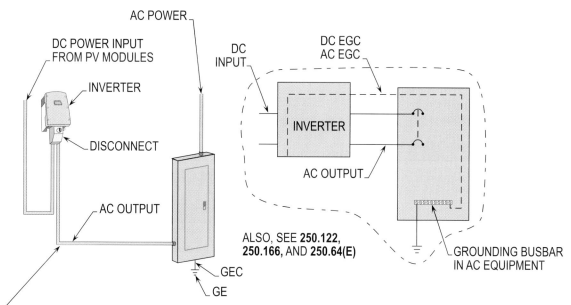

COMBINED DIRECT-CURRENT GROUNDING ELECTRODE CONDUCTOR AND ALTERNATING-CURRENT EQUIPMENT GROUNDING CONDUCTOR ROUTED WITH CIRCUIT CONDUCTORS, IRREVERSIBLY SPLICED, BONDED AT ENTRY AND EXIT OF ALL METAL RACEWAYS AND ENCLOSURES AND SIZED TO THE GREATER OF **250.166** OR **250.122**.
• **690.47(C)(3)**

COMBINED DIRECT-CURRENT GROUNDING ELECTRODE CONDUCTOR AND ALTERNATING-CURRENT EQUIPMENT GROUNDING CONDUCTOR
NEC 690.47(C)(3)

Purpose of Change: To provide requirements pertaining to direct-current grounding electrode conductor and alternating-current equipment grounding conductor and their connections in PV systems.

Type of Change	New Subsection		Committee Change	Accept in Principle		2008 NEC	-
ROP	pg. 884	# 4-250	log: 2513	ROC	pg. 487	# 4-107	log: 281	UL	1703
Submitter: John Wiles			Submitter: Technical Correlating Committee			OSHA	-
NFPA 70B	-		NFPA 70E	-		NFPA 79	-

690.72 Charge Control.

(C) Buck/Boost Direct-Current Converters. When buck/boost charge controllers and other dc power converters that increase or decrease the output current or output voltage with respect to the input current or input voltage are installed, the requirements shall comply with 690.72(C)(1) and (C)(2).

(1) The ampacity of the conductors in output circuits shall be based on the maximum rated continuous output current of the charge controller or converter for the selected output voltage range.

Stallcup's Comment: A new subsection has been added to address the requirements of buck/boost direct-current converters for solar photovoltaic systems.

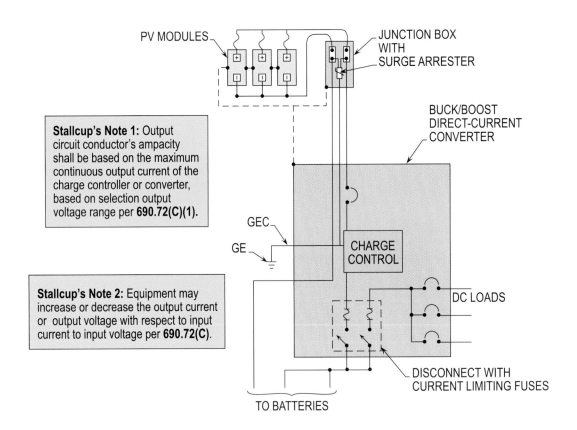

BUCK/BOOST DIRECT-CURRENT CONVERTERS
NEC 690.72(C)(1)

Purpose of Change: To provide requirements for the use of buck/boost direct-current charge controllers or converters.

Type of Change	New Subsection			Committee Change	Accept in Principle			2008 NEC	-
ROP	pg. 884	# 4-250	log: 2513	ROC	pg. 487	# 4-107	log: 281	UL	1703
Submitter: John Wiles				Submitter: Technical Correlating Committee				OSHA	-
NFPA 70B	-			NFPA 70E	-			NFPA 79	-

690.72 Charge Control.

(C) Buck/Boost Direct-Current Converters. When buck/boost charge controllers and other dc power converters that increase or decrease the output current or output voltage with respect to the input current or input voltage are installed, the requirements shall comply with 690.72(C)(1) and (C)(2).

(2) The voltage rating of the output circuits shall be based on the maximum voltage output of the charge controller or converter for the selected output voltage range.

Stallcup's Comment: A new subsection has been added to address the requirements of buck/boost direct-current converters for solar photovoltaic systems.

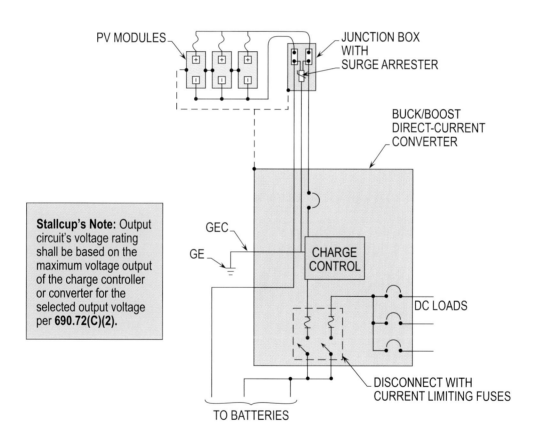

BUCK/BOOST DIRECT-CURRENT CONVERTERS
NEC 690.72(C)(2)

Purpose of Change: To provide requirements for the use of buck/boost direct-current charge controllers or converters.

Type of Change	New Article			Committee Change	Accept			2008 NEC	-
ROP	pg. 888	# 4-262	log: 3818	ROC	pg. 490	# 4-121	log: 2224	UL	-
Submitter: Thomas J. Baker				Submitter: Robert H. WIllis				OSHA	-
NFPA 70B	-			NFPA 70E	-			NFPA 79	-

Article 694 Small Wind Electric Systems.

Part I General

Part II Circuit Requirements

Part III Disconnecting Means

Part IV Wiring Methods

Part V Grounding

Part VI Marking

Part VII Connection to Other Sources

Part VIII Storage Batteries

Part IX Systems over 600 Volts

Stallcup's Comment: A new **Article 694** has been added, numbered as **694.1 through 694.85,** and divided into nine parts for general requirements, circuit requirements, disconnecting means, wiring methods, grounding, marking, connection to other sources, storage batteries, and systems over 600 volts.

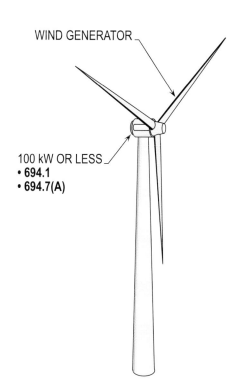

WIND GENERATOR

100 kW OR LESS
- **694.1**
- **694.7(A)**

SMALL WIND ELECTRIC SYSTEMS	
PART I	GENERAL • **694.1 THRU 694.7**
PART II	CIRCUIT REQUIREMENTS • **694.10 THRU 694.18**
PART III	DISCONNECTING MEANS • **694.20 THRU 694.28**
PART IV	WIRING METHODS • **694.30(A) THRU (C)**
PART V	GROUNDING • **694.40(A) THRU (C)**
PART VI	MARKING • **694.50 THRU 694.56**
PART VII	CONNECTION TO OTHER SOURCES • **694.60 THRU 694.68**
PART VIII	STORAGE BATTERIES • **694.70 THRU 694.75**
PART IX	SYSTEMS OVER 600 VOLTS • **694.80 THRU 694.85**

**SMALL WIND ELECTRIC SYSTEMS
ARTICLE 694**

Purpose of Change: To introduce new **Article 694.**

Type of Change	New Article			Committee Change	Accept			2008 NEC	-
ROP	pg. 888	# 4-262	log: 3818	ROC	pg. 490	# 4-121	log: 2224	UL	-
Submitter: Thomas J. Baker				Submitter: Robert H. WIllis				OSHA	-
NFPA 70B	-			NFPA 70E	-			NFPA 79	-

Article 694 Small Wind Electric Systems.

Part I General

- Scope – 694.1

- Definitions – 694.2

- Other Articles – 694.3

- Installation – 694.7

Stallcup's Comment: A new of **Article 694** has been added, numbered as **694.1 through 694.85,** and divided into nine parts for general requirements, circuit requirements, disconnecting means, wiring methods, grounding, marking, connection to other sources, storage batteries, and systems over 600 volts.

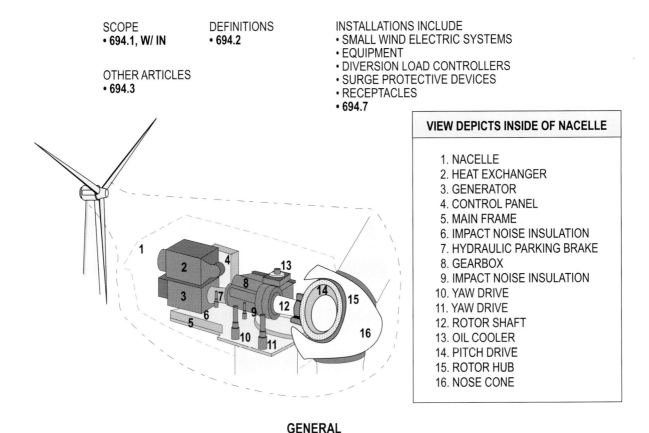

SCOPE
- **694.1, W/ IN**

DEFINITIONS
- **694.2**

INSTALLATIONS INCLUDE
- SMALL WIND ELECTRIC SYSTEMS
- EQUIPMENT
- DIVERSION LOAD CONTROLLERS
- SURGE PROTECTIVE DEVICES
- RECEPTACLES
- **694.7**

OTHER ARTICLES
- **694.3**

VIEW DEPICTS INSIDE OF NACELLE

1. NACELLE
2. HEAT EXCHANGER
3. GENERATOR
4. CONTROL PANEL
5. MAIN FRAME
6. IMPACT NOISE INSULATION
7. HYDRAULIC PARKING BRAKE
8. GEARBOX
9. IMPACT NOISE INSULATION
10. YAW DRIVE
11. YAW DRIVE
12. ROTOR SHAFT
13. OIL COOLER
14. PITCH DRIVE
15. ROTOR HUB
16. NOSE CONE

**GENERAL
ARTICLE 694, PART I**

Purpose of Change: Part I to the new **Article 694** covers the scope, definitions, and installation requirements.

Type of Change	New Article			Committee Change	Accept			2008 NEC	-
ROP	pg. 888	# 4-262	log: 3818	ROC	pg. 490	# 4-121	log: 2224	UL	-
Submitter: Thomas J. Baker				Submitter: Robert H. WIllis				OSHA	-
NFPA 70B	-			NFPA 70E	-			NFPA 79	-

Article 694 Small Wind Electric Systems.

Part II Circuit Requirements

- Maximum Voltage – 694.10(A) through (C)

- Circuit Sizing and Current – 694.12(A) and (B)

- Overcurrent Protection – 694.15(A) through (C)

- Stand-Alone Systems – 694.18(A) through (D)

Stallcup's Comment: A new of **Article 694** has been added, numbered as **694.1 through 694.85,** and divided into nine parts for general requirements, circuit requirements, disconnecting means, wiring methods, grounding, marking, connection to other sources, storage batteries, and systems over 600 volts.

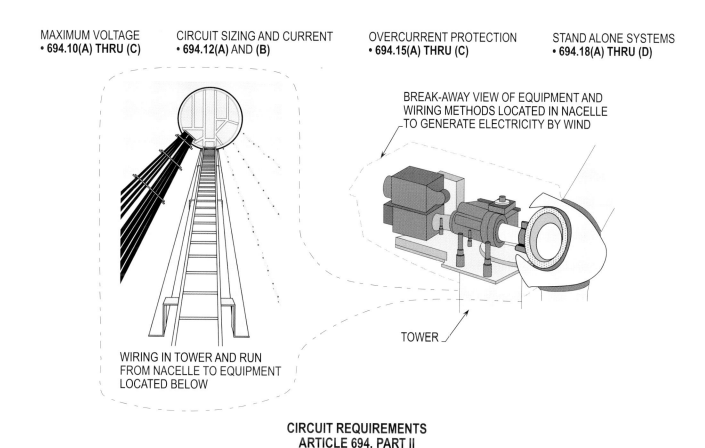

MAXIMUM VOLTAGE
• **694.10(A) THRU (C)**

CIRCUIT SIZING AND CURRENT
• **694.12(A)** AND **(B)**

OVERCURRENT PROTECTION
• **694.15(A) THRU (C)**

STAND ALONE SYSTEMS
• **694.18(A) THRU (D)**

BREAK-AWAY VIEW OF EQUIPMENT AND WIRING METHODS LOCATED IN NACELLE TO GENERATE ELECTRICITY BY WIND

TOWER

WIRING IN TOWER AND RUN FROM NACELLE TO EQUIPMENT LOCATED BELOW

**CIRCUIT REQUIREMENTS
ARTICLE 694, PART II**

Purpose of Change: Part II to new **Article 694** covers circuit requirements pertaining to installation of wind generators.

Type of Change	New Article			Committee Change	Accept			2008 NEC	-
ROP	pg. 888	# 4-262	log: 3818	ROC	pg. 490	# 4-121	log: 2224	UL	-
Submitter: Thomas J. Baker				Submitter: Robert H. WIllis				OSHA	-
NFPA 70B	-			NFPA 70E	-			NFPA 79	-

Article 694 Small Wind Electric Systems.

Part III Disconnecting Means

- • All Conductors – 694.20

- • Additional Provisions – 694.22(A) through (D)

- • Disconnection of Small Wind Electric System Equipment – 694.24

- • Fuses – 694.26

- • Installation and Service of a Wind Turbine – 694.28

Stallcup's Comment: A new of **Article 694** has been added, numbered as **694.1 through 694.85,** and divided into nine parts for general requirements, circuit requirements, disconnecting means, wiring methods, grounding, marking, connection to other sources, storage batteries, and systems over 600 volts.

ALL CONDUCTORS
• **694.20**

ADDITIONAL PROVISIONS
• **694.22(A) THRU (D)**

DISCONNECTION OF SMALL WIND ELECTRIC SYSTEM EQUIPMENT
• **694.24**

FUSES
• **694.26**

INSTALLATION AND SERVICE OF A WIND TURBINE
• **694.28**

Stallcup's Note: Personnel should wear PPE and perform PM as outlined in NFPA 70B and 70E.

DISCONNECT IN TURBINE TOWER
• **694.22(C)(1)**

DISCONNECT ADJACENT TO TURBINE TOWER
• **694.22(C)(1)**

(a) INSIDE TURBINE TOWER

(b) ADJACENT TO TURBINE TOWER

**DISCONNECTING MEANS
ARTICLE 694, PART III**

Purpose of Change: Part III to new **Article 694** covers requirements for disconnecting means.

Type of Change	New Article			Committee Change	Accept		2008 NEC	-
ROP pg. 888	# 4-262	log: 3818		ROC pg. 490	# 4-121	log: 2224	UL	-
Submitter: Thomas J. Baker				Submitter: Robert H. WIllis			OSHA	-
NFPA 70B -				NFPA 70E -			NFPA 79	-

Article 694 Small Wind Electric Systems.

Part IV Wiring Methods

- • Permitted Methods – 694.30(A) through (C)

Stallcup's Comment: A new of **Article 694** has been added, numbered as **694.1 through 694.85,** and divided into nine parts for general requirements, circuit requirements, disconnecting means, wiring methods, grounding, marking, connection to other sources, storage batteries, and systems over 600 volts.

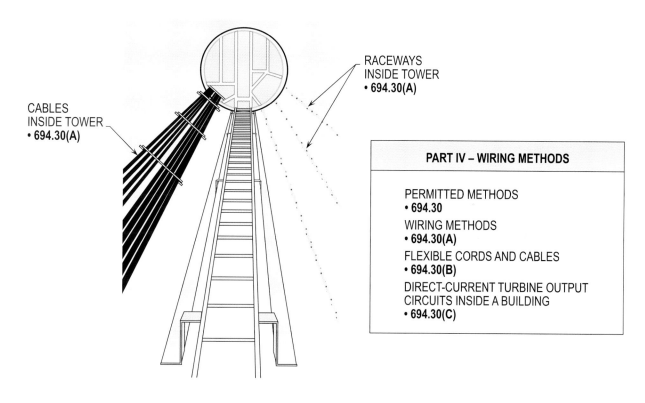

CABLES
INSIDE TOWER
• 694.30(A)

RACEWAYS
INSIDE TOWER
• 694.30(A)

PART IV – WIRING METHODS

PERMITTED METHODS
• 694.30
WIRING METHODS
• 694.30(A)
FLEXIBLE CORDS AND CABLES
• 694.30(B)
DIRECT-CURRENT TURBINE OUTPUT
CIRCUITS INSIDE A BUILDING
• 694.30(C)

**WIRING METHODS
ARTICLE 694, PART IV**

Purpose of Change: Part IV to new **Article 694** covers requirements for installation and use of wiring methods.

Type of Change	New Article			Committee Change	Accept			2008 NEC	-
ROP pg. 888		# 4-262	log: 3818	**ROC** pg. 490		# 4-121	log: 2224	**UL**	-
Submitter: Thomas J. Baker				Submitter: Robert H. WIllis				**OSHA**	-
NFPA 70B	-			**NFPA 70E**	-			**NFPA 79**	-

Article 694 Small Wind Electric Systems.

Part V Grounding

- Equipment Grounding – 694.40(A) through (C)

Stallcup's Comment: A new of **Article 694** has been added, numbered as **694.1 through 694.85,** and divided into nine parts for general requirements, circuit requirements, disconnecting means, wiring methods, grounding, marking, connection to other sources, storage batteries, and systems over 600 volts.

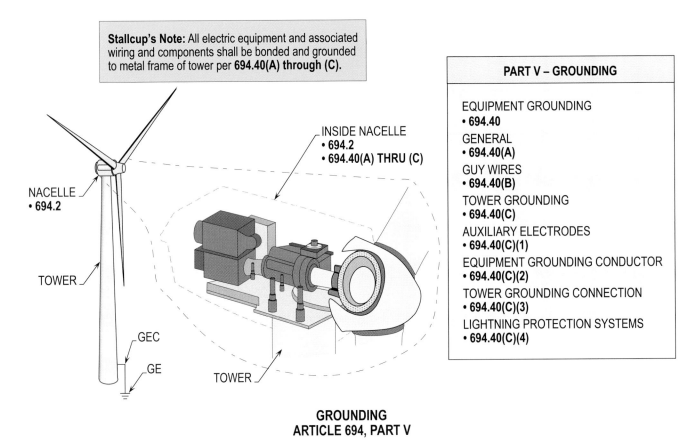

Stallcup's Note: All electric equipment and associated wiring and components shall be bonded and grounded to metal frame of tower per **694.40(A) through (C).**

INSIDE NACELLE
• **694.2**
• **694.40(A) THRU (C)**

NACELLE
• 694.2

TOWER

GEC

GE

TOWER

PART V – GROUNDING

EQUIPMENT GROUNDING
• **694.40**
GENERAL
• **694.40(A)**
GUY WIRES
• **694.40(B)**
TOWER GROUNDING
• **694.40(C)**
AUXILIARY ELECTRODES
• **694.40(C)(1)**
EQUIPMENT GROUNDING CONDUCTOR
• **694.40(C)(2)**
TOWER GROUNDING CONNECTION
• **694.40(C)(3)**
LIGHTNING PROTECTION SYSTEMS
• **694.40(C)(4)**

GROUNDING
ARTICLE 694, PART V

Purpose of Change: Part V to new **Article 694** covers requirements for the grounding tower and equipment related to the wind generator.

Type of Change	New Article		Committee Change	Accept		2008 NEC	-
ROP	pg. 888	# 4-262	log: 3818	ROC	pg. 490	# 4-121	log: 2224	UL	-
Submitter: Thomas J. Baker			Submitter: Robert H. WIllis			OSHA	-
NFPA 70B	-		NFPA 70E	-		NFPA 79	-

Article 694 Small Wind Electric Systems.

Part VI Marking

- Interactive System Point of Interconnection – 694.50

- Power Systems Employing Energy Storage – 694.52

- Identification of Power Sources – 694.54(A) and (B)

- Instructions for Disabling Turbine – 694.56

Stallcup's Comment: A new of **Article 694** has been added, numbered as **694.1 through 694.85,** and divided into nine parts for general requirements, circuit requirements, disconnecting means, wiring methods, grounding, marking, connection to other sources, storage batteries, and systems over 600 volts.

Stallcup's Note: Small wind electric systems shall be marked as described in **694.50 through 694.56.**

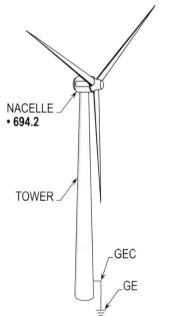

NACELLE
• **694.2**

TOWER

GEC

GE

PART VI – MARKING
INTERACTIVE SYSTEM POINT OF INTERCONNECTION • **694.50**
POWER SYSTEMS EMPLOYING ENERGY STORAGE • **694.52**
IDENTIFICATION OF POWER SOURCES • **694.54**
FACILITIES WITH STAND-ALONE SYSTEMS • **694.54(A)**
FACILITIES WITH UTILITY SERVICES AND SMALL WIND ELECTRIC SYSTEMS • **694.54(B)**
INSTRUCTIONS FOR DISABLING TURBINE • **694.56**

MARKING
ARTICLE 694, PART VI

Purpose of Change: Part VI to new **Article 694** covers requirements for marking small wind electric system installations.

Type of Change	New Article			Committee Change	Accept			2008 NEC	-
ROP	pg. 888	# 4-262	log: 3818	ROC	pg. 490	# 4-121	log: 2224	UL	-
Submitter: Thomas J. Baker				Submitter: Robert H. Willis				OSHA	-
NFPA 70B	-			NFPA 70E	-			NFPA 79	-

Article 694 Small Wind Electric Systems.

Part VII Connection to Other Sources

- Identified Interactive Equipment – 694.60

- Installation – 694.62

- Operating Voltage Range – 694.66

- Point of Connection – 694.68

Stallcup's Comment: A new of **Article 694** has been added, numbered as **694.1 through 694.85,** and divided into nine parts for general requirements, circuit requirements, disconnecting means, wiring methods, grounding, marking, connection to other sources, storage batteries, and systems over 600 volts.

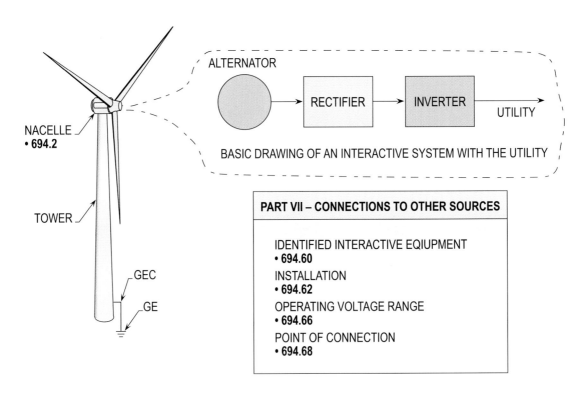

ALTERNATOR

RECTIFIER → INVERTER → UTILITY

BASIC DRAWING OF AN INTERACTIVE SYSTEM WITH THE UTILITY

NACELLE
• **694.2**

TOWER

GEC

GE

PART VII – CONNECTIONS TO OTHER SOURCES

IDENTIFIED INTERACTIVE EQIPUMENT
• **694.60**
INSTALLATION
• **694.62**
OPERATING VOLTAGE RANGE
• **694.66**
POINT OF CONNECTION
• **694.68**

CONNECTIONS TO OTHER SOURCES
ARTICLE 694, PART VII

Purpose of Change: Part VII to new **Article 694** covers requirements for connections to other sources such as the utility.

Type of Change	New Article			Committee Change	Accept			2008 NEC	-
ROP	pg. 888	# 4-262	log: 3818	ROC	pg. 490	# 4-121	log: 2224	UL	-
Submitter: Thomas J. Baker				Submitter: Robert H. WIllis				OSHA	-
NFPA 70B	-			NFPA 70F	-			NFPA 79	-

Article 694 Small Wind Electric Systems.

Part VIII Storage Batteries

- Installation – 694.70(A) through (G)

- Charge Control – 694.75(A) and (B)

Stallcup's Comment: A new of **Article 694** has been added, numbered as **694.1 through 694.85,** and divided into nine parts for general requirements, circuit requirements, disconnecting means, wiring methods, grounding, marking, connection to other sources, storage batteries, and systems over 600 volts.

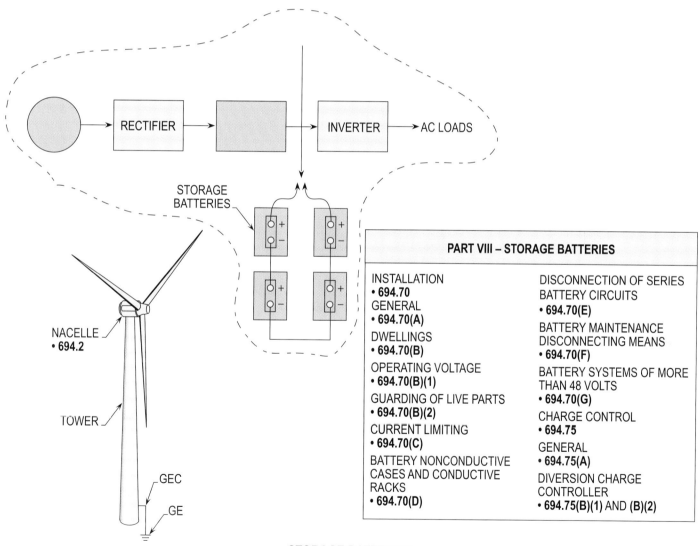

STORAGE BATTERIES
ARTICLE 694, PART VIII

Purpose of Change: Part VIII to new **Article 694** covers requirements for storage batteries.

Type of Change	New Article			Committee Change	Accept			2008 NEC	-
ROP	pg. 888	# 4-262	log: 3818	ROC	pg. 490	# 4-121	log: 2224	UL	-
Submitter: Thomas J. Baker				Submitter: Robert H. WIllis				OSHA	-
NFPA 70B	-			NFPA 70E	-			NFPA 79	-

Article 694 Small Wind Electric Systems.

Part IX Systems over 600 Volts

- General – 694.80

- Cable and Equipment Ratings – 694.85(A) and (B)

Stallcup's Comment: A new of **Article 694** has been added, numbered as **694.1 through 694.85,** and divided into nine parts for general requirements, circuit requirements, disconnecting means, wiring methods, grounding, marking, connection to other sources, storage batteries, and systems over 600 volts.

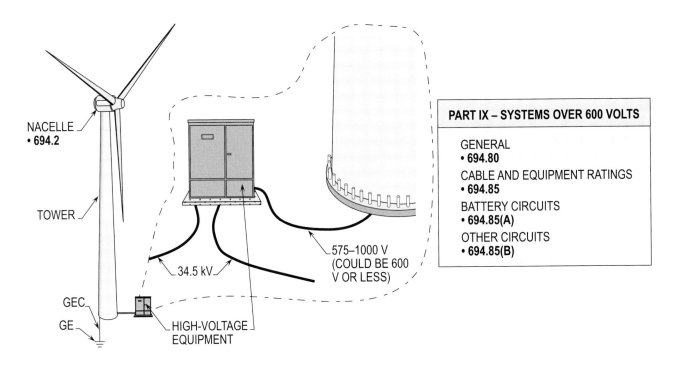

**SYSTEMS OVER 600 VOLTS
ARTICLE 694, PART IX**

Purpose of Change: Part IX to new **Article 694** covers equipment and wiring used for systems over 600 volts.

Type of Change	Revision			Committee Change	Accept			2008 NEC	695.3(B)
ROP	pg. 909	# 13-60a	log: CP 1300	ROC	pg. -	# -	log: -	UL	1004-5
Submitter: Code-Making Panel 13				Submitter: -				OSHA	-
NFPA 70B	-			NFPA 70E	-			NFPA 79	-

695.3 Power Source(s) for Electric Motor-Driven Fire Pumps.

(B) Multiple Sources. If reliable power cannot be obtained from a source described in 695.3(A), power shall be supplied by one of the following:

(1) Individual Sources. An approved combination of two or more of the sources from 695.3(A).

Exception to (B)(1) and (B)(2): An alternate source of power shall not be required where a back-up engine-driven or back-up steam turbine-driven fire pump is installed.

Stallcup's Comment: A revision has been made to clarify the requirements of multiple sources for fire pumps.

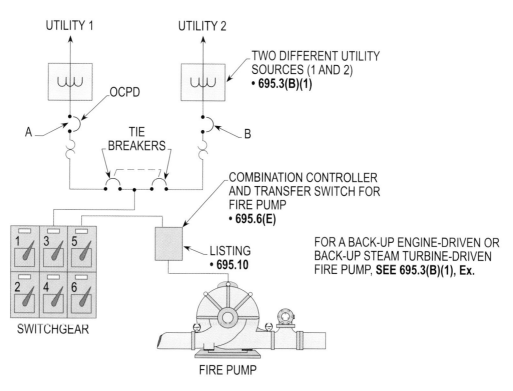

INDIVIDUAL SOURCES
NEC 695.3(B)(1) and Ex.

Purpose of Change: To clarify that two different utility power sources can be considered reliable power for a fire pump.

Type of Change	Revision			Committee Change	Accept			2008 NEC	695.3(B)
ROP	pg. 909	# 13-60a	log: CP 1300	ROC	pg. -	# -	log: -	UL	1004-5
Submitter: Code Making Panel 13				Submitter: -				OSHA	-
NFPA 70B	-			NFPA 70E	-			NFPA 79	-

695.3 Power Source(s) for Electric Motor-Driven Fire Pumps.

(B) Multiple Sources. If reliable power cannot be obtained from a source described in 695.3(A), power shall be supplied by one of the following:

(2) Individual Source and On-site Standby Generator. An approved combination of one or more of the sources in 695.3(A) and an on-site standby generator complying with 695.3(D).

Exception to (B)(1) and (B)(2): An alternate source of power shall not be required where a back-up engine-driven or back-up steam turbine-driven fire pump is installed.

Stallcup's Comment: A revision has been made to clarify the requirements of multiple sources for fire pumps.

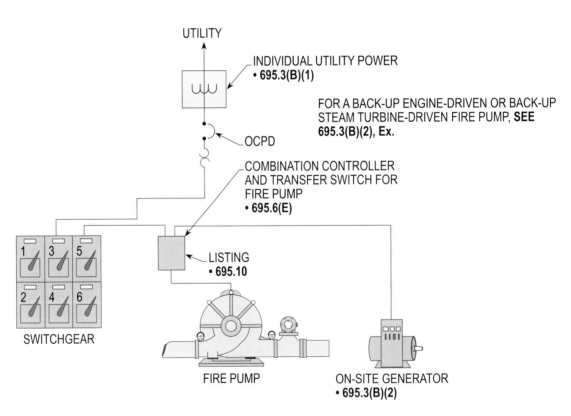

INDIVIDUAL SOURCE AND ON-SITE STANDBY GENERATOR
NEC 695.3(B)(2) and Ex.

Purpose of Change: To clarify that an individual power source and an on-site generator is considered reliable power for a fire pump.

Type of Change	New Subsection			Committee Change	Accept			2008 NEC	-
ROP	pg. 909	# 13-60a	log: CP 1300	ROC	pg. 501	# 13-83	log: 1802	UL	1004-5
Submitter: Code-Making Panel 13				Submitter: Michael P. Walls				OSHA	-
NFPA 70B	-			NFPA 70E	-			NFPA 79	-

695.3 Power Source(s) for Electric Motor-Driven Fire Pumps.

(C) Multibuilding Campus-Style Complexes. If the sources in 695.3(A) are not practicable and the installation is part of a multibuilding campus-style complex, feeder sources shall be permitted if approved by the authority having jurisdiction and installed in accordance with either (C)(1) and (C)(3) or (C)(2) and (C)(3).

(1) Feeder Sources. Two or more feeders shall be permitted as more than one power source if such feeders are connected to, or derived from, separate utility services. The connection(s), overcurrent protective device(s), and disconnecting means for such feeders shall meet the requirements of 695.4(B).

Stallcup's Comment: A new subsection has been added to address the requirements of multibuilding campus-style complexes for fire pumps.

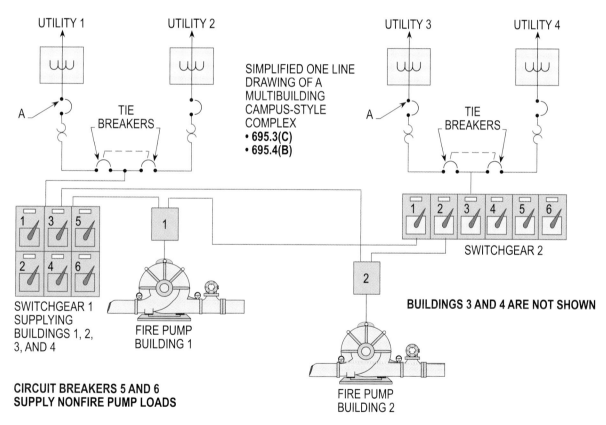

FEEDER SOURCES
NEC 695.3(C)(1)

Purpose of Change: To clarify that two or more feeders from different utility sources can supply reliable power to fire pumps located on multibuilding campus-style complexes.

Type of Change	New Subsection		Committee Change	Accept		2008 NEC	-
ROP	pg. 909	# 13-60a	log: CP 1300	ROC	pg. 501	# 13-83	log: 1802	UL	1004-5
Submitter: Code-Making Panel 13			Submitter: Michael P. Walls			OSHA	-
NFPA 70B	-		NFPA 70E	-		NFPA 79	-

695.3 Power Source(s) for Electric Motor-Driven Fire Pumps.

(C) Multibuilding Campus-Style Complexes. If the sources in 695.3(A) are not practicable and the installation is part of a multibuilding campus-style complex, feeder sources shall be permitted if approved by the authority having jurisdiction and installed in accordance with either (C)(1) and (C)(3) or (C)(2) and (C)(3).

(2) Feeder and Alternate Source. A feeder shall be permitted as a normal source of power if an alternate source of power independent from the feeder is provided. The connection(s), overcurrent protective device(s), and disconnecting means for such feeders shall meet the requirements of 695.4(B).

(3) Selective Coordination. The overcurrent protective device(s) in each disconnecting means shall be selectively coordinated with any other supply-side overcurrent protective device(s).

Stallcup's Comment: A new subsection has been added to address the requirements of multibuilding campus-style complexes for fire pumps.

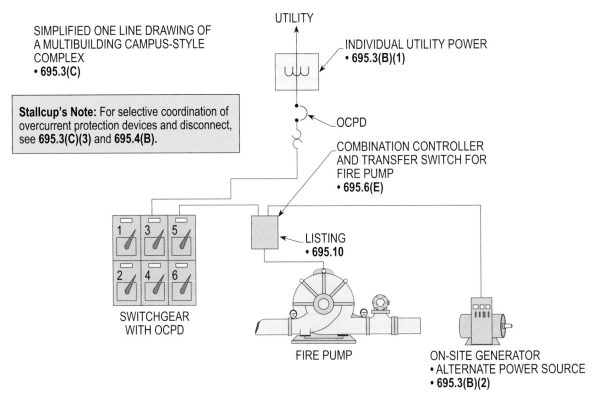

SIMPLIFIED ONE LINE DRAWING OF A MULTIBUILDING CAMPUS-STYLE COMPLEX
• 695.3(C)

Stallcup's Note: For selective coordination of overcurrent protection devices and disconnect, see **695.3(C)(3)** and **695.4(B).**

UTILITY

INDIVIDUAL UTILITY POWER
• 695.3(B)(1)

OCPD

COMBINATION CONTROLLER AND TRANSFER SWITCH FOR FIRE PUMP
• 695.6(E)

LISTING
• 695.10

SWITCHGEAR WITH OCPD

FIRE PUMP

ON-SITE GENERATOR
• ALTERNATE POWER SOURCE
• 695.3(B)(2)

**FEEDER AND ALTERNATE SOURCE AND SELECTIVE COORDINATION
NEC 695.3(C)(2) AND (C)(3)**

Purpose of Change: To clarify that a feeder shall be permitted to be used as a normal power supply if an alternate source is provided.

Type of Change	New Subsection			Committee Change	Accept			2008 NEC	-
ROP	pg. 909	# 13-60a	log: CP 1300	ROC	pg. -	# -	log: -	UL	1004-5
Submitter: Code-Making Panel 13				Submitter: -				OSHA	-
NFPA 70B	-			NFPA 70E	-			NFPA 79	-

695.3 Power Source(s) for Electric Motor-Driven Fire Pumps.

(D) On-Site Standby Generator as Alternate Source. An on-site standby generator(s) used as an alternate source of power shall comply with (D)(1) through (D)(3).

(1) Capacity. The generator shall have sufficient capacity to allow normal starting and running of the motor(s) driving the fire pump(s) while supplying all other simultaneously operated load(s).

Automatic shedding of one or more optional standby loads in order to comply with this capacity requirement shall be permitted.

(2) Connection. A tap ahead of the generator disconnecting means shall not be required.

(3) Adjacent Disconnects. The requirements of 430.113 shall not apply.

Stallcup's Comment: A new subsection has been added to address the requirements of an on-site standby generator as alternate source for fire pumps.

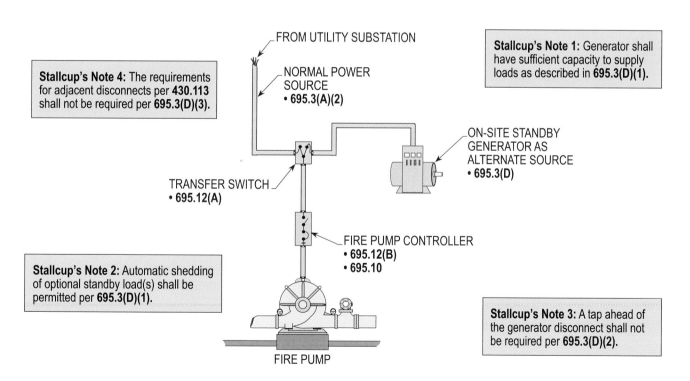

**ON-SITE STANDBY GENERATOR AS ALTERNATE SOURCE
NEC 695.3(D)(1) THRU (D)(3)**

Purpose of Change: To clarify the requirements for using an on-site standby generator as an alternate source of power for fire pumps.

Type of Change	New Subsection			Committee Change	Accept			2008 NEC	-
ROP	pg. 909	# 13-60a	log: CP 1300	ROC	pg. -	# -	log: -	UL	1004-5
Submitter: Code-Making Panel 13				Submitter: -				OSHA	-
NFPA 70B	-			NFPA 70E	-			NFPA 79	-

695.3 Power Source(s) for Electric Motor-Driven Fire Pumps.

(E) Arrangement. All power supplies shall be located and arranged to protect against damage by fire from within the premises and exposing hazards.

Multiple power sources shall be arranged so that a fire at one source does not cause an interruption at the other source.

(F) Phase Converters. Phase converters shall not be permitted to be used for fire pump service.

Stallcup's Comment: A new subsection has been added to address the arrangement and phase converter requirements for fire pumps.

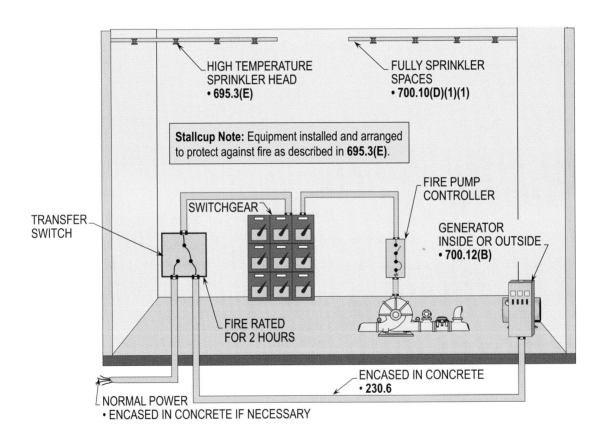

ARRANGEMENT AND PHASE CONVERTERS
NEC 695.3(E) AND (F)

Purpose of Change: To provide requirements for protection from fire and hazards as well as prohibiting phase converters for fire pump installations.

Type of Change	New Subdivision			Committee Change	Accept			**2008 NEC**	-
ROP	pg. 915	# 13-77a	log: CP 1301	**ROC**	pg. -	# -	log: -	**UL**	1004-5
Submitter: Code-Making Panel 13				Submitter: -				**OSHA**	-
NFPA 70B	-			**NFPA 70E**	-			**NFPA 79**	-

695.4 Continuity of Power.

(B) Connection Through Disconnecting Means and Overcurrent Device.

(2) Overcurrent Device Selection. Overcurrent devices shall comply with (a) or (b).

(b) *On-Site Standby Generators.* Overcurrent protective devices between an on-site standby generator and a fire pump controller shall be selected and sized to allow for instantaneous pickup of the full pump room load, but shall not be larger than the value selected to comply with 430.62 to provide short-circuit protection only.

Stallcup's Comment: A new subdivision has been added to address the overcurrent device selection requirements of on-site standby generators.

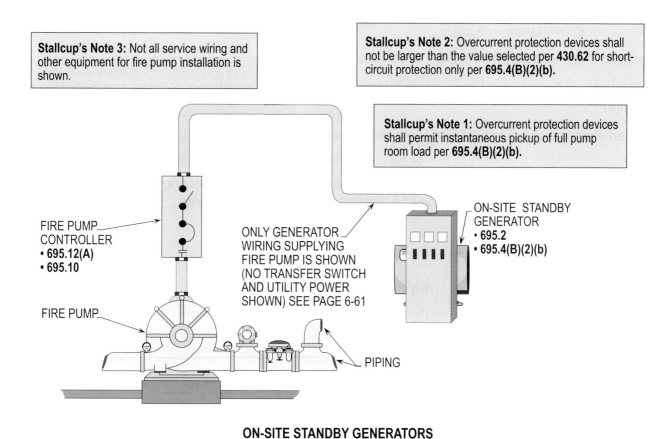

Stallcup's Note 3: Not all service wiring and other equipment for fire pump installation is shown.

Stallcup's Note 2: Overcurrent protection devices shall not be larger than the value selected per **430.62** for short-circuit protection only per **695.4(B)(2)(b).**

Stallcup's Note 1: Overcurrent protection devices shall permit instantaneous pickup of full pump room load per **695.4(B)(2)(b).**

FIRE PUMP CONTROLLER
• **695.12(A)**
• **695.10**

FIRE PUMP

ONLY GENERATOR WIRING SUPPLYING FIRE PUMP IS SHOWN (NO TRANSFER SWITCH AND UTILITY POWER SHOWN) SEE PAGE 6-61

ON-SITE STANDBY GENERATOR
• **695.2**
• **695.4(B)(2)(b)**

PIPING

ON-SITE STANDBY GENERATORS
NEC 695.4(B)(2)(b)

Purpose of Change: To provide requirements for sizing and selecting overcurrent protection devices for on-site standby generators.

Type of Change	New Subsection			Committee Change	Accept in Principle		2008 NEC	-
ROP	pg. 922	# 13-97	log: 4392	ROC	pg. -	# -	log: -	UL	1004-5
Submitter: John R. Kovacik				Submitter: -			OSHA	-
NFPA 70B	-			NFPA 70E	-		NFPA 79	-

695.6 Power Wiring.

(E) Loads Supplied by Controllers and Transfer Switches. A fire pump controller and fire pump power transfer switch, if provided, shall not serve any load other than the fire pump for which it is intended.

Stallcup's Comment: A new subsection has been added to address the requirements of loads supplied by controllers and transfer switches for fire pumps.

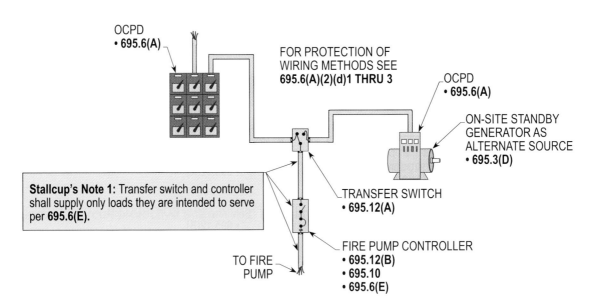

LOADS SUPPLIED BY CONTROLLERS AND TRANSFER SWITCHES
NEC 695.6(E)

Purpose of Change: To add requirements for fire-pump loads supplied by transfer switches and controllers.

Type of Change	New Subsection			Committee Change	Accept in Principle		2008 NEC	-
ROP pg. 922	# 13-97	log: 4392		ROC pg. -	# -	log: -	UL	1004-5
Submitter: John R. Kovacik				Submitter: -			OSHA	-
NFPA 70B	▪			NFPA 70E			NFPA 79	-

695.6 Power Wiring.

(H) Listed Electrical Circuit Protective System to Controller Wiring. Electrical circuit protective system installation shall comply with any restrictions provided in the listing of the electrical circuit protective system used and the following also shall apply:

(1) A junction box shall be installed ahead of the fire pump controller a minimum of 300 mm (12 in.) beyond the fire-rated wall or floor bounding the fire zone.

(2) Where required by the manufacturer of a listed electrical circuit protective system or by the listing, or as required elsewhere in this *Code*, the raceway between a junction box and the fire pump controller shall be sealed at the junction box end as required and in accordance with the instructions of the manufacturer.

(3) Standard wiring between the junction box and the controller shall be permitted.

Stallcup's Comment: A new subsection has been added to address electrical circuit protective system installation and comply with any restrictions provided in the listing of the electrical circuit protective system.

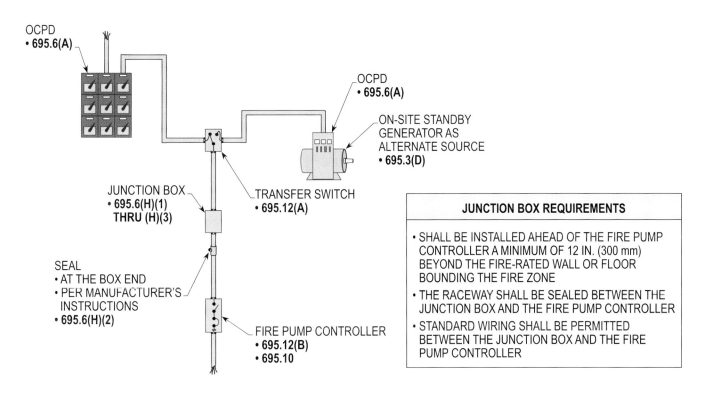

**LISTED ELECTRICAL CIRCUIT PROTECTIVE SYSTEM TO CONTROLLER WIRING
NEC 695.6(H)(1) THRU (H)(3)**

Purpose of Change: To clarify the installation of an electrical circuit protective system and comply with any restrictions provided in the listing of the electrical circuit protective system.

_____

Name   Date

Chapter 6   Section   Answer
Special Equipment

1. The switch or circuit breaker shall open all _____ conductors simultaneously on multi-wire branch circuits for sign and outline lighting systems.   _____ _____

(A) bonding   **(B)** equipment grounding
(C) grounded   **(D)** ungrounded

2. The conductors for the secondary wiring of LED sign illumination systems shall have an ampacity not less than the load to be supplied and shall not be sized smaller than _____ AWG.   _____ _____

(A) 22   **(B)** 18
(C) 16   **(D)** 14

3. Information technology equipment shall be permitted to be connected to a branch circuit by a power-supply cord that does not exceed _____ ft.   _____ _____

(A) 6   **(B)** 10
(C) 15   **(D)** 20

4. The remote disconnect controls for the control of electronic equipment power and HVAC systems shall be grouped and _____.   _____ _____

(A) approved   **(B)** identified
(C) labeled   **(D)** listed

5. Outlets supplying pool pump motors connected to single-phase, 120 volt through 240 volt branch circuits, rated 15 or 20 amps, whether by receptacle or by direct connection, shall be provided with _____ protection for personnel.   _____ _____

(A) AFCI   **(B)** GFCI
(C) both (A) and (B)   **(D)** neither (A) nor (B)

6. Perimeter surfaces less than _____ ft separated by a permanent wall or building 5 ft in height or more shall require equipotential bonding on the pool side of the permanent wall or building.   _____ _____

(A) 3   **(B)** 5
(C) 10   **(D)** 15

7. Where the hydromassage bathtub is cord-and-plug connected with the supply receptacle accessible only through a service access opening, the receptacle shall be installed so that its face is within direct view and not more than _____ ft of the opening.   _____ _____

(A) 1   **(B)** 3
(C) 5   **(D)** 6

8. The disconnecting means for submersible or floating equipment shall be readily accessible on land, located not more than _____ in. from the receptacle it controls, and shall be located in the supply circuit ahead of the receptacle.   _____ _____

(A) 12   **(B)** 18
(C) 24   **(D)** 30

Section   Answer

9. Where the conductors of more than one photovoltaic system occupy the same junction box, raceway, or equipment, the conductors of each system shall be _____ at all termination, connection, and splice points.

(A) approved   **(B)** listed
(C) identified   **(D)** marked

10. Where the conductors of more than one photovoltaic system occupy the same junction box or raceway with a removable cover(s), the ac and dc conductors or each system shall be grouped separately by wire ties or similar means at least once, and then shall be grouped at intervals not to exceed _____ ft.

(A) 1   **(B)** 3
(C) 5   **(D)** 6

11. Circuit breakers for stand-alone solar photovoltaic systems that are marked _____ shall not be backfed.

(A) line   **(B)** load
(C) both (A) and (B)   **(D)** neither (A) nor (B)

12. Disconnecting means shall be installed on photovoltaic output circuits where overcurrent devices (fuses) must be serviced that cannot be isolated from energized circuits. Where the disconnecting means are located more than _____ ft from the overcurrent device, a directory showing the location of each disconnect shall be installed at the overcurrent device location.

(A) 6   **(B)** 10
(C) 12   **(D)** 15

13. Wiring methods shall not be installed within _____ in. of the roof decking or sheathing except where directly below the roof surface covered by photovoltaic modules and associated equipment.

(A) 6   **(B)** 10
(C) 12   **(D)** 24

14. Spacing between labels or markings, or between a label and a marking, shall not be more than _____ ft for direct-current photovoltaic source and output circuits inside a building.

(A) 3   **(B)** 5
(C) 6   **(D)** 10

15. Circuit conductors and overcurrent devices shall be sized to carry not less than _____ percent of the maximum current for small wind electric systems.

(A) 100   **(B)** 125
(C) 135   **(D)** 150

16. The turbine disconnecting means for small wind electric systems shall consist of not more than _____ switches or circuit breakers mounted in a single enclosure, in a group of separate enclosures, or in or on a switchboard.

(A) 6   **(B)** 8
(C) 10   **(D)** 12

Section   Answer

17. Storage batteries for small wind electric systems in dwellings shall have the cells connected to operate at less than _____ volts.

_____ _____

(A) 12   (B) 24
(C) 50   (D) 100

18. Where routed through a building, the feeder conductors shall be encased in a minimum _____ in. of concrete.

_____ _____

(A) 1   (B) 2
(C) 4   (D) 6

19. A junction box shall be installed ahead of the fire pump controller a minimum of _____ in. beyond the fire-rated wall or floor bounding the fire zone for an electrical circuit protective system.

_____ _____

(A) 2   (B) 6
(C) 10   (D) 12

20. Control conductors installed between the fire pump power transfer switch and the standby generator supplying the fire pump during normal power loss shall be kept entirely independent of all other wiring and shall be permitted to be encased in a minimum of _____ in. of concrete.

_____ _____

(A) 2   (B) 4
(C) 6   (D) 8

Type of Change	New Subsection			Committee Change	Accept in Principle			2008 NEC	770.110
ROP	pg. 1052	# 16-47	log: 2083	ROC	pg. -	# -	log: -	UL	1651
Submitter: Ron L. Janikowski				Submitter: -				OSHA	-
NFPA 70B	-			NFPA 70E	-			NFPA 79	-

770.110 Raceways for Optical Fiber Cables.

(B) Raceway Fill for Optical Fiber Cables. Raceway fill for optical fibers cables shall comply with either (B)(1) or (B)(2).

(1) Without Electric Light or Power Conductors. Where optical fiber cables are installed in raceway without electric light or power conductors, the raceway fill requirements of Chapters 3 and 9 shall not apply.

(2) Nonconductive Optical Fiber Cables with Electric Light or Power Conductors. Where nonconductive optical fiber cables are installed with electric light or power conductors in a raceway, the raceway fill requirements of Chapters 3 and 9 shall apply.

Stallcup's Comment: A revision has been made to clarify the raceway fill for optical fiber cables.

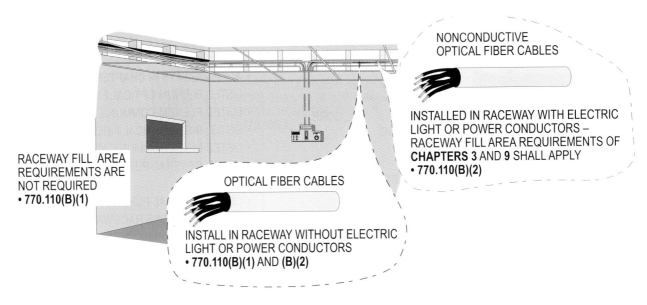

NONCONDUCTIVE OPTICAL FIBER CABLES

INSTALLED IN RACEWAY WITH ELECTRIC LIGHT OR POWER CONDUCTORS – RACEWAY FILL AREA REQUIREMENTS OF **CHAPTERS 3** AND **9** SHALL APPLY
• **770.110(B)(2)**

RACEWAY FILL AREA REQUIREMENTS ARE NOT REQUIRED
• **770.110(B)(1)**

OPTICAL FIBER CABLES

INSTALL IN RACEWAY WITHOUT ELECTRIC LIGHT OR POWER CONDUCTORS
• **770.110(B)(1)** AND **(B)(2)**

RACEWAY FILL FOR OPTICAL FIBER CABLES
NEC 770.110(B)(1) AND (B)(2)

Purpose of Change: To provide and clarify the requirements for the fill area for optical fiber cables installed in raceways or without electric light and power conductors.

Type of Change	New Subsections			Committee Change	Accept in Princple			2008 NEC	770.113
ROP	pg. 1053	# 16-48	log: 2084	ROC	pg. 575	# 16-37	log: 1627	UL	1651
Submitter: Ron L. Janikowski				Submitter: Craig Sato				OSHA	-
NFPA 70B	-			NFPA 70E	-			NFPA 79	-

770.113 Installation of Optical Fiber Cables and Raceways, and Cable Routing Assemblies.

(A) Listing. Optical fiber cables and raceways, and cable routing assemblies installed in buildings shall be listed.

Exception: Optical fiber cables that comply with 770.48 shall not be required to be listed.

(B) Fabricated Ducts Used for Environmental Air. The following cables shall be permitted in ducts, as described in 300.22(B) if they are directly associated with the air distribution system:

(1) Up to 1.22 m (4 ft) of Types OFNP and OFCP cables

(2) Types OFNP, OFCP, OFNR, OFCR, OFNG, OFCG, OFN, and OFC cables installed in raceways that are installed in compliance with 300.22(B)

Informational Note: For information on fire protection of wiring installed in fabricated ducts see 4.3.4.1 and 4.3.11.3.3 in NFPA 90A-2009, *Standard for the Installation of Air-Conditioning and Ventilating Systems*.

Stallcup's Comment: A revision has been made to clarify the installation requirements of optical fiber cables and raceways and cable routing assembllies.

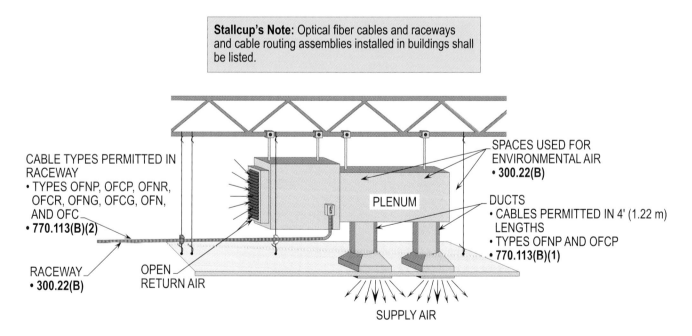

> **Stallcup's Note:** Optical fiber cables and raceways and cable routing assemblies installed in buildings shall be listed.

LISTING AND FABRICATED DUCTS USED FOR ENVIRONMENTAL AIR
NEC 770.113(A) AND (B)

Purpose of Change: To provide requirements for installing optical fiber cables and raceways and cable routing assemblies in spaces used for environmental air circulation.

Type of Change	New Subsection			Committee Change	Accept in Principle			2008 NEC	770.113
ROP	pg. 1053	# 16-48	log: 2084	ROC	pg. 575	# 16-37	log: 1627	UL	1651
Submitter: Ron L. Janikowski				Submitter: Craig Sato				OSHA	-
NFPA 70B	-			NFPA 70E	-			NFPA 79	-

770.113 Installation of Optical Fiber Cables and Raceways, and Cable Routing Assemblies.

(C) Other Spaces Used For Environmental Air (Plenums). The following cables and raceways shall be permitted in other spaces used for environmental air as described in 300.22(C):

(1) Types OFNP and OFCP cables

(2) Plenum optical fiber raceway

(3) Types OFNP and OFCP cables installed in plenum optical fiber raceway or plenum communications raceway

(4) Types OFNP and OFCP cables and plenum optical fiber raceways supported by open metallic cable trays or cable tray systems

Stallcup's Comment: A new subsection has been added to address the installation requirements of optical fiber cables and raceways in other spaces used for environmental air (plenums).

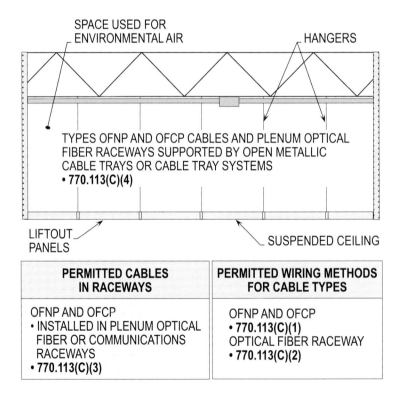

OTHER SPACES USED FOR ENVIRONMENTAL AIR (PLENUMS)
NEC 770.113(C)(1) THRU (C)(4)

Purpose of Change: To outline requirements for installing optical fiber cables and raceways in other spaces used for environmental air.

Type of Change	New Subsection			Committee Change	Accept in Principle			2008 NEC	770.113
ROP	pg. 1053	# 16-48	log: 2084	ROC	pg. 575	# 16-37	log: 1627	UL	1651
Submitter: Ron L. Janikowski				Submitter: Craig Sato				OSHA	-
NFPA 70B				NFPA 70E	-			NFPA 79	-

770.113 Installation of Optical Fiber Cables and Raceways, and Cable Routing Assemblies.

(C) Other Spaces Used For Environmental Air (Plenums). The following cables and raceways shall be permitted in other spaces used for environmental air as described in 300.22(C):

(5) Types OFNP, OFCP, OFNR, OFCR, OFNG, OFCG, OFN, and OFC cables installed in raceways that are installed in compliance with 300.22(C)

(6) Types OFNP, OFCP, OFNR, OFCR, OFNG, OFCG, OFN, OFC cables and plenum optical fiber raceways, riser optical fiber raceways and general-purpose optical fiber raceways supported by solid bottom metal cable trays with solid metal covers in other spaces used for environmental air (plenums) as described in 300.22(C)

Informational Note: For information on fire protection of wiring installed in other spaces used for environmental air see 4.3.11.2, 4.3.11.4, and 4.3.11.5 of NFPA 90A-2009, *Standard for the Installation of Air-Conditioning and Ventilating Systems*.

Stallcup's Comment: A new subsection has been added to address the installation requirements of optical fiber cables and raceways in other spaces used for environmental air (plenums).

> **Stallcup's Note 1:** Types OFNP, OFCP, OFNR, OFCR, OFNG, OFCG, OFN, and OFC cables shall be installed in raceways that comply with **770.113(C)(5)** and **300.22(C).**

> **Stallcup's Note 2:** Types OFNP, OFCP, OFNR, OFCR, OFNG, OFCG, OFN, and OFC cables, plenum optical fiber raceways, and general-purpose optical fiber raceways shall be installed per **770.113(C)(6).**

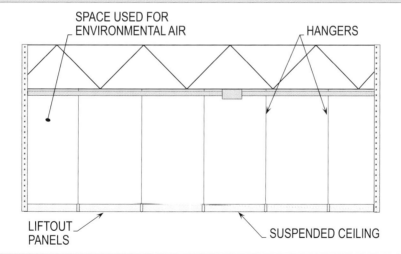

> **Stallcup's Note 3:** General-purpose optical fiber raceways supported by solid bottom metal cable trays with solid metals covers as outlined in **770.113(C)(6)** and **300.22(C).**

OTHER SPACES USED FOR ENVIRONMENTAL AIR (PLENUMS)
NEC 770.113(C)(5) AND (C)(6)

Purpose of Change: To clarify the requirements for installing optical fiber cables and raceways in other spaces used for environmental air.

Type of Change	New Subsection			Committee Change	Accept in Principle			2008 NEC	770.113
ROP	pg. 1053	# 16-48	log: 2084	ROC	pg. 575	# 16-37	log: 1627	UL	1651
Submitter: Ron L. Janikwoski				Submitter: Craig Sato				OSHA	-
NFPA 70B	-			NFPA 70E	-			NFPA 79	-

770.113 Installation of Optical Fiber Cables and Raceways, and Cable Routing Assemblies.

(D) Risers — Cables, Raceways and Cable Routing Assemblies in Vertical Runs. The following cables, raceways and cable routing assemblies shall be permitted in vertical runs penetrating one or more floors and in vertical runs in a shaft:

(1) Types OFNP, OFCP, OFNR, and OFCR cables

(2) Plenum and riser optical fiber raceways

(3) Riser cable routing assemblies

Stallcup's Comment: A new subsection has been added to address the installation requirements of optical fiber cables, raceways and cable routing assemblies in vertical runs.

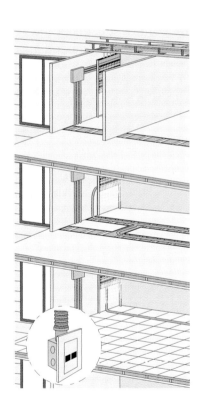

CABLES, RACEWAYS AND CABLE ROUTING ASSEMBLIES IN VERTICAL RUNS

TYPES OFNP, OFCP, OFNR, AND OFCR CABLES
• **770.113(D)(1)**
PLENUM AND RISER OPTICAL FIBER RACEWAYS
• **770.113(D)(2)**
RISER CABLE ROUTING ASSEMBLIES
• **770.113(D)(3)**

RISERS – CABLES, RACEWAYS AND CABLE ROUTING ASSEMBLIES IN VERTICAL RUNS
NEC 770.113(D)(1) THRU (D)(3)

Purpose of Change: To provide and clarify the installation requirements for optical fiber cables, raceways and cable routing assemblies used in vertical runs.

Type of Change	New Subsection			Committee Change	Accept in Principle		2008 NEC	770.113
ROP	pg. 1053	# 16-48	log: 2084	ROC	pg. 575	# 16-37	log: 1627	UL	1651
Submitter: Ron L. Janikwoski				Submitter: Craig Sato			OSHA	-
NFPA 70B	-			NFPA 70E	-		NFPA 79	-

770.113 Installation of Optical Fiber Cables and Raceways, and Cable Routing Assemblies.

(D) Risers — Cables, Raceways and Cable Routing Assemblies in Vertical Runs. The following cables, raceways and cable routing assemblies shall be permitted in vertical runs penetrating one or more floors and in vertical runs in a shaft:

(4) Types OFNP, OFCP, OFNR, and OFCR cables installed in:

a. Plenum optical fiber raceway

b. Plenum communications raceway

c. Riser optical fiber raceway

d. Riser communications raceway

e. Riser cable routing assembly

Informational Note: See 770.26 for firestop requirements for floor penetrations.

Stallcup's Comment: A new subsection has been added to address the installation requirements of optical fiber cables, raceways and cable routing assemblies in vertical runs.

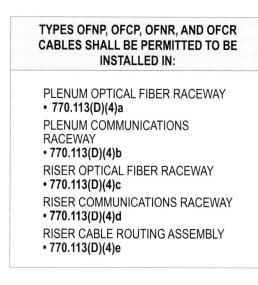

TYPES OFNP, OFCP, OFNR, AND OFCR CABLES SHALL BE PERMITTED TO BE INSTALLED IN:

PLENUM OPTICAL FIBER RACEWAY
• **770.113(D)(4)a**
PLENUM COMMUNICATIONS RACEWAY
• **770.113(D)(4)b**
RISER OPTICAL FIBER RACEWAY
• **770.113(D)(4)c**
RISER COMMUNICATIONS RACEWAY
• **770.113(D)(4)d**
RISER CABLE ROUTING ASSEMBLY
• **770.113(D)(4)e**

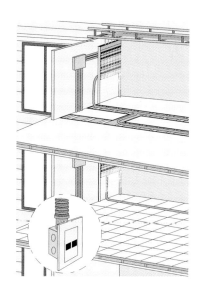

**RISERS – CABLES, RACEWAYS AND CABLE ROUTING ASSEMBLIES IN VERTICAL RUNS
NEC 770.113(D)(4)**

Purpose of Change: To provide and clarify the installation requirements for optical fiber cables, raceways and cable routing assemblies used in vertical runs in a shaft.

Type of Change	New Subsection			Committee Change	Accept in Principle			2008 NEC	770.113
ROP	pg. 1053	# 16-48	log: 2084	ROC	pg. 575	# 16-37	log: 1627	UL	1651
Submitter: Ron L. Janikowski				Submitter: Craig Sato				OSHA	-
NFPA 70B	-			NFPA 70E	-			NFPA 79	-

770.113 Installation of Optical Fiber Cables and Raceways, and Cable Routing Assemblies.

(E) Risers — Cables and Raceways in Metal Raceways. The following cables and raceways shall be permitted in metal raceways in a riser having firestops at each floor:

(1) Types OFNP, OFCP, OFNR, OFCR, OFNG, OFCG, OFN, and OFC cables

(2) Plenum, riser, and general-purpose optical fiber raceways

Stallcup's Comment: A new subsection has been added to address the installation requirements of optical cables and raceways in metal raceways.

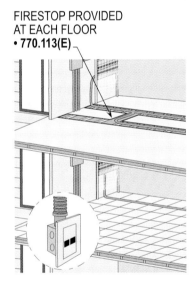

CABLES AND RACEWAYS THAT SHALL BE
PERMITTED IN METAL RACEWAYS IN A RISER
HAVING A FIRESTOP AT EACH FLOOR

TYPES OFNP, OFCP, OFNR, OFCR, OFNG,
OFCG, OFN, AND OFC CABLES
• 770.113(E)(1)
PLENUM, RISER, AND GENERAL-PURPOSE
OPTICAL FIBER RACEWAYS
• 770.113(E)(2)

FIRESTOP PROVIDED
AT EACH FLOOR
• 770.113(E)

RISERS – CABLES AND RACEWAYS IN METAL RACEWAYS
NEC 770.113(E)(1) AND (E)(2)

Purpose of Change: To clarify the installation requirements for optical fiber cables and raceways routing through floors with a firestop.

Type of Change	New Subsection			Committee Change	Accept in Principle		2008 NEC	770.113
ROP	pg. 1053	# 16-48	log: 2084	ROC	pg. 575	# 16-37	log: 1627	UL	1651
Submitter: Ron L. Janikowski				Submitter: Craig Sato			OSHA	-
NFPA 70B				NFPA 70E	-		NFPA 79	-

770.113 Installation of Optical Fiber Cables and Raceways, and Cable Routing Assemblies.

(E) Risers — Cables and Raceways in Metal Raceways. The following cables and raceways shall be permitted in metal raceways in a riser having firestops at each floor:

(3) Types OFNP, OFCP, OFNR, OFCR, OFNG, OFCG, OFN, and OFC cables installed in:

a. Plenum optical fiber raceway
b. Plenum communications raceway
c. Riser optical fiber raceway
d. Riser communications raceway
e. General-purpose optical fiber raceway
 f. General-purpose communications raceway

Informational Note: See 770.26 for firestop requirements for floor penetrations.

Stallcup's Comment: A new subsection has been added to address the installation requirements of optical cables and raceways in metal raceways.

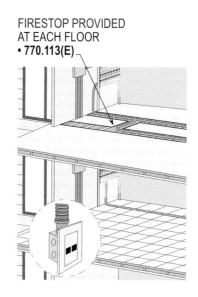

TYPES OFNP, OFCP, OFNR, OFCR, OFNG, OFCG, OFN, AND OFC CABLES SHALL BE PERMITTED TO BE INSTALLED IN:

PLENUM OPTICAL FIBER RACEWAY
• 770.113(E)(3)a
PLENUM COMMUNICATIONS RACEWAY
• 770.113(E)(3)b
RISER OPTICAL FIBER RACEWAY
• 770.113(E)(3)c
RISER COMMUNICATIONS RACEWAY
• 770.113(E)(3)d
GENERAL-PURPOSE OPTICAL FIBER RACEWAY
• 770.113(E)(3)e
GENERAL-PURPOSE COMMUNICATIONS RACEWAY
• 770.113(E)(3)f

FIRESTOP PROVIDED
AT EACH FLOOR
• 770.113(E)

RISERS – CABLES AND RACEWAYS IN METAL RACEWAYS
NEC 770.113(E)(3)

Purpose of Change: To provide and clarify the installation requirements for optical fiber cables and raceways in metal raceways.

Type of Change	New Subsection			Committee Change	Accept in Principle			2008 NEC	770.113
ROP	pg. 1053	# 16-48	log: 2084	ROC	pg. 575	# 16-37	log: 1627	UL	1651
Submitter: Ron L. Janikowski				Submitter: Craig Sato				OSHA	-
NFPA 70B	-			NFPA 70E	-			NFPA 79	-

770.113 Installation of Optical Fiber Cables and Raceways, and Cable Routing Assemblies.

(F) Risers — Cables, Raceways, and Cable Routing Assemblies in Fireproof Shafts. The following cables, raceways, and cable routing assemblies shall be permitted to be installed in fireproof riser shafts having firestops at each floor:

(1) Types OFNP, OFCP, OFNR, OFCR, OFNG, OFCG, OFN, and OFC cables

(2) Plenum, riser, and general-purpose optical fiber raceways

(3) Riser and general-purpose cable routing assemblies

Stallcup's Comment: A new subsection has been added to address the installation requirements of optical cables, raceways, and cable routing assemblies in fireproof riser shafts having firestops at each floor.

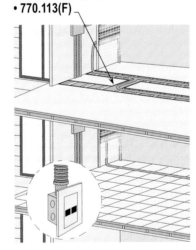

CABLES AND RACEWAYS THAT SHALL BE PERMITTED IN METAL RACEWAYS IN A RISER HAVING A FIRESTOP AT EACH FLOOR:

TYPES OFNP, OFCP, OFNR, OFCR, OFNG, OFCG, OFN, AND OFC CABLES
•770.113(F)(1)

PLENUM, RISER, AND GENERAL-PURPOSE OPTICAL FIBER RACEWAYS
• 770.113(F)(2)

RISER AND GENERAL-PURPOSE CABLE ROUTING ASSEMBLIES
• 770.113(F)(3)

FIRESTOP PROVIDED AT EACH FLOOR IN FIREPROOF SHAFT
• 770.113(F)

**RISERS – CABLES, RACEWAYS, AND CABLE ROUTING ASSEMBLIES IN FIREPROOF SHAFTS
NEC 770.113(F)(1) THRU (F)(3)**

Purpose of Change: To provide requirements for the installation of optical cables, raceways, and cable routing assemblies in fireproof riser shafts having firestops at each floor.

Type of Change	New Subsection		Committee Change	Accept in Principle		2008 NEC	770.113
ROP	pg. 1053	# 16-48	log: 2084	ROC	pg. 575	# 16-37	log: 1627	UL	1651

Type of Change	New Subsection		Committee Change	Accept in Principle		2008 NEC	770.113
ROP	pg. 1053	# 16-48 log: 2084	ROC	pg. 575	# 16-37 log: 1627	UL	1651
Submitter: Ron L. Janikowski			Submitter: Craig Sato		OSHA	-
NFPA 70B	-		NFPA 70E	-		NFPA 79	-

770.113 Installation of Optical Fiber Cables and Raceways, and Cable Routing Assemblies.

(F) Risers — Cables, Raceways, and Cable Routing Assemblies in Fireproof Shafts. The following cables, raceways, and cable routing assemblies shall be permitted to be installed in fireproof riser shafts having firestops at each floor:

(4) Types OFNP, OFCP, OFNR, OFCR, OFNG, OFCG, OFN, and OFC cables installed in:

a. Plenum optical fiber raceway
b. Plenum communications raceway
c. Riser optical fiber raceway
d. Riser communications raceway
e. General-purpose optical fiber raceway
f. General-purpose communications raceway
g. Riser cable routing assembly
h. General-purpose cable routing assembly

Informational Note: See 770.26 for firestop requirements for floor penetrations.

Stallcup's Comment: A new subsection has been added to address the installation requirements of optical cables, raceways, and cable routing assemblies in fireproof riser shafts having firestops at each floor.

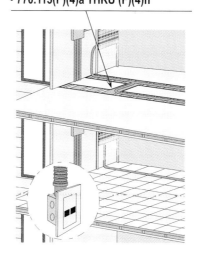

TYPES OFNP, OFCP, OFNR, OFCR, OFNG, OFCG, OFN, AND OFC CABLES SHALL BE PERMITTED TO BE INSTALLED IN:

PLENUM OPTICAL FIBER RACEWAY
•**770.113(F)(4)a**
PLENUM COMMUNICATIONS RACEWAY
• **770.113(F)(4)b**
RISER OPTICAL FIBER RACEWAY
• **770.113(F)(4)c**
RISER COMMUNICATIONS RACEWAY
• **770.113(F)(4)d**
GENERAL-PURPOSE OPTICAL FIBER RACEWAY
• **770.113(F)(4)e**
GENERAL-PURPOSE COMMUNICATIONS RACEWAY
• **770.113(F)(4)f**
RISER CABLE ROUTING ASSEMBLY
• **770.113(F)(4)g**
GENERAL-PURPOSE CABLE ROUTING ASSEMBLY
• **770.113(F)(4)h**

FIRESTOP PROVIDED AT EACH FLOOR IN FIREPROOF RISER SHAFT
• **770.113(F)(4)a THRU (F)(4)h**

**RISERS – CABLES, RACEWAYS, AND CABLE ROUTING ASSEMBLIES IN FIREPROOF SHAFTS
NEC 770.113(F)(4)**

Purpose of Change: To provide requirements for the installation of optical cables, raceways, and cable routing assemblies in fireproof riser shafts having firestops at each floor.

Type of Change	New Subsection		Committee Change	Accept in Principle		2008 NEC	770.113
ROP	pg. 1053	# 16-48	log: 2084	ROC	pg. 575	# 16-37	log: 1627	UL	1651
Submitter: Ron L. Janikowski			Submitter: Craig Sato			OSHA	-
NFPA 70B	-		NFPA 70E	-		NFPA 79	-

770.113 Installation of Optical Fiber Cables and Raceways, and Cable Routing Assemblies.

(G) Risers — One- and Two-Family Dwellings. The following cables, raceways, and cable routing assemblies shall be permitted in one- and two-family dwellings:

(1) Types OFNP, OFCP, OFNR, OFCR, OFNG, OFCG, OFN, and OFC cables

(2) Plenum, riser, and general-purpose optical fiber raceways

(3) Riser and general-purpose cable routing assemblies

Stallcup's Comment: A new subsection has been added to address the installation requirements of optical cables, raceways, and cable routing assemblies in one- and two-family dwellings.

CABLES AND RACEWAYS AND CABLE ROUTING ASSEMBLIES SHALL BE PERMITTED IN ONE- AND TWO-FAMILY DWELLINGS

TYPES OFNP, OFCP, OFNR, OFCR, OFNG, OFCG, OFN, AND OFC CABLES
•770.113(G)(1)

PLENUM, RISER, AND GENERAL-PURPOSE OPTICAL FIBER RACEWAYS
• 770.113(G)(2)

RISER AND GENERAL-PURPOSE CABLE ROUTING ASSEMBLIES
• 770.113(G)(3)

**RISERS – ONE- AND TWO-FAMILY DWELLINGS
NEC 770.113(G)(1) THRU (G)(3)**

Purpose of Change: To provide requirements for the installation of optical cables, raceways, and routing assemblies in one- and two-family dwellings.

Type of Change	New Subsection		Committee Change	Accept in Principle		2008 NEC	770.113
ROP	pg. 1053	# 16-48	log: 2084	ROC	pg. 575	# 16-37	log: 1627	UL	1651
Submitter: Ron L. Janikowski			Submitter: Craig Sato			OSHA	-
NFPA 70B	-		NFPA 70E	-		NFPA 79	-

770.113 Installation of Optical Fiber Cables and Raceways, and Cable Routing Assemblies.

(G) Risers — One- and Two-Family Dwellings. The following cables, raceways, and cable routing assemblies shall be permitted in one- and two-family dwellings:

(4) Types OFNP, OFCP, OFNR, OFCR, OFNG, OFCG, OFN, and OFC cables installed in:

a. Plenum optical fiber raceway
b. Plenum communications raceway
c. Riser optical fiber raceway
d. Riser communications raceway
e. General-purpose optical fiber raceway
 f. General-purpose communications raceway
g. Riser cable routing assembly
h. General-purpose cable routing assembly

Stallcup's Comment: A new subsection has been added to address the installation requirements of optical cables, raceways, and cable routing assemblies in one- and two-family dwellings.

TYPES OFNP, OFCP, OFNR, OFCR, OFNG, OFCG, OFN, AND OFC CABLES SHALL BE PERMITTED TO BE INSTALLED IN:

PLENUM OPTICAL FIBER RACEWAY
• **770.113(G)(4)a**
PLENUM COMMUNICATIONS RACEWAY
• **770.113(G)(4)b**
RISER OPTICAL FIBER RACEWAY
• **770.113(G)(4)c**
RISER COMMUNICATIONS RACEWAY
• **770.113(G)(4)d**
GENERAL-PURPOSE OPTICAL FIBER RACEWAY
• **770.113(G)(4)e**
GENERAL-PURPOSE COMMUNICATIONS RACEWAY
• **770.113(G)(4)f**
RISER CABLE ROUTING ASSEMBLY
• **770.113(G)(4)g**
GENERAL-PURPOSE CABLE ROUTING ASSEMBLY
• **770.113(G)(4)h**

RISERS – ONE- AND TWO-FAMILY DWELLINGS
NEC 770.113(G)(4)

Purpose of Change: To provide requirements for the installation of optical cables, raceways, and routing assemblies in one- and two-family dwellings.

Type of Change	New Subsection			Committee Change	Accept in Principle		2008 NEC	770.113
ROP	pg. 1053	# 16-48	log: 2084	ROC	pg. 575	# 16-37	log: 1627	UL	1651
Submitter: Ron L. Janikowski				Submitter: Craig Sato			OSHA	-
NFPA 70B	-			NFPA 70E	-		NFPA 79	-

770.113 Installation of Optical Fiber Cables and Raceways, and Cable Routing Assemblies.

(H) Cable Trays. The following cables and raceways shall be permitted to be supported by cable trays:

(1) Types OFNP, OFCP, OFNR, OFCR, OFNG, OFCG, OFN, and OFC cables

(2) Plenum, riser, and general-purpose optical fiber raceways

Stallcup's Comment: A new subsection has been added to address the installation requirements of optical cables and raceways that are supported by cable trays.

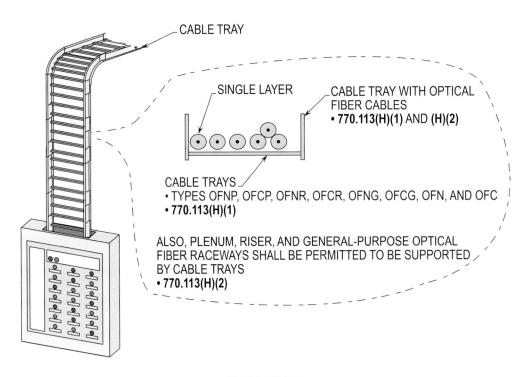

CABLE TRAY

SINGLE LAYER

CABLE TRAY WITH OPTICAL FIBER CABLES
• **770.113(H)(1) AND (H)(2)**

CABLE TRAYS
• TYPES OFNP, OFCP, OFNR, OFCR, OFNG, OFCG, OFN, AND OFC
• **770.113(H)(1)**

ALSO, PLENUM, RISER, AND GENERAL-PURPOSE OPTICAL FIBER RACEWAYS SHALL BE PERMITTED TO BE SUPPORTED BY CABLE TRAYS
• **770.113(H)(2)**

**CABLE TRAYS
NEC 770.113(H)(1) AND (H)(2)**

Purpose of Change: To add requirements pertaining to cable trays used to support optical cables and raceways.

Type of Change	New Subsection		Committee Change	Accept in Principle		2008 NEC	770.113
ROP	pg. 1053	# 16-48	log: 2084	ROC	pg. 575	# 16-37	log: 1627	UL	1651
Submitter: Ron L. Janikowski			Submitter: Craig Sato			OSHA	-
NFPA 70B	-		NFPA /UE	-		NFPA 79	-

770.113 Installation of Optical Fiber Cables and Raceways, and Cable Routing Assemblies.

(H) Cable Trays. The following cables and raceways shall be permitted to be supported by cable trays:

(3) Types OFNP, OFCP, OFNR, OFCR, OFNG, OFCG, OFN, and OFC cables installed in:

a. Plenum optical fiber raceway

b. Plenum communications raceway

c. Riser optical fiber raceway

d. Riser communications raceway

e. General-purpose optical fiber raceway

f. General-purpose communications raceway

Stallcup's Comment: A new subsection has been added to address the installation requirements of optical cables and raceways that are supported by cable trays.

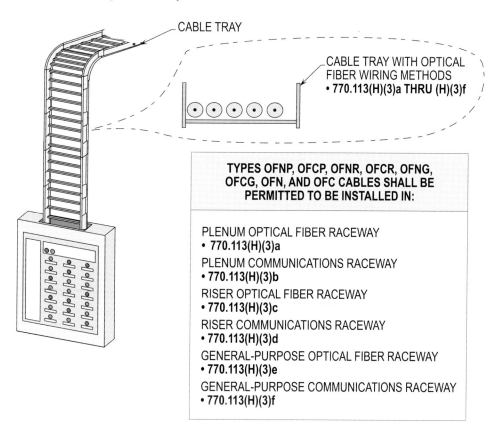

CABLE TRAY

CABLE TRAY WITH OPTICAL
FIBER WIRING METHODS
• 770.113(H)(3)a THRU (H)(3)f

TYPES OFNP, OFCP, OFNR, OFCR, OFNG, OFCG, OFN, AND OFC CABLES SHALL BE PERMITTED TO BE INSTALLED IN:

PLENUM OPTICAL FIBER RACEWAY
• 770.113(H)(3)a

PLENUM COMMUNICATIONS RACEWAY
• 770.113(H)(3)b

RISER OPTICAL FIBER RACEWAY
• 770.113(H)(3)c

RISER COMMUNICATIONS RACEWAY
• 770.113(H)(3)d

GENERAL-PURPOSE OPTICAL FIBER RACEWAY
• 770.113(H)(3)e

GENERAL-PURPOSE COMMUNICATIONS RACEWAY
• 770.113(H)(3)f

CABLE TRAYS
NEC 770.113(H)(3)

Purpose of Change: To add requirements pertaining to cable trays used to support optical cables and raceways.

Type of Change	New Subsection		Committee Change	Accept in Principle		2008 NEC	770.113
ROP	pg. 1053	# 16-48	log: 2084	ROC	pg. 575	# 16-37	log: 1627	UL	1651
Submitter: Ron L. Janikowski			Submitter: Craig Sato			OSHA	-
NFPA 70B	-		NFPA 70E	-		NFPA 79	-

770.113 Installation of Optical Fiber Cables and Raceways, and Cable Routing Assemblies.

(I) Distributing Frames and Cross-Connect Arrays. The following cables, raceways, and cable routing assemblies shall be permitted to be installed in distributing frames and cross-connect arrays:

(1) Types OFNP, OFCP, OFNR, OFCR, OFNG, OFCG, OFN, and OFC cables

(2) Plenum, riser, and general-purpose optical fiber raceways

(3) Riser or general-purpose cable routing assemblies

Stallcup's Comment: A new subsection has been added to address the installation requirements of optical cables, raceways, and cable routing assemblies in distributing frames and cross-connect arrays.

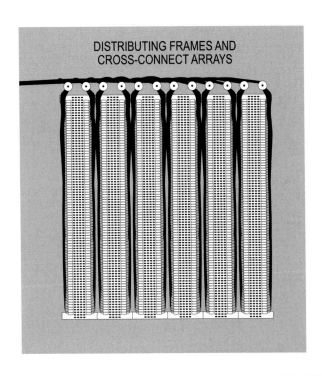

DISTRIBUTING FRAMES AND CROSS-CONNECT ARRAYS

CABLES AND RACEWAYS AND CABLE ROUTING ASSEMBLIES SHALL BE PERMITTED IN DISTRIBUTING FRAMES AND CROSS-CONNECT ARRAYS

TYPES OFNP, OFCP, OFNR, OFCR, OFNG, OFCG, OFN, AND OFC CABLES
• **770.113(I)(1)**
PLENUM, RISER, AND GENERAL-PURPOSE OPTICAL FIBER RACEWAYS
• **770.113(I)(2)**
RISER OR GENERAL-PURPOSE CABLE ROUTING ASSEMBLIES
• **770.113(I)(3)**

DISTRIBUTING FRAMES AND CROSS-CONNECT ARRAYS
NEC 770.113(I)(1) THRU (I)(3)

Purpose of Change: To address requirements for installing optical cables, raceways, and cable routing assemblies in distributing frames and cross-connect arrays.

Type of Change	New Subsection			Committee Change	Accept in Principle			2008 NEC	770.113
ROP	pg. 1053	# 16-48	log: 2084	ROC	pg. 575	# 16-37	log: 1627	UL	1651
Submitter: Ron L. Janikowski				Submitter: Craig Sato				OSHA	-
NFPA 70B	-			NFPA 70C	-			NFPA 79	-

770.113 Installation of Optical Fiber Cables and Raceways, and Cable Routing Assemblies.

(I) Distributing Frames and Cross-Connect Arrays. The following cables, raceways, and cable routing assemblies shall be permitted to be installed in distributing frames and cross-connect arrays:

(4) Types OFNP, OFCP, OFNR, OFCR, OFNG, OFCG, OFN, and OFC cables installed in:

a. Plenum optical fiber raceway

b. Plenum communications raceway

c. Riser optical fiber raceway

d. Riser communications raceway

e. General-purpose optical fiber raceway

 f. General-purpose communications raceway

g. Riser cable routing assembly

h. General-purpose cable routing assembly

Stallcup's Comment: A new subsection has been added to address the installation requirements of optical cables, raceways, and cable routing assemblies in distributing frames and cross-connect arrays.

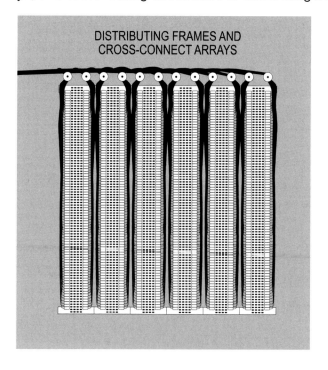

DISTRIBUTING FRAMES AND CROSS-CONNECT ARRAYS
NEC 770.113(I)(4)

Purpose of Change: To address requirements for installing optical cables, raceways, and cable routing assemblies in distributing frames and cross-connect arrays.

Type of Change	Subsection			Committee Change	Accept in Principle			2008 NEC	770.113
ROP	pg. 1053	# 16-48	log: 2084	ROC	pg. 575	# 16-37	log: 1627	UL	1651
Submitter: Ron L. Janikowski				Submitter: Craig Sato				OSHA	-
NFPA 70B	-			NFPA 70E	-			NFPA 79	-

770.113 Installation of Optical Fiber Cables and Raceways, and Cable Routing Assemblies.

(J) Other Building Locations. The following cables, raceways, and cable routing assemblies shall be permitted to be installed in building locations other than the locations covered in 770.113(B) through (I):

(1) Types OFNP, OFCP, OFNR, OFCR, OFNG, OFCG, OFN, and OFC cables

(2) Plenum, riser, and general-purpose optical fiber raceways

(3) Riser and general-purpose cable routing assemblies

Stallcup's Comment: A new subsection has been added to address the installation requirements of optical cables, raceways, and cable routing assemblies in building locations other than those specified for fabricated ducts used for environmental air; other spaces used for environmental air (plenums); cables, raceways, and cable routing assemblies in vertical runs; cables and raceways in metal raceways; cables, raceways, and cable routing assemblies in fireproof shafts; one- and two-family dwellings; cable trays; and distributing frames and cross-connect arrays.

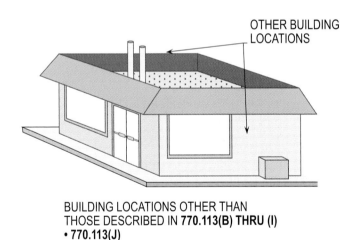

OTHER BUILDING LOCATIONS

BUILDING LOCATIONS OTHER THAN THOSE DESCRIBED IN **770.113(B) THRU (I)**
• **770.113(J)**

CABLES AND RACEWAYS AND CABLE ROUTING ASSEMBLIES SHALL BE PERMITTED TO BE INSTALLED IN OTHER BUILDING LOCATIONS

TYPES OFNP, OFCP, OFNR, OFCR, OFNG, OFCG, OFN, AND OFC CABLES
• **770.113(J)(1)**
PLENUM, RISER, AND GENERAL-PURPOSE OPTICAL FIBER RACEWAYS
• **770.113(J)(2)**
RISER AND GENERAL-PURPOSE CABLE ROUTING ASSEMBLIES
• **770.113(J)(3)**

**OTHER BUILDING LOCATIONS
NEC 770.113(J)(1) THRU (J)(3)**

Purpose of Change: To add requirements for routing optical fiber systems in building locations other than those outlined in **770.113(B) through (I)**.

Type of Change	Subsection			Committee Change	Accept in Principle			2008 NEC	770.113
ROP	pg. 1053	# 16-48	log: 2084	ROC	pg. 575	# 16-37	log: 1627	UL	1651
Submitter: Ron L. Janikowski				Submitter: Craig Sato				OSHA	-
NFPA 70B	•			NFPA 70E	-			NFPA 79	-

770.113 Installation of Optical Fiber Cables and Raceways, and Cable Routing Assemblies.

(J) Other Building Locations. The following cables, raceways, and cable routing assemblies shall be permitted to be installed in building locations other than the locations covered in 770.113(B) through (I):

(4) Types OFNP, OFCP, OFNR, OFCR, OFNG, OFCG, OFN, and OFC cables installed in:

a. Plenum optical fiber raceway
b. Plenum communications raceway
c. Riser optical fiber raceway
d. Riser communications raceway
e. General-purpose optical fiber raceway
f. General-purpose communications raceway
g. Riser cable routing assembly
h. General-purpose cable routing assembly

Stallcup's Comment: A new subsection has been added to address the installation requirements of optical cables, raceways, and cable routing assemblies in building locations other than those specified for fabricated ducts used for environmental air; other spaces used for environmental air (plenums); cables, raceways, and cable routing assemblies in vertical runs; cables and raceways in metal raceways; cables, raceways, and cable routing assemblies in fireproof shafts; one- and two-family dwellings; cable trays; and distributing frames and cross-connect arrays.

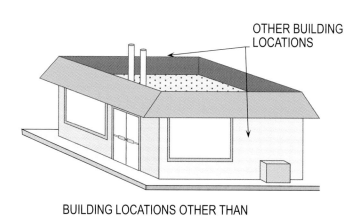

OTHER BUILDING LOCATIONS
BUILDING LOCATIONS OTHER THAN THOSE DESCRIBED IN **770.113(B) THRU (I)** • **770.113(J)**

TYPES OFNP, OFCP, OFNR, OFCR, OFNG, OFCG, OFN, AND OFC CABLES SHALL BE PERMITTED TO BE INSTALLED IN:

PLENUM OPTICAL FIBER RACEWAY • **770.113(J)(4)a**
PLENUM COMMUNICATIONS RACEWAY • **770.113(J)(4)b**
RISER OPTICAL FIBER RACEWAY • **770.113(J)(4)c**
RISER COMMUNICATIONS RACEWAY • **770.113(J)(4)d**
GENERAL-PURPOSE OPTICAL FIBER RACEWAY • **770.113(J)(4)e**
GENERAL-PURPOSE COMMUNICATIONS RACEWAY • **770.113(J)(4)f**
RISER CABLE ROUTING ASSEMBLY • **770.113(J)(4)g**
GENERAL-PURPOSE CABLE ROUTING ASSEMBLY • **770.113(J)(4)h**

**OTHER BUILDING LOCATIONS
NEC 770.113(J)(4)**

Purpose of Change: To add requirements for routing optical fiber systems in building locations other than those outlined in **770.113(B) through (I).**

Type of Change	Subsection			Committee Change	Accept in Principle			2008 NEC	770.113
ROP	pg. 1053	# 16-48	log: 2084	ROC	pg. 575	# 16-37	log: 1627	UL	1651
Submitter: Ron L. Janikowski				Submitter: Craig Sato				OSHA	-
NFPA 70B	-			NFPA 70E	-			NFPA 79	-

770.113 Installation of Optical Fiber Cables and Raceways, and Cable Routing Assemblies.

(J) Other Building Locations. The following cables, raceways, and cable routing assemblies shall be permitted to be installed in building locations other than the locations covered in 770.113(B) through (I):

(5) Types OFNP, OFCP, OFNR, OFCR, OFNG, OFCG, OFN, and OFC cables installed in a raceway of a type recognized in Chapter 3

Stallcup's Comment: A new subsection has been added to address the installation requirements of optical cables, raceways, and cable routing assemblies in building locations other than those specified for fabricated ducts used for environmental air; other spaces used for environmental air (plenums); cables, raceways, and cable routing assemblies in vertical runs; cables and raceways in metal raceways; cables, raceways, and cable routing assemblies in fireproof shafts; one- and two-family dwellings; cable trays; and distributing frames and cross-connect arrays.

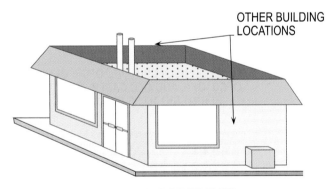

OTHER BUILDING LOCATIONS

BUILDING LOCATIONS OTHER THAN
THOSE DESCRIBED IN **770.113(B) THRU (I)**
• **770.113(J)**

CABLES AND RACEWAYS AND CABLE
ROUTING ASSEMBLIES SHALL BE PERMITTED
TO BE INSTALLED IN OTHER BUILDING
LOCATIONS

TYPES OFNP, OFCP, OFNR, OFCR, OFNG, OFCG,
OFN, AND OFC CABLES
• **770.113(J)(5)**

OTHER BUILDING LOCATIONS
NEC 770.113(J)(5)

Purpose of Change: To add requirements for routing optical fiber systems in building locations other than those outlined in **770.113(B) through (I).**

Type of Change	New Section			Committee Change	Accept			2008 NEC	-
ROP	pg. 1052	# 16-41	log: 1123	ROC	pg. -	# -	log: -	UL	1651
Submitter: James E. Brunssen				Submitter: -				OSHA	-
NFPA 70B	-			NFPA 70E	-			NFPA 79	-

770.114 Grounding.

Non-current-carrying conductive members of optical fiber cables shall be bonded to a grounded equipment rack or enclosure, or grounded in accordance with the grounding methods specified by 770.100(B)(2).

Stallcup's Comment: A new section has been added to address the grounding requirements for noncurrent-carrying conductive members of optical fiber cables.

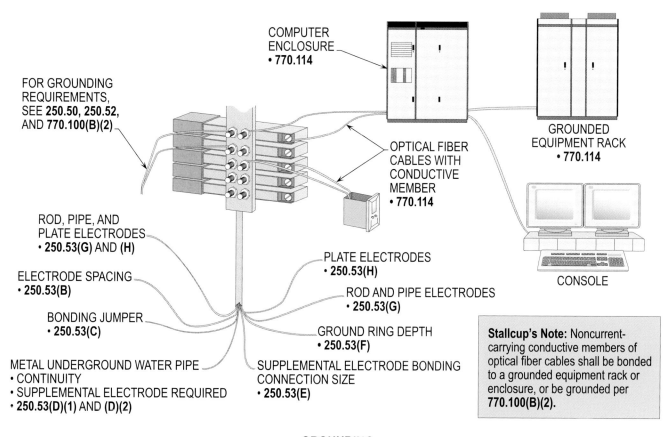

GROUNDING
NEC 770.114

Purpose of Change: To add and address the installation requirements of noncurrent-carrying conductive members of optical fiber cables.

Name   Date

Chapter 7
Special Conditions

Section   Answer

1. For installations under single management, where conditions of maintenance and supervision ensure that only qualified persons will monitor and service the installation, and where documented safe switching procedures are established and maintained for disconnection, the generator set disconnecting means for emergency system shall not be required to be located within _____ of the building or structure served.

(A) sight   **(B)** 6 ft
(C) 12 ft   **(D)** 20 ft

2. The sensor for the ground-fault signal devices shall be located at, or ahead of, the main system disconnecting means for the legally required standby source, and the maximum setting of the signal devices shall be for a ground-fault current of _____ amps.

(A) 1000   **(B)** 1200
(C) 1500   **(D)** 2000

3. The circuit conductors and overcurrent devices shall be sized to carry not less than _____ percent of the maximum currents for interconnected electric power production sources.

(A) 95   **(B)** 100
(C) 125   **(D)**135

4. The circuit disconnecting means for NPLFA circuits shall have _____ identification, shall be accessible to qualified personnel, and shall be identified as "Fire Alarm Circuit."

(A) blue   **(B)** red
(C) orange   **(D)** yellow

5. Where there is no mobile home service equipment located within _____ ft of the exterior wall of the mobile home it serves, the noncurrent-carrying metallic members of optical fiber cables entering the mobile home shall be grounded.

(A) 6   **(B)** 10
(C) 20   **(D)** 30

6. Where there is no mobile home disconnecting means grounded and located within _____ ft of the exterior wall of the mobile home it serves, the noncurrent-carrying metallic members or optical fiber cables entering the mobile home shall be grounded.

(A) 20   **(B)** 30
(C) 40   **(D)** 50

7. Types OFNP and OFCP cables shall be permitted in fabricated ducts used for environmental air up to _____ ft in length.

(A) 4   **(B)** 6
(C) 10   **(D)** 12

Section   Answer

8. Which of the following cables or raceways shall be permitted in other spaces used for environmental air:

(A) OFNR cable
(C) Plenum optical fiber raceway

(B) OFNG cable
(D) General-purpose optical fiber raceways

9. Which of the following cables shall be permitted in vertical runs penetrating one or more floors and in vertical runs in a shaft:

(A) OFNR cable
(C) OFCG cable

(B) OFNG cable
(D) All of the above

10. Which of the following cables shall be permitted in metal raceways in a riser having firestops at each floor:

(A) OFNP cable
(C) OFCG cable

(B) OFCR cable
(D) All of the above

Communications Systems

Chapter 8 of the *National Electrical Code* covers communications systems. Articles and sections of this chapter stand alone and are independent from **Chapters 1 through 7**, except in specific cases where they are referenced. Communications systems include such systems as telephone, cable TV, radio, and broadband communications.

The rules and regulations in this article are mainly applied to those systems that are connected to a central station and operate as elements of such systems. When communications are involved, one of the articles in the 800 series must be selected, based on the system utilized.

Chapter 8 also includes **Article 810**, which deals with radio, television, receiving equipment, and amateur radio transmitting and receiving equipment, but not equipment and antenna used for coupling carrier current to power line conductors.

Article 820 covers coaxial cable distribution of radio frequency signals typically employed in community antenna television (CATV) systems.

Article 830 covers network-powered broadband communications where a carrier frequency has multiple signals impressed on the carrier. At the network interface unit, the signals are converted into individual signals and then distributed for phone, TV, burglar alarm, and other similar uses.

Type of Change	Revision			Committee Change	Accept in Principle		2008 NEC	800.100(A)(3)
ROP	pg. 1072	# 16-91	log: 4190	ROC	pg. 598	# 16-125	log: 1886	UL	444
Submitter: Paul Dobrowsky				Submitter: Phil Simmons			OSHA	1910.308(e)(4)
NFPA 70B	-			NFPA 70E	-		NFPA 79	-

800.100 Cable and Primary Protector Bonding and Grounding.

(A) Bonding Conductor or Grounding Electrode Conductor.

(3) Size. The bonding conductor or grounding electrode conductor shall not be smaller than 14 AWG. It shall have a current-carrying capacity not less than the grounded metallic sheath member(s) and protected conductor(s) of the communications cable. The bonding conductor or grounding electrode conductor shall not be required to exceed 6 AWG.

Stallcup's Comment: A revision has been made to clarify the sizing requirements of the bonding conductor or grounding electrode conductor.

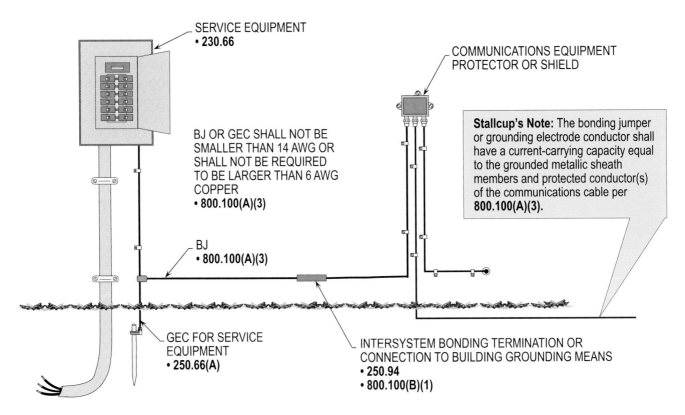

BONDING CONDUCTOR OR GROUNDING ELECTRODE CONDUCTOR – SIZE
NEC 800.100(A)(3)

Purpose of Change: To add requirements for the sizing and installation of the bonding conductor and grounding electrode conductor used to earth ground communications circuits.

Type of Change	Revision			Committee Change	Accept in Principle		2008 NEC	800.100(A)(3)
ROP	pg. 1072	# 16-91	log: 4190	ROC	pg. 598	# 16-125	log: 1886	UL	444
Submitter: Paul Dobrowsky				Submitter: Phil Simmons			OSHA	1910.308(e)(4)
NFPA 70B	-			NFPA 70E	-		NFPA 79	-

800.100 Cable and Primary Protector Bonding and Grounding.

(A) Bonding Conductor or Grounding Electrode Conductor.

(3) Size. The bonding conductor or grounding electrode conductor shall not be smaller than 14 AWG. It shall have a current-carrying capacity not less than the grounded metallic sheath member(s) and protected conductor(s) of the communications cable. The bonding conductor or grounding electrode conductor shall not be required to exceed 6 AWG.

Stallcup's Comment: A revision has been made to clarify the sizing requirements of the bonding conductor or grounding electrode conductor.

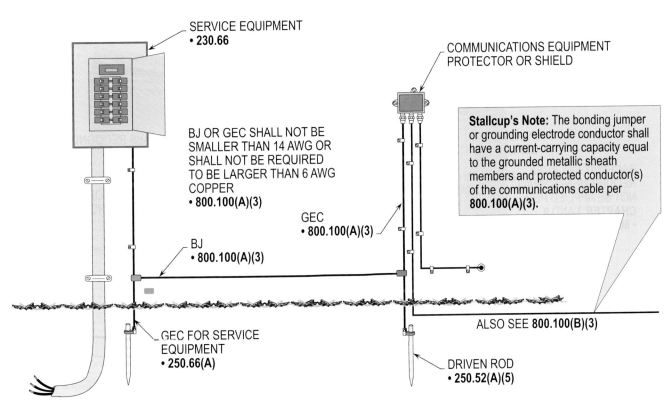

BONDING CONDUCTOR OR GROUNDING ELECTRODE CONDUCTOR – SIZE
NEC 800.100(A)(3)

Purpose of Change: To add requirements for the sizing and installation of the bonding conductor and grounding electrode conductor used to earth ground communications circuits.

Type of Change	New Subsection		Committee Change	Accept in Principle		2008 NEC	800.113
ROP	pg. 1093	# 16-160	log: 2102	ROC	pg. 603	# 16-138	log: 1631	UL	444
Submitter: Ron L. Janikowski			Submitter: Craig Sato			OSHA	-
NFPA 70B	-			NFPA 70E	-		NFPA 79	-

800.113 Installation of Communications Wires, Cables, and Raceways.

(H) Cable Trays. The following wires, cables, and raceways shall be permitted to be supported by cable trays:

(3) Communications wires and Types CMP, CMR, CMG, and CM cables installed in:

a. Plenum communications raceway

b. Riser communications raceway

c. General-purpose communications raceway

Stallcup's Comment: A new subsection has been added to address the installation requirements of communications wires, cables, and raceways that are supported by cable trays.

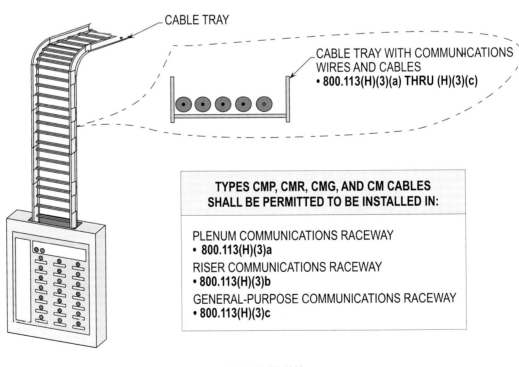

CABLE TRAY

CABLE TRAY WITH COMMUNICATIONS WIRES AND CABLES
• 800.113(H)(3)(a) THRU (H)(3)(c)

TYPES CMP, CMR, CMG, AND CM CABLES SHALL BE PERMITTED TO BE INSTALLED IN:

PLENUM COMMUNICATIONS RACEWAY
• 800.113(H)(3)a
RISER COMMUNICATIONS RACEWAY
• 800.113(H)(3)b
GENERAL-PURPOSE COMMUNICATIONS RACEWAY
• 800.113(H)(3)c

CABLE TRAYS
NEC 800.113(H)(3)

Purpose of Change: To add requirements pertaining to cable trays used to support communications cables and raceways.

Type of Change	New Subsection			Committee Change	Accept in Principle			2008 NEC	800.113
ROP	pg. 1093	# 16-160	log: 2102	ROC	pg. 603	# 16-138	log: 1631	UL	444
Submitter: Ron L. Janikowski				Submitter: Craig Sato				OSHA	-
NFPA 70B	-			NFPA 70E	-			NFPA 79	-

800.113 Installation of Communications Wires, Cables, and Raceways.

(I) Distributing Frames and Cross-Connect Arrays. The following wires, cables, and raceways shall be permitted to be installed in distributing frames and cross-connect arrays:

(1) Types CMP, CMR, CMG, and CM cables and communications wires

(2) Plenum, riser, and general-purpose communications raceways

Stallcup's Comment: A new subsection has been added to address the installation requirements of communications wires, cables, and raceways in distributing frames and cross-connect arrays.

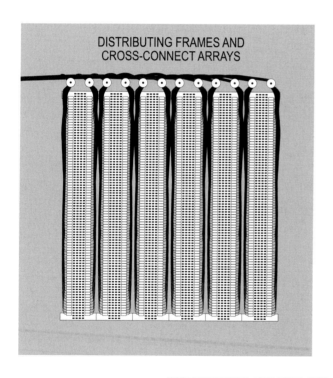

CABLES AND RACEWAYS AND CABLE ROUTING ASSEMBLIES SHALL BE PERMITTED IN DISTRIBUTING FRAMES AND CROSS-CONNECT ARRAYS

TYPES CMP, CMR, CMG, AND CM CABLES AND COMMUNICATIONS WIRES
• **800.113(I)(1)**

PLENUM, RISER, AND GENERAL-PURPOSE COMMUNICATIONS RACEWAYS
• **800.113(I)(2)**

**DISTRIBUTING FRAMES AND CROSS-CONNECT ARRAYS
NEC 800.113(I)(1) AND (I)(2)**

Purpose of Change: To add requirements for installing communications wires, cables, and raceways in distributing frames and cross-connect arrays.

Type of Change	New Subsection			Committee Change	Accept in Principle			2008 NEC	800.113
ROP	pg. 1093	# 16-160	log: 2102	ROC	pg. 603	# 16-138	log: 1631	UL	444
Submitter: Ron L. Janikowski				Submitter: Craig Sato				OSHA	-
NFPA 70B	-			NFPA 70E	-			NFPA 79	-

800.113 Installation of Communications Wires, Cables, and Raceways.

(I) Distributing Frames and Cross-Connect Arrays. The following wires, cables, and raceways shall be permitted to be installed in distributing frames and cross-connect arrays:

(3) Communications wires and Types CMP, CMR, CMG, and CM cables installed in:

a. Plenum communications raceway

b. Riser communications raceway

c. General-purpose communications raceway

d. Riser cable routing assembly

e. General-purpose cable routing assembly

Stallcup's Comment: A new subsection has been added to address the installation requirements of communications wires, cables, and raceways in distributing frames and cross-connect arrays.

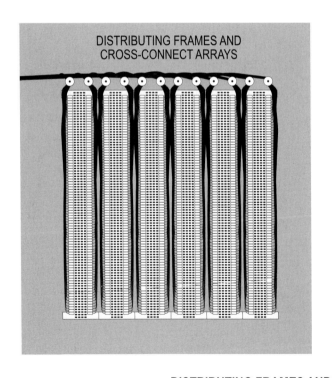

**DISTRIBUTING FRAMES AND CROSS-CONNECT ARRAYS
NEC 800.113(I)(3)**

Purpose of Change: To address requirements for installing communications wires, cables, raceways, and cable routing assemblies in distributing frames and cross-connect arrays.

Type of Change	New Subsection			Committee Change	Accept in Principle		2008 NEC	800.113
ROP	pg. 1093	# 16-160	log: 2102	ROC	pg. 603	# 16-138	log: 1631	UL	444
Submitter: Ron L. Janikowski				Submitter: Craig Sato			OSHA	-
NFPA 70B				NFPA 70E	-		NFPA 79	-

800.113 Installation of Communications Wires, Cables, and Raceways.

(J) Other Building Locations. The following wires, cables, and raceways shall be permitted to be installed in building locations other than the locations covered in 800.113(B) through (I):

(1) Types CMP, CMR, CMG, and CM cables

(2) A maximum of 3 m (10 ft) of exposed Type CMX in nonconcealed spaces

(3) Plenum, riser, and general-purpose communications raceways

Stallcup's Comment: A new subsection has been added to address the installation requirements of communications cables, raceways, and cable routing assemblies in building locations other than specified for fabricated ducts used for environmental air; other spaces used for environmental air (plenums); cables, raceways, and cable routing assemblies in vertical runs; cables and raceways in metal raceways; cables, raceways, and cable routing assemblies in fireproof shafts; one- and two-family dwellings; cable trays; and distributing frames and cross-connect arrays.

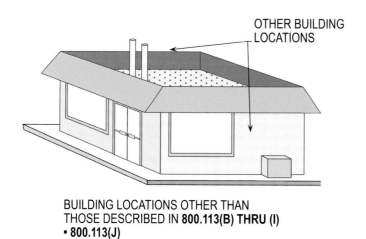

OTHER BUILDING LOCATIONS

BUILDING LOCATIONS OTHER THAN THOSE DESCRIBED IN **800.113(B) THRU (I)**
• **800.113(J)**

CABLES AND RACEWAYS AND GENERAL-PURPOSE COMMUNICATIONS RACEWAYS SHALL BE PERMITTED TO BE INSTALLED IN OTHER BUILDING LOCATIONS

TYPES CMP, CMR, CMG, AND CM CABLES
• **800.113(J)(1)**

A MAXIMUM OF 10' (3 m) OF EXPOSED CMX IN NONCONCEALED SPACES
• **800.113(J)(2)**

PLENUM, RISER, AND GENERAL-PURPOSE COMMUNICATIONS RACEWAYS
• **800.113(J)(3)**

OTHER BUILDING LOCATIONS
NEC 800.113(J)(1) THRU (J)(3)

Purpose of Change: To add requirements for routing communications systems in building locations other than those outlined in **800.113(B) through (I).**

Type of Change	New Subsection			Committee Change	Accept in Principle			2008 NEC	800.113
ROP	pg. 1093	# 16-160	log: 2102	ROC	pg. 603	# 16-138	log: 1631	UL	444
Submitter: Ron L. Janikowski				Submitter: Craig Sato				OSHA	-
NFPA 70B	-			NFPA 70E	-			NFPA 79	-

800.113 Installation of Communications Wires, Cables, and Raceways.

(J) Other Building Locations. The following wires, cables, and raceways shall be permitted to be installed in building locations other than the locations covered in 800.113(B) through (I):

(4) Communications wires and Types CMP, CMR, CMG, and CM cables installed in:

a. Plenum communications raceway

b. Riser communications raceway

c. General-purpose communications raceway

Stallcup's Comment: A new subsection has been added to address the installation requirements of communications cables, raceways, and cable routing assemblies in building locations other than specified for fabricated ducts used for environmental air; other spaces used for environmental air (plenums); cables, raceways, and cable routing assemblies in vertical runs; cables and raceways in metal raceways; cables, raceways, and cable routing assemblies in fireproof shafts; one- and two-family dwellings; cable trays; and distributing frames and cross-connect arrays.

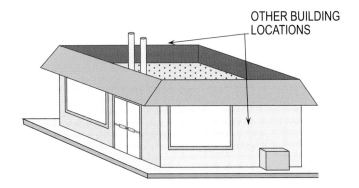

OTHER BUILDING LOCATIONS

TYPES CMP, CMR, CMG, AND CM CABLES SHALL BE PERMITTED TO BE INSTALLED IN:

PLENUM COMMUNICATIONS RACEWAY
• 800.113(J)(4)a
RISER COMMUNICATIONS RACEWAY
• 800.113(J)(4)b
GENERAL-PURPOSE COMMUNICATIONS RACEWAY
• 800.113(J)(4)c

OTHER BUILDING LOCATIONS
NEC 800.113(J)(4)

Purpose of Change: To add requirements for routing communications systems in building locations other than those outlined in **800.113(B) through (I).**

Type of Change	New Subsection			Committee Change	Accept in Principle		2008 NEC	800.113
ROP	pg. 1093	# 16-160	log: 2102	ROC	pg. 603	# 16-138	log: 1631	UL	444
Submitter: Ron L. Janikowski				Submitter: Craig Sato			OSHA	-
NFPA 70B	-			NFPA 70C			NFPA 79	-

800.113 Installation of Communications Wires, Cables, and Raceways.

(J) Other Building Locations. The following wires, cables, and raceways shall be permitted to be installed in building locations other than the locations covered in 800.113(B) through (I):

(5) Types CMP, CMR, CMG, and CM cables installed in:

a. Riser cable routing assembly

b. General-purpose cable routing assembly

(6) Communications wire and Types CMP, CMR, CMG, CM, and CMX cables installed in a raceway of a type recognized in Chapter 3

(7) Type CMUC undercarpet communications wires and cables installed under carpet

Stallcup's Comment: A new subsection has been added to address the installation requirements of communications cables, raceways, and cable routing assemblies in building locations other than specified for fabricated ducts used for environmental air; other spaces used for environmental air (plenums); cables, raceways, and cable routing assemblies in vertical runs; cables and raceways in metal raceways; cables, raceways, and cable routing assemblies in fireproof shafts; one- and two-family dwellings; cable trays; and distributing frames and cross-connect arrays.

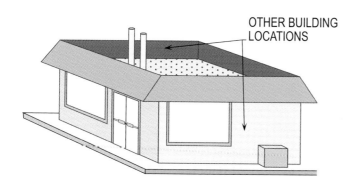

OTHER BUILDING LOCATIONS

CABLES AND RACEWAYS AND CABLE ROUTING ASSEMBLIES SHALL BE PERMITTED TO BE INSTALLED IN OTHER BUILDING LOCATIONS

TYPES CM, CMR, CMG, AND CM CABLES
• 800.113(J)(5)
RISER CABLE ROUTING ASSEMBLY
• 800.113(J)(5)a
GENERAL-PURPOSE CABLE ROUTING ASSEMBLY
• 800.113(J)(5)b
TYPES CMP, CMR, CMG, CM, AND CMX CABLES
• 800.113(J)(6)
UNDER CARPET COMMUNICATIONS WIRES AND CABLES
• 800.113(J)(7)

OTHER BUILDING LOCATIONS
NEC 800.113(J)(5) THRU (J)(7)

Purpose of Change: To add requirements for routing communications systems in building locations other than those outlined in **800.113(B) through (I).**

Type of Change	New Subsection			Committee Change	Accept in Principle			2008 NEC	800.113
ROP	pg. 1093	# 16-160	log: 2102	ROC	pg. 603	# 16-138	log: 1631	UL	444
Submitter: Ron L. Janikowski				Submitter: Craig Sato				OSHA	-
NFPA 70B	-			NFPA 70E	-			NFPA 79	-

800.113 Installation of Communications Wires, Cables, and Raceways.

(K) Multifamily Dwellings. The following cables, raceways, and wiring assemblies shall be permitted to be installed in multifamily dwellings in locations other than the locations covered in 800.113(B) through (G):

(1) Types CMP, CMR, CMG, and CM cables

(2) Type CMX cable less than 6 mm (0.25 in.) in diameter in nonconcealed spaces

(3) Plenum, riser, and general-purpose communications raceways

Stallcup's Comment: A new subsection has been added to address the installation requirements of communications cables, raceways, and wiring assemblies in multifamily dwellings.

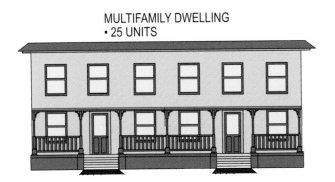

MULTIFAMILY DWELLING
• 25 UNITS

CABLES AND RACEWAYS AND WIRING ASSEMBLIES SHALL BE PERMITTED IN MULTIFAMILY DWELLINGS

TYPES CMP, CMR, CMG, AND CM CABLES
• 800.113(K)(1)
TYPE CMX CABLES LESS THAN 0.25" (6 mm) IN DIAMETER IN NONCONCEALED SPACES
• 8000.113(K)(2)
PLENUM, RISER, AND GENERAL-PURPOSE COMMUNICATIONS RACEWAYS
• 800.113(K)(3)

MULTIFAMILY DWELLINGS
NEC 800.113(K)(1) THRU (K)(3)

Purpose of Change: To add requirements for routing communications systems in multifamily dwellings other than those outlined in **800.113(B) through (G)**.

Type of Change	New Subsection			Committee Change	Accept in Principle		2008 NEC	800.113
ROP	pg. 1093	# 16-160	log: 2102	ROC	pg. 603	# 16-138 log: 1631	UL	444
Submitter: Ron L. Janikowski				Submitter: Craig Sato			OSHA	-
NFPA 70D				NFPA 70E	-		NFPA 79	-

800.113 Installation of Communications Wires, Cables, and Raceways.

(K) Multifamily Dwellings. The following cables, raceways, and wiring assemblies shall be permitted to be installed in multifamily dwellings in locations other than the locations covered in 800.113(B) through (G):

(4) Communications wires and Types CMP, CMR, CMG, and CM cables installed in:

a. Plenum communications raceway

b. Riser communications raceway

c. General-purpose communications raceway

Stallcup's Comment: A new subsection has been added to address the installation requirements of communications cables, raceways, and wiring assemblies in multifamily dwellings.

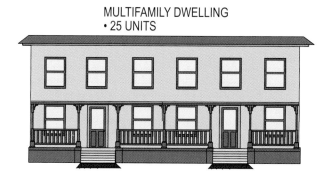

MULTIFAMILY DWELLING
• 25 UNITS

TYPES CMP, CMR, CMG, AND CM CABLES SHALL BE PERMITTED TO BE INSTALLED IN:

PLENUM COMMUNICATIONS RACEWAY
• **800.113(K)(4)a**
RISER COMMUNICATIONS RACEWAY
• **800.113(K)(4)b**
GENERAL-PURPOSE COMMUNICATIONS RACEWAY
• **800.113(K)(4)c**

MULTIFAMILY DWELLINGS
NEC 800.113(K)(4)

Purpose of Change: To add requirements for routing communications systems in multifamily dwellings other than those outlined in **800.113(B) through (G).**

Type of Change	New Subsection			Committee Change	Accept in Principle			2008 NEC	800.113
ROP	pg. 1093	# 16-160	log: 2102	ROC	pg. 603	# 16-138	log: 1631	UL	444
Submitter: Ron L. Janikowski				Submitter: Craig Sato				OSHA	-
NFPA 70B	-			NFPA 70E	-			NFPA 79	-

800.113 Installation of Communications Wires, Cables, and Raceways.

(K) Multifamily Dwellings. The following cables, raceways, and wiring assemblies shall be permitted to be installed in multifamily dwellings in locations other than the locations covered in 800.113(B) through (G):

(5) Types CMP, CMR, CMG, and CM cables installed in:

a. Riser cable routing assembly

b. General-purpose cable routing assembly

(6) Communications wire and Types CMP, CMR, CMG, CM, and CMX cables installed in a raceway of a type recognized in Chapter 3

(7) Type CMUC undercarpet communications wires and cables installed under carpet

Stallcup's Comment: A new subsection has been added to address the installation requirements of communications cables, raceways, and wiring assemblies in multifamily dwellings.

MULTIFAMILY DWELLING
• 25 UNITS

CABLES AND RACEWAYS AND CABLE ROUTING ASSEMBLIES SHALL BE PERMITTED TO BE INSTALLED IN MULTIFAMILY DWELLINGS

TYPES CMP, CMR, CMG, AND CM CABLES
• **800.113(K)(5)**
RISER CABLE ROUTING ASSEMBLY
• **800.113(K)(5)a**
GENERAL-PURPOSE CABLE ROUTING ASSEMBLY
• **800.113(K)(5)b**
TYPES CMP, CMR, CMG, CM, AND CMX CABLES
• **800.113(K)(6)**
UNDER CARPET COMMUNICATIONS WIRES AND CABLES
• **800.113(K)(7)**

**MULTIFAMILY DWELLINGS
NEC 800.113(K)(5) THRU (K)(7)**

Purpose of Change: To add requirements for routing communications systems in multifamily dwellings other than those outlined in **800.113(B) through (G)**.

Type of Change	New Subsection			Committee Change	Accept in Principle			2008 NEC	800.113
ROP	pg. 1093	# 16-160	log: 2102	ROC	pg. 603	# 16-138	log: 1631	UL	444
Submitter: Ron L. Janikowski				Submitter: Craig Sato				OSHA	-
NFPA 70B				NFPA 70F	-			NFPA 79	-

800.113 Installation of Communications Wires, Cables, and Raceways.

(L) One- and Two-Family Dwellings. The following cables and raceways shall be permitted to be installed in one- and two-family dwellings in locations other than the locations covered in 800.113(B) through (F):

(1) Types CMP, CMR, CMG, and CM cables

(2) Type CMX cable less than 6 mm (0.25 in.) in diameter

(3) Plenum, riser, and general-purpose communications raceways

Stallcup's Comment: A new subsection has been added to address the installation requirements for communications cables and raceways in one- and two-family dwellings.

TWO-FAMILY DWELLING

CABLES AND RACEWAYS SHALL BE PERMITTED IN ONE- AND TWO-FAMILY DWELLINGS

TYPES CMP, CMR, CMG, AND CM CABLES
• **800.113(L)(1)**

TYPE CMX CABLES LESS THAN 0.25" (6 mm) IN DIAMETER
• **8000.113(L)(2)**

PLENUM, RISER, AND GENERAL-PURPOSE COMMUNICATIONS RACEWAYS
• **800.113(L)(3)**

ONE- AND TWO-FAMILY DWELLINGS
NEC 800.113(L)(1) THRU (L)(3)

Purpose of Change: To add requirements for routing communications systems in one- and two-family dwellings in locations other than those outlined in **800.113(B) through (F).**

Type of Change	New Subsection			Committee Change	Accept in Principle			2008 NEC	800.113
ROP	pg. 1093	# 16-160	log: 2102	ROC	pg. 603	# 16-138	log: 1631	UL	444
Submitter: Ron L. Janikowski				Submitter: Craig Sato				OSHA	-
NFPA 70B	-			NFPA 70E	-			NFPA 79	-

800.113 Installation of Communications Wires, Cables, and Raceways.

(L) One- and Two-Family Dwellings. The following cables and raceways shall be permitted to be installed in one- and two-family dwellings in locations other than the locations covered in 800.113(B) through (F):

(4) Communications wires and Types CMP, CMR, CMG, and CM cables installed in:

a. Plenum communications raceway

b. Riser communications raceway

c. General-purpose communications raceway

Stallcup's Comment: A new subsection has been added to address the installation requirements for communications cables and raceways in one- and two-family dwellings.

TWO-FAMILY DWELLING

TYPES CMP, CMR, CMG, AND CM CABLES SHALL BE PERMITTED TO BE INSTALLED IN:

PLENUM COMMUNICATIONS RACEWAY
• **800.113(L)(4)a**
RISER COMMUNICATIONS RACEWAY
• **800.113(L)(4)b**
GENERAL-PURPOSE COMMUNICATIONS RACEWAY
• **800.113(L)(4)c**

**ONE- AND TWO-FAMILY DWELLINGS
NEC 800.113(L)(4)**

Purpose of Change: To add requirements for routing communications systems in one- and two-family dwellings in locations other than those outlined in **800.113(B) through (F).**

Type of Change	New Subsection			Committee Change	Accept in Principle			2008 NEC	800.113
ROP	pg. 1093	# 16-160	log: 2102	ROC	pg. 603	# 16-138	log: 1631	UL	444
Submitter: Ron L. Janikowski				Submitter: Craig Sato				OSHA	-
NFPA 70B	-			NFPA 70E	-			NFPA 79	-

800.113 Installation of Communications Wires, Cables, and Raceways.

(L) One- and Two-Family Dwellings. The following cables and raceways shall be permitted to be installed in one- and two-family dwellings in locations other than the locations covered in 800.113(B) through (F):

(5) Types CMP, CMR, CMG, and CM cables installed in:

a. Riser cable routing assembly

b. General-purpose cable routing assembly

(6) Communication wires and Types CMP, CMR, CMG, CM, and CMX cables installed in a raceway of a type recognized in Chapter 3

(7) Type CMUC undercarpet communications wires and cables installed under carpet

(8) Hybrid power and communications cable listed in accordance with 800.179(I)

Stallcup's Comment: A new subsection has been added to address the installation requirements for communications cables and raceways in one- and two-family dwellings.

TWO-FAMILY DWELLING

CABLES AND RACEWAYS AND CABLE ROUTING ASSEMBLIES SHALL BE PERMITTED TO BE INSTALLED IN ONE- AND TWO-FAMILY DWELLINGS

TYPES CMP, CMR, CMG, AND CM CABLES
• **800.113(L)(5)**

RISER CABLE ROUTING ASSEMBLY
• **800.113(L)(5)a**

GENERAL-PURPOSE CABLE ROUTING ASSEMBLY
• **800.113(L)(5)b**

TYPES CMP, CMR, CMG, CM, AND CMX CABLES
• **800.113(L)(6)**

TYPE CMUC UNDER CARPET COMMUNICATIONS WIRES AND CABLES
• **800.113(L)(7)**

HYBRID POWER AND COMMUNICATIONS CABLE COMPLYING WITH **800.179(I)**
• **800.113(L)(8)**

ONE- AND TWO-FAMILY DWELLINGS
NEC 800.113(L)(5) THRU (L)(8)

Purpose of Change: To add requirements for routing communications systems in one- and two-family dwellings in locations other than those outlined in **800.113(B) through (F)**.

Type of Change	Revision			Committee Change	Accept			2008 NEC	820.44(A)
ROP	pg. 1078	# 16-126	log: 2098	ROC	pg. -	# -	log: -	UL	1655
Submitter: Ron L. Janikowski				Submitter: -				OSHA	-
NFPA 70B	-			NFPA 70E	-			NFPA 79	-

820.44 Overhead (Aerial) Coaxial Cables.

Overhead (aerial) coaxial cables, prior to the point of grounding, as specified in 820.93, shall comply with 820.44(A) through (E).

(A) On Poles and In-Span. Where coaxial cables and electric light or power conductors are supported by the same pole or are run parallel to each other in-span, the conditions described in 820.44(A)(1) through (A)(4) shall be met.

(1) Relative Location. Where practicable, the coaxial cables shall be located below the electric light or power conductors.

(2) Attachment to Cross-Arms. Coaxial cables shall not be attached to cross-arm that carries electric light or power conductors.

(3) Climbing Space. The climbing space through coaxial cables shall comply with the requirements of 225.14(D).

Stallcup's Comment: A revision has been made to clarify the requirements for coaxial cables and electric light or power conductors that are supported by the same pole or are run parallel to each other in-span.

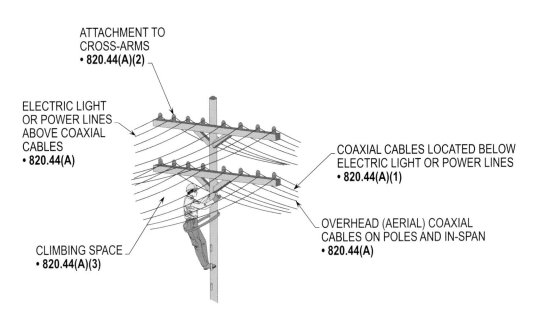

ON POLES AND IN-SPAN
NEC 820.44(A)(1) THRU (A)(3)

Purpose of Change: To add requirements that address coaxial cables and electric light and power conductors (lines) that are supported by the same pole or run parallel to each other in-span situations.

Type of Change	Revision			Committee Change	Accept			2008 NEC	820.44(A)
ROP	pg. 1078	# 16-126	log: 2098	ROC	pg. -	# -	log: -	UL	1655
Submitter: Ron L. Janikowski				Submitter: -				OSHA	-
NFPA 70B	-			NFPA 70E	-			NFPA 79	-

820.44 Overhead (Aerial) Coaxial Cables.

Overhead (aerial) coaxial cables, prior to the point of grounding, as specified in 820.93, shall comply with 820.44(A) through (E).

(4) Clearance. Lead-in or overhead (aerial) -drop coaxial cables from a pole or other support, including the point of initial attachment to a building or structure, shall be kept away from electric light, power, Class 1, or non–power-limited fire alarm circuit conductors so as to avoid the possibility of accidental contact.

Exception: Where proximity to electric light, power, Class 1, or non-power-limited fire alarm circuit conductors cannot be avoided, the installation shall provide clearances of not less than 300 mm (12 in.) from electric light, power, Class 1, or non–power-limited fire alarm circuit conductors. The clearance requirement shall apply at all points along the drop, and it shall increase to 1.0 m (40 in.) at the pole.

Stallcup's Comment: A revision has been made to clarify the requirements for coaxial cables and electric light or power conductors that are supported by the same pole or are run parallel to each other in-span.

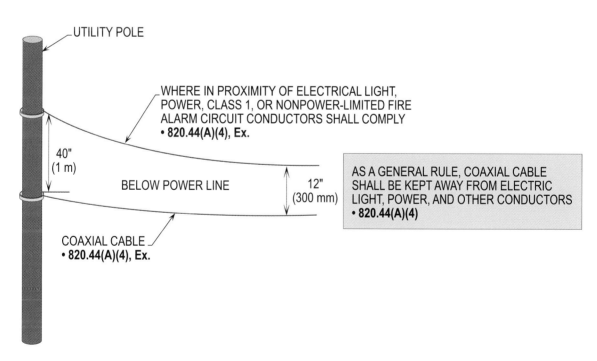

CLEARANCE
NEC 820.44(A)(4) and Ex.

Purpose of Change: To add requirements that address coaxial cables and electric light and power conductors (lines) that are supported by the same pole or run parallel to each other in-span situations.

Type of Change	New Exception			Committee Change	Accept in Principle			2008 NEC	-
ROP	pg. 1117	# 16-225	log: 4187	ROC	pg. 628	# 16-224	log: 1902	UL	1655
Submitter: Paul Dobrowsky				Submitter: Phil Simmons				OSHA	-
NFPA 70B	-			NFPA 70E	-			NFPA 79	-

820.100 Cable Bonding and Grounding.

The shield of the coaxial cable shall be bonded or grounded as specified in 820.100(A) through (D).

Exception: For communications systems using coaxial cable confined within the premises and isolated from outside cable plant, the shield shall be permitted to be grounded by a connection to an equipment grounding conductor as described in 250.118. Connecting to an equipment grounding conductor through a grounded receptacle using a dedicated grounding conductor and permanently connected listed device shall be permitted. Use of a cord and plug for the connection to an equipment grounding conductor shall not be permitted.

Stallcup's Comment: A new exception has been added to address the grounding requirements for communications systems using coaxial cable confined within the premises and isolated from an outside cable plant.

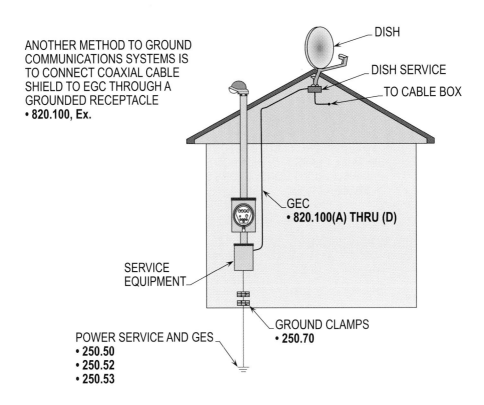

ANOTHER METHOD TO GROUND COMMUNICATIONS SYSTEMS IS TO CONNECT COAXIAL CABLE SHIELD TO EGC THROUGH A GROUNDED RECEPTACLE
• **820.100, Ex.**

DISH

DISH SERVICE

TO CABLE BOX

GEC
• **820.100(A) THRU (D)**

SERVICE EQUIPMENT

GROUND CLAMPS
• **250.70**

POWER SERVICE AND GES
• **250.50**
• **250.52**
• **250.53**

CABLE BONDING AND GROUNDING
NEC 820.100, Ex.

Purpose of Change: To add requirements for addressing the grounding for communications systems using coaxial cable confined within the premises.

Type of Change	New Subsections			Committee Change	Accept in Principle			2008 NEC	820.110
ROP	pg. 1126	# 16-265	log: 2111	ROC	pg. 649	# 16-292	log: 320	UL	1655
Submitter: Ron L. Janikowski				Submitter: Technical Correlating Committee				OSHA	-
NFPA 70B	-			NFPA 70E	-			NFPA 79	-

820.110 Raceways for Coaxial Cables.

(A) Types of Raceways. Coaxial cables shall be permitted to be installed in any raceway that complies with either (A)(1) or (A)(2).

(1) Raceways Recognized in Chapter 3. Coaxial cables shall be permitted to be installed in any raceway included in Chapter 3. The raceways shall be installed in accordance with the requirements of Chapter 3.

(2) Other Permitted Raceways. Coaxial cables shall be permitted to be installed in listed plenum communications raceway, listed riser communications raceway, or listed general-purpose communications raceway selected in accordance with the provisions of 820.113, and installed in accordance with 362.24 through 362.56, where the requirements applicable to electrical nonmetallic tubing apply.

(B) Raceway Fill for Coaxial Cables. The raceway fill requirements of Chapters 3 and 9 shall not apply to coaxial cables.

Stallcup's Comment: A new subsection has been added to address the raceway and raceway fill requirements for coaxial cables.

See **800.110(A)(1), (A)(2),** and **(B)** for similar illustrations.

Type of Change	New Subsections			Committee Change	Accept in Principle			2008 NEC	800.113
ROP	pg. 1127	# 16-267	log: 2112	ROC	pg. 631	# 16-233	log: 1635	UL	1655
Submitter: Ron L. Janikowski				Submitter: Criag Sato				OSHA	-
NFPA 70B	-			NFPA 70E	-			NFPA 79	-

820.113 Installation of Coaxial Cables.

(A) Listing. Coaxial cables installed in buildings shall be listed.

Exception: Coaxial cables that comply with 820.48 shall not be required to be listed.

(B) Fabricated Ducts Used for Environmental Air. The following cables shall be permitted in ducts as described in 300.22(B) if they are directly associated with the air distribution system:

(1) Up to 1.22 m (4 ft) of Type CATVP cable

(2) Types CATVP, CATVR, CATV, and CATVX cables installed in raceways that are installed in compliance with 300.22(B)

Informational Note: For information on fire protection of wiring installed in fabricated ducts see 4.3.4.1 and 4.3.11.3.3 in NFPA 90A-2009, *Standard for the Installation of Air-Conditioning and Ventilating Systems.*

Stallcup's Comment: A revision has been made to clarify the installation requirements of coaxial cables in buildings and fabricated ducts used for environmental air.

See **800.110(A)(1)**, **(A)(2)**, and **(B)** for a similar illustration.

Type of Change	New Subdivision			Committee Change	Accept in Principle			2008 NEC	820.113
ROP	pg. 1127	# 16-267	log: 2112	ROC	pg. 631	# 16-233	log: 1635	UL	1655
Submitter: Ron L. Janikowski				Submitter: Craig Sato				OSHA	-
NFPA 70B	-			NFPA 70E				NFPA 79	-

820.113 Installation of Coaxial Cables.

(C) Other Spaces Used For Environmental Air (Plenums). The following cables shall be permitted in other spaces used for environmental air as described in 300.22(C):

(1) Type CATVP cable

(2) Type CATVP cable installed in plenum communications raceway

(3) Type CATVP cable supported by open metallic cable trays or cable tray systems

(4) Types CATVP, CATVR, CATV, and CATVX cables installed in raceways that are installed in compliance with 300.22(C)

(5) Types CATVP, CATVR, CATV, and CATVX cables supported by solid bottom metal cable trays with solid metal covers in other spaces used for environmental air (plenums) as described in 300.22(C)

Informational Note: For information on fire protection of wiring installed in other spaces used for environmental air see 4.3.11.2, 4.3.11.4, and 4.3.11.5 of NFPA 90A-2009, *Standard for the Installation of Air-Conditioning and Ventilating Systems.*

Stallcup's Comment: A new subsection has been added to address the installation requirements of coaxial cables in other spaces used for environmental air (plenums).

See **800.113(C)(1) through (C)(6)** for similar illustrations.

Type of Change	New Subdivision			Committee Change	Accept in Principle			2008 NEC	820.113
ROP	pg. 1127	# 16-267	log: 2112	ROC	pg. 631	# 16-233	log: 1635	UL	1655
Submitter: Ron L. Janikowski				Submitter: Craig Sato				OSHA	-
NFPA 70B	-			NFPA 70E	-			NFPA 79	-

820.113 Installation of Coaxial Cables.

(D) Risers — Cables in Vertical Runs. The following cables shall be permitted in vertical runs penetrating one or more floors and in vertical runs in a shaft:

(1) Types CATVP and CATVR cables

(2) Types CATVP and CATVR cables installed in:

a. Plenum communications raceway

b. Riser communications raceway

c. Riser cable routing assembly

Informational Note: See 820.26 for firestop requirements for floor penetrations.

Stallcup's Comment: A new subsection has been added to address the installation requirements of coaxial cables in vertical runs.

See **800.113(D)(1) through (D)(3)** for similar illustrations.

Type of Change	New Subsection			Committee Change	Accept in Principle			2008 NEC	820.113
ROP	pg. 1127	# 16-267	log: 2112	ROC	pg. 631	# 16-233	log: 1635	UL	1655
Submitter: Ron L. Janikowski				Submitter: Craig Sato				OSHA	-
NFPA 70B	-			NFPA 70E	-			NFPA 79	-

820.113 Installation of Coaxial Cables.

(E) Risers — Cables in Metal Raceways. The following cables shall be permitted in metal raceways in a riser having firestops at each floor:

(1) Types CATVP, CATVR, CATV, and CATVX cables

(2) Types CATVP, CATVR, CATV, and CATVX cables installed in:

a. Plenum communications raceway

b. Riser communications raceway

c. General-purpose communications raceway

Informational Note: See 820.26 for firestop requirements for floor penetrations.

Stallcup's Comment: A new subsection has been added to address the installation requirements of coaxial cables in metal raceways.

See **800.113(E)(1) through (E)(3)** for similar illustrations.

Type of Change	New Subsection			Committee Change	Accept in Principle			2008 NEC	820.113
ROP	pg. 1127	# 16-267	log: 2112	ROC	pg. 631	# 16-233	log: 1635	UL	1655
Submitter: Ron L. Janikowski				Submitter: Craig Sato				OSHA	-
NFPA 70B	-			NFPA 70E	-			NFPA 79	-

820.113 Installation of Coaxial Cables.

(F) Risers — Cables in Fireproof Shafts. The following cables shall be permitted to be installed in fireproof riser shafts with firestops at each floor:

(1) Types CATVP, CATVR, CATV, and CATVX cables

(2) Types CATVP, CATVR, and CATV cables installed in:

a. Plenum communications raceway

b. Riser communications raceway

c. General-purpose communications raceway

d. Riser cable routing assembly

e. General-purpose cable routing assembly

Informational Note: See 820.26 for firestop requirements for floor penetrations.

Stallcup's Comment: A new subsection has been added to address the installation requirements of coaxial cables in fireproof riser shafts having firestops at each floor.

See **800.113(F)(1) through (F)(3)** for similar illustrations.

Type of Change	New Subsection			Committee Change	Accept in Principle		2008 NEC	820.113
ROP	pg. 1127	# 16-267	log: 2112	ROC	pg. 631	# 16-233	log: 1635	UL	1655
Submitter: Ron L. Janikowski				Submitter: Craig Sato			OSHA	-
NFPA 70B	-			NFPA 70E	-		NFPA 79	-

820.113 Installation of Coaxial Cables.

(G) Risers — One- and Two-Family Dwellings. The following cables shall be permitted in one- and two-family dwellings:

(1) Types CATVP, CATVR, and CATV cables

(2) Type CATVX cable less than 10 mm (0.375 in.) in diameter

(3) Types CATVP, CATVR, and CATV cables installed in:

a. Plenum communications raceway

b. Riser communications raceway

c. General-purpose communications raceway

d. Riser cable routing assembly

e. General-purpose cable routing assembly

Informational Note: See 820.26 for firestop requirements for floor penetrations.

Stallcup's Comment: A new subsection has been added to address the installation requirements of coaxial cables in one- and two-family dwellings.

See **800.113(G)(1) through (G)(4)** for similar illustrations.

Type of Change	New Subsection		Committee Change	Accept in Principle		2008 NEC	820.113
ROP	pg. 1127	# 16-267	log: 2112	ROC	pg. 631	# 16-233	log: 1635	UL	1655
Submitter: Ron L. Janikowski			Submitter: Craig Sato			OSHA	-
NFPA 70B	-		NFPA 70E	-		NFPA 79	-

820.113 Installation of Coaxial Cables.

(H) Cable Trays. The following cables shall be permitted to be supported by cable trays:

(1) Types CATVP, CATVR, and CATV cables

(2) Types CATVP, CATVR, and CATV cables installed in:

a. Plenum communications raceway

b. Riser communications raceway

c. General-purpose communications raceway

Stallcup's Comment: A new subsection has been added to address the installation requirements of coaxial cables that are supported by cable trays.

See **800.113(H)(1) through (H)(3)** for similar illustrations.

Type of Change	New Subsection		Committee Change	Accept in Principle		2008 NEC	820.113
ROP	pg. 1127	# 16-267	log: 2112	ROC	pg. 631	# 16-233	log: 1635	UL	1655
Submitter: Ron L. Janikowski			Submitter: Craig Sato			OSHA	-
NFPA 70B	-		NFPA 70E			NFPA 79	-

820.113 Installation of Coaxial Cables.

(I) Distributing Frames and Cross-Connect Arrays. The following cables shall be permitted to be installed in distributing frames and cross-connect arrays:

(1) Types CATVP, CATVR, and CATV cables

(2) Types CATVP, CATVR, and CATV cables installed in:

a. Plenum communications raceway

b. Riser communications raceway

c. General-purpose communications raceway

d. Riser cable routing assembly

e. General-purpose cable routing assembly

Stallcup's Comment: A new subsection has been added to address the installation requirements of coaxial cables in distributing frames and cross-connect arrays.

See **800.113(I)(1) through (I)(3)** for similar illustrations.

Type of Change	New Subsection			Committee Change	Accept in Principle			2008 NEC	820.113
ROP	pg. 1127	# 16-267	log: 2112	ROC	pg. 631	# 16-233	log: 1635	UL	1655
Submitter: Ron L. Janikowski				Submitter: Craig Sato				OSHA	-
NFPA 70B	-			NFPA 70E	-			NFPA 79	-

820.113 Installation of Coaxial Cables.

(J) Other Building Locations. The following cables and cable routing assemblies shall be permitted to be installed in building locations other than the locations covered in 820.113(B) through (I):

(1) Types CATVP, CATVR, and CATV cables

(2) A maximum of 3 m (10 ft) of exposed Type CATVX cable in nonconcealed spaces

(3) Types CATVP, CATVR, and CATV cables installed in:

a. Plenum communications raceway

b. Riser communications raceway

c. General-purpose communications raceway

d. Riser cable routing assembly

e. General-purpose cable routing assembly

(4) Types CATVP, CATVR, CATV, and Type CATVX cables installed in a raceway of a type recognized in Chapter 3

Stallcup's Comment: A new subsection has been added to address the installation requirements of coaxial cables and cable routing assemblies in building locations other than specified for fabricated ducts used for environmental air; other spaces used for environmental air (plenums); vertical runs; metal raceways; fireproof shafts; one- and two-family dwellings; cable trays; and distributing frames and cross-connect arrays.

See **800.113(J)(1) through (J)(7)** for similar illustrations.

Type of Change	New Subsection			Committee Change		Accept in Principle		2008 NEC	820.113
ROP	pg. 1127	# 16-267	log: 2112	ROC	pg. 631	# 16-233	log: 1635	UL	1655
Submitter: Ron L. Janikowski				Submitter: Craig Sato				OSHA	-
NFPA 70B	-			NFPA 70E	■			NFPA 79	-

820.113 Installation of Coaxial Cables.

(K) One- and Two-Family and Multifamily Dwellings. The following cables and cable routing assemblies shall be permitted to be installed in one- and two-family and multifamily dwellings in locations other than those locations covered in 820.113(B) through (I):

(1) Types CATVP, CATVR, and CATV cables

(2) Type CATVX cable less than 10 mm (0.375 in.) in diameter

(3) Types CATVP, CATVR, and CATV cables installed in:

a. Plenum communications raceway

b. Riser communications raceway

c. General-purpose communications raceway

d. Riser cable routing assembly

e. General-purpose cable routing assembly

(4) Types CATVP, CATVR, CATV, and Type CATVX cables installed in a raceway of a type recognized in Chapter 3

Stallcup's Comment: A new subsection has been added to address the installation requirements of coaxial cables and cable routing assemblies in one- and two-family and multifamily dwellings.

See **800.113(K)(1) through (K)(7)** and **800.113(L)(1) through (L)(8)** for similar illustrations.

Type of Change	New Subsection			Committee Change	Accept			2008 NEC	830.44
ROP	pg. 1078	# 16-126	log: 2098	ROC	pg. -	# -	log: -	UL	444
Submitter: Ron L. Janikowski				Submitter: -				OSHA	-
NFPA 70B	-			NFPA 70E	-			NFPA 79	-

830.44 Overhead (Aerial) Cables.

(A) On Poles and In-Span. Where network-powered broadband communications cables and electric light or power conductors are supported by the same pole or are run parallel to each other in-span, the conditions described in 830.44(A)(1) through (A)(4) shall be met.

(1) Relative Location. Where practicable, the network-powered broadband communications cables shall be located below the electric light or power conductors.

(2) Attachment to Cross-Arms. Network-powered broadband communications cables shall not be attached to a cross-arm that carries electric light or power conductors.

(3) Climbing Space. The climbing space through network-powered broadband communications wires and cables shall comply with the requirements of 225.14(D).

(4) Clearance. Lead-in or overhead (aerial) -drop network-powered broadband communications cables from a pole or other support, including the point of initial attachment to a building or structure, shall be kept away from electric light, power, Class 1, or non-power-limited fire alarm circuit conductors so as to avoid the possibility of accidental contact.

Stallcup's Comment: A revision has been made to clarify the requirements for network-powered broadband communications cables and electric light or power conductors that are supported by the same pole or are run parallel to each other in-span.

See **820.44(A)(1) through (A)(4)** for similar illustrations.

Type of Change	New Subdivisions			Committee Change	Accept			2008 NEC	830.110
ROP	pg. 1147	# 16-330	log: 2121	ROC	pg. 649	# 16-292	log: 320	UL	444
Submitter: Ron L. Janikowski				Submitter: Technical Correlating Committee				OSHA	-
NFPA 70B	-			NFPA 70C				NFPA 79	-

830.110 Raceways for Low- and Medium-Power Network-Powered Broadband Communications Cables.

(A) Raceways Recognized in Chapter 3. Low- and medium-power network-powered broadband communications cables shall be permitted to be installed in any raceway included in Chapter 3. The raceways shall be installed in accordance with the requirements of Chapter 3.

(B) Raceway Fill for Network-Powered Broadband Communications Cables. Raceway fill for network-powered broadband communications cables shall comply with either (B)(1) or (B)(2).

(1) Low-Power Network-Powered Broadband Communications Cables. The raceway fill requirements of Chapters 3 and 9 shall not apply to low-power network-powered broadband communications cables.

(2) Medium-Power Network-Powered Broadband Communications Cables. Where medium-power network-powered broadband communications cables are installed in a raceway, the raceway fill requirements of Chapters 3 and 9 shall apply.

Stallcup's Comment: A new subsection has been added to address the raceway and raceway fill requirements for low- and medium-power network-powered broadband communications cables.

See **800.110(A)(1)**, **(A)(2)**, and **(B)** for similar illustrations.

Type of Change	New Subsections			Committee Change	Accept in Principle		2008 NEC	-
ROP	pg. 1148	# 16-331	log: 2122	ROC	pg. 649	# 16-293	log: 1637	UL	444
Submitter: Ron L. Janikowski				Submitter: Craig Sato			OSHA	-
NFPA 70B	-			NFPA 70E	-		NFPA 79	-

830.113 Installation of Network-Powered Broadband Communications Cables.

(A) Listing. Network-powered broadband communications cables installed in buildings shall be listed.

(B) Fabricated Ducts Used for Environmental Air. The following cables shall be permitted in ducts and as described in 300.22(B) if they are directly associated with the air distribution system:

(1) Up to 1.22 m (4 ft) of Type BLP cable

(2) Types BLP, BMR, BLR, BM, BL, and BLX cables installed in raceways that are installed in compliance with 300.22(B)

Informational Note: For information on fire protection of wiring installed in fabricated ducts see 4.3.4.1 and 4.3.11.3.3 in NFPA 90A-2009, *Standard for the Installation of Air-Conditioning and Ventilating Systems.*

Stallcup's Comment: A revision has been made to clarify the installation requirements of network-powered broadband communications cables in buildings and fabricated ducts used for environmental air.

See **800.113(A)** and **(B)** for a similar illustration.

Type of Change	New Subsection			Committee Change	Accept in Principle			2008 NEC	-
ROP	pg. 1148	# 16-331	log: 2122	ROC	pg. 649	# 16-293	log: 1637	UL	444
Submitter: Ron L. Janikowski				Submitter: Craig Sato				OSHA	-
NFPA 70B	-			NFPA 70C				NFPA 79	-

830.113 Installation of Network-Powered Broadband Communications Cables.

(C) Other Spaces Used For Environmental Air (Plenums). The following cables shall be permitted in other spaces used for environmental air as described in 300.22(C):

(1) Type BLP cable

(2) Type BLP cable installed in plenum communications raceway

(3) Type BLP cable supported by open metallic cable trays or cable tray systems

(4) Types BLP, BMR, BLR, BM, BL, and BLX cables installed in raceways that are installed in compliance with 300.22(C)

(5) Types BLP, BMR, BLR, BM, BL, and BLX cables supported by solid bottom metal cable trays with solid metal covers in other spaces used for environmental air (plenums) as described in 300.22(C)

Informational Note: For information on fire protection of wiring installed in other spaces used for environmental air see 4.3.11.2, 4.3.11.4, and 4.3.11.5 of NFPA 90A-2009, *Standard for the Installation of Air-Conditioning and Ventilating Systems.*

Stallcup's Comment: A new subsection has been added to address the installation requirements of network-powered broadband communications cables in other spaces used for environmental air (plenums).

See **800.113(C)(1) through (C)(6)** for similar illustrations.

Type of Change	New Subsection			Committee Change	Accept in Principle		2008 NEC	-
ROP	pg. 1148	# 16-331	log: 2122	ROC	pg. 649	# 16-293	log: 1637	UL	444
Submitter: Ron L. Janikowski				Submitter: Craig Sato			OSHA	-
NFPA 70B	-			NFPA 70E	-		NFPA 79	-

830.113 Installation of Network-Powered Broadband Communications Cables.

(D) Risers — Cables in Vertical Runs. The following cables shall be permitted in vertical runs penetrating one or more floors and in vertical runs in a shaft:

(1) Types BLP, BMR, and BLR cables

(2) Types BLP and BLR cables installed in:

a. Plenum communications raceway

b. Riser communications raceway

c. Riser cable routing assembly

Informational Note: See 830.26 for firestop requirements for floor penetrations.

Stallcup's Comment: A new subsection has been added to address the installation requirements of network-powered broadband communications cables in vertical runs.

See **800.113(D)(1) through (D)(3)** for similar illustrations.

Type of Change	New Subsection			Committee Change	Accept in Principle			2008 NEC	-
ROP	pg. 1148	# 16-331	log: 2122	ROC	pg. 649	# 16-293	log: 1637	UL	444
Submitter: Ron L. Janikowski				Submitter: Craig Sato				OSHA	-
NFPA 70B	-			NFPA 70E				NFPA 79	-

830.113 Installation of Network-Powered Broadband Communications Cables.

(E) Risers — Cables in Metal Raceways. The following cables shall be permitted in a metal raceway in a riser with firestops at each floor:

(1) Types BLP, BMR, BLR, BM, BL, and BLX cables

(2) Types BLP, BLR, and BL cables installed in:

a. Plenum communications raceway

b. Riser communications raceway

c. General-purpose communications raceway

Informational Note: See 830.26 for firestop requirements for floor penetrations.

Stallcup's Comment: A new subsection has been added to address the installation requirements of network-powered broadband communications cables in metal raceways.

See **800.113(E)(1) through (E)(3)** for similar illustrations.

Type of Change	New Subsection			Committee Change	Accept in Principle		2008 NEC	-
ROP	pg. 1148	# 16-331	log: 2122	ROC	pg. 649	# 16-293	log: 1637	UL	444
Submitter: Ron L. Janikowski				Submitter: Craig Sato			OSHA	-
NFPA 70B	-			NFPA 70E	-		NFPA 79	-

830.113 Installation of Network-Powered Broadband Communications Cables.

(F) Risers — Cables in Fireproof Shafts. The following cables shall be permitted to be installed in fireproof riser shafts with firestops at each floor:

(1) Types BLP, BMR, BLR, BM, BL, and BLX cables

(2) Types BLP, BLR, and BL cables installed in:

a. Plenum communications raceway

b. Riser communications raceway

c. General-purpose communications raceway

d. Riser cable routing assembly

e. General-purpose cable routing assembly

Informational Note: See 830.26 for firestop requirements for floor penetrations.

Stallcup's Comment: A new subsection has been added to address the installation requirements of network-powered broadband communications cables in fireproof riser shafts having firestops at each floor.

See **800.113(F)(1) through (F)(3)** for similar illustrations.

Type of Change	New Subsection			Committee Change	Accept in Principle		2008 NEC	-
ROP	pg. 1148	# 16-331	log: 2122	ROC	pg. 649	# 16-293	log: 1637	UL	444
Submitter: Ron L. Janikowski				Submitter: Craig Sato			OSHA	-
NFPA 70B	-			NFPA TOC			NFPA 79	-

830.113 Installation of Network-Powered Broadband Communications Cables.

(G) Risers — One- and Two-Family Dwellings. The following cables shall be permitted in one- and two-family dwellings:

(1) Types BLP, BMR, BLR, BM, BL cables and Types BL and BLX cables less than 10 mm (0.375 in.) in diameter

(2) Types BLP, BLR, and BL cables installed in:

a. Plenum communications raceway

b. Riser communications raceway

c. General-purpose communications raceway

d. Riser cable routing assembly

e. General-purpose cable routing assembly

Informational Note: See 830.26 for firestop requirements for floor penetrations.

Stallcup's Comment: A new subsection has been added to address the installation requirements of network-powered broadband communications cables and cable routing assemblies in one- and two-family dwellings.

See **800.113(G)(1) through (G)(4)** for similar illustrations.

Type of Change	New Subdivision		Committee Change	Accept in Principle		2008 NEC	-
ROP	pg. 1148	# 16-331	log: 2122	ROC	pg. 649	# 16-293	log: 1637	UL	444
Submitter: Ron L. Janikowski			Submitter: Craig Sato			OSHA	-
NFPA 70B	-		NFPA 70E	-		NFPA 79	-

830.113 Installation of Network-Powered Broadband Communications Cables.

(H) Other Building Locations. The following cables and raceways shall be permitted to be installed in building locations other than those covered in 830.113(B) through (G):

(1) Types BLP, BMR, BLR, BM, and BL cables

(2) Types BLP, BMR, BLR, BM, BL, and BLX cables installed in a raceway

(3) Types BLP, BLR, and BL cables

a. Plenum communications raceway

b. Riser communications raceway

c. General-purpose communications raceway

d. Riser cable routing assembly

e. General-purpose cable routing assembly

(4) Types BLX and BL cables less than 10 mm (0.375 in.) in diameter in one- and two-family dwellings

(5) Types BMU and BLU cables entering the building from outside and run in rigid metal conduit or intermediate metal conduit where the conduit is connected by a bonding conductor or grounding electrode conductor in accordance with 830.100(B)

Informational Note: This provision limits the length of Type BLX cable to 15 m (50 ft), while 830.90(B) requires that the primary protector, or NIU with integral protection, be located as close as practicable to the point at which the cable enters the building. Therefore, in installations requiring a primary protector, or NIU with integral protection, Type BLX cable may not be permitted to extend 15 m (50 ft), into the building, if it is practicable to place the primary protector closer than 15 m (50 ft) to the entrance point.

(6) A maximum length of 15 m (50 ft), within the building, of Type BLX cable entering the building from outside and terminating at an NIU or a primary protection location.

Stallcup's Comment: A new subsection has been added to address the installation requirements of network-powered broadband communications cables in fabricated ducts used for environmental air; other spaces used for environmental air (plenums); vertical runs; metal raceways; fireproof shafts; one- and two-family dwellings; and cable trays.

See **800.113(J)(1) through (J)(7)** for similar illustrations.

_____

Name   Date

Chapter 8
Communications Systems

Section   Answer

1. The bonding conductor or grounding electrode conductor shall not be smaller than _____   _____   _____
AWG for communications circuits.

(A) 18   **(B)** 16
(C) 14   **(D)** 12

2. The bonding or grounding electrode conductor shall not be required to exceed _____ AWG for   _____   _____
communications circuits.

(A) 8   **(B)** 6
(C) 4   **(D)** 2

3. Type CMP communications cable shall be permitted in ducts used for environmental air up to   _____   _____
_____ ft in length.

(A) 2   **(B)** 3
(C) 4   **(D)** 6

4. Which of the following communications wires and cables shall permitted in other spaces used for   _____   _____
environmental air:

(A) CMR cables   **(B)** CMG cables
(C) CMX cables   **(D)** Plenum communications raceway

5. Which of the following communications cables shall be permitted in metal raceways in a riser having   _____   _____
firestops at each floor:

(A) CMP cables   **(B)** CMR cables
(C) Plenum, riser, and general-purpose communications raceway   **(D)** All of the above

6. Type CMX communications cable less than _____ in. diameter shall be permitted in one- and   _____   _____
two-family dwellings.

(A) 0.25   **(B)** 0.33
(C) 0.40   **(D)** 0.50

7. Which of the following communications cables shall be permitted to be supported by cable trays: _____   _____

(A) CM cables   **(B)** CMP cables
(C) CMX cables   **(D)** None of the above

8. A maximum of _____ ft of exposed Type CMX communications cable shall be permitted to be   _____   _____
installed in nonconcealed spaces.

(A) 4   **(B)** 6
(C) 10   **(D)** 12

Section   Answer

9. Where proximity to electric light, power, Class 1, or nonpower-limited fire alarm circuit conductors cannot be avoided, the installation shall provide clearance of not less than _____ in. from electric light, power, Class 1, or nonpower-limited fire alarm circuit conductors.

(A) 4   **(B)** 6
(C) 12   **(D)** 18

10. For communications systems using coaxial cable confined within the premises and isolated from outside cable plant, the shield shall be permitted to be grounded by a connection to a(n) _____ conductor.

(A) equipment grounding   **(B)** bonding
(C) grounded   **(D)** ungrounded

Annex A
Abbreviations

A

A – amps
AC – alternating current
A/C – air-conditioning
AEGCP – assured equipment grounding conductor program
AFCI – arc-fault circuit-interrupter
AFCIP – arc-fault circuit-interrupter protection
AHJ – authority having jurisdiction
Alu. – aluminum
ASCC – available short-circuit current
AWG – American Wire Gage

B

BC – branch circuit
BCSC – branch-circuit selection current
BJ – bonding jumper
BK – black
BL – blue
BR – brown

C

°C – Celsius
CB – circuit breaker
CEE – concrete-encased electrode
CL – code letter
CM – circular mils
CMP – code-making panel
Comp. – compressor
Cond. – condenser
Cont. – continuous
cu. – copper
cu. in. – cubic inches

D

DC – direct-current
dia. – diameter
DPCB – double pole circuit breaker

E

EBJ – equipment bonding jumper
EExde – increase safety
Eexe – flameproof/increased safety components
Eff. – efficiency
EGFCP – effective ground-fault current path
EGB – equipment grounding bar
EGC – equipment grounding conductor
EMT – electrical metallic tubing
ENT or ENMT – electrical nonmetallic tubing
Epf – explosionproof
Ex. – exception

F

°F – Fahrenheit
FLA – full-load amperage
FLC – full-load current
FMC – flexible metal conduit
ft – foot

G

G – ground
GE – grounding electrode
GEC – grounding electrode conductor
GES – grounding electrode system
GFC – ground-fault current
GFCI – ground-fault circuit-interrupter
GFL – ground fault limiter
GFP – ground-fault protection
GFPE – ground-fault protection of equipment
GR – green
GRY – gray
GSC – grounded service conductor
GS-EC – grounded service-entrance conductor
GSE – grounded service enclosure

H

H – ungrounded (phase) conductor
HACR – heating, air conditioning, cooling, and refrigeration
HP – horsepower
Htg. – heating
Hz – hertz

I

I – amperage or current
IEC – International Electrotechnical Commission
IG – isolated ground
in. – inches
IN – Informational Note
INST. CB – instantaneous trip circuit breaker
INVT – inverse-time circuit breaker
IRA – inrush amps
ISC – intrinsically safe circuits

L

L – length of conductor
LD. – load
LFMC – liquidtight flexible metal conduit
LFNC – liquidtight flexible nonmetallic conduit
LPB – lighting panelboard
LRA – locked rotor amps
LRC – locked rotor current
LTR – long time rated

M

mA – milliamperes
MAX. – maximum
MEL – maximum energy level
mf – microfarads
MGFA – maximum ground fault available
MIN. – minimum
MIN. – minute
MR – momentary rated
MT. – motor
MWP – metal water pipe

N

N – neutral
NACB – nonautomatic circuit breaker
NB – neutral bar
NEC – *National Electrical Code*
NEMA – National Electrical Manufacturers Association
NFD – nonfused disconnect
NFPA – National Fire Protection Association
NLTFMC – nonmetallic liquidtight flexible metal conduit
NPC – nameplate current of motor
NTDF – nontime-delay fuse

O

OFC – optical fiber cable
OCP – overcurrent protection
OCPD – overcurrent protection device
OL – overload
OLP – overload protection
OL – overload
OR – orange
OSHA – Occupational Safety and Health Administration

P

PDS – power distribution system
PF – power factor
PFFC – path for fault current
PH. – phases
PRI. – primary
PPS – power production source
PSA – power supply assembly
PU – purple
PV – photovoltaic

R

R – ohms or resistance
RD – red
RMC – rigid metal conduit

S

SBS – structural building steel
SCC – short-circuit current
SDS – separately derived system
SEC. – secondary
SF – service factor
SIA – seal-in amps
SMM – structural metal member
SP – single-pole
SPCB – single-pole circuit breaker
sq. ft – square foot (feet)
sq. in. – square inches
STR – short-time rated
SWD – switched disconnect
SWG – switchgear
S-SBJ – supply-side bonding jumper

T

TDC – time-delay cycle
TDF – time-delay fuse
TDL – time-delay limiter
TP – thermal protector
TR – temperature rise
TS – trip setting
TV – touch voltage

U

UF – underground feeder

V

V – volts
VA – volt-amps
VD – voltage drop

W

W – watts
WT – white
WP – weatherproof

X

XFMR – transformer
X0 – neutral point

Y

YEL – yellow

Article 100 – Definitions

Arc-Fault Circuit Interrupter (AFCI). A device intended to provide protection from the effects of arc faults by recognizing characteristics unique to arcing and by functioning to de-energize the circuit when an arc fault is detected.

Automatic. Performing a function without the necessity of human intervention.

Bathroom. An area including a basin with one or more of the following: a toilet, a urinal, a tub, a shower, a bidet, or similar plumbing fixtures.

Bonding Jumper, System. The connection between the grounded circuit conductor and the supply-side bonding jumper, or the equipment grounding conductor, or both, at a separately derived system.

Ground Fault. An unintentional, electrically conducting connection between an ungrounded conductor of an electrical circuit and the normally non-current-carrying conductors, metallic enclosures, metallic raceways, metallic equipment, or earth.

Nonautomatic. Requiring human intervention to perform a function.

Separately Derived System. A premises wiring system whose power is derived from a source of electric energy or equipment other than a service. Such systems have no direct connection from circuit conductors of one system to circuit conductors of another system, other than connection through the earth, metal enclosures, metallic raceways, or equipment grounding conductors.

Service Conductors, Overhead. The overhead conductors between the service point and the first point of connection to the service-entrance conductors at the building or their structure.

Service Conductors, Underground. The underground conductors between the service point and the first point of connection to the service-entrance conductors in a terminal box, meter, or other enclosure, inside or outside the building wall.

Informational Note: Where there is no terminal box, meter, or other enclosure, the point of connection is considered to be the point of entrance of the service conductors into the building.

Service Lateral. The underground conductors between the utility electric supply system and the service point.

Service Point. The point of connection between the facilities of the serving utility and the premises wiring.

Informational Note: The service point can be described as the point of demarcation between where the serving utility ends and the premises wiring begins. The serving utility generally specifies the location of the service point based on the conditions of service.

Uninterruptible Power Supply. A power supply used to provide alternating-current power to a load for some period of time in the event of a power failure.

Informational Note: In addition, it may provide a more constant voltage and frequency supply to the load, reducing the effects of voltage and frequency variations.

Article 225 – Outside Branch Circuits and Feeders

Substation. An enclosed assemblage of equipment (e.g., switches, circuit breakers, buses, and transformers) under the control of qualified persons, through which electric energy is passed for the purpose of switching or modifying its characteristics.

Article 250 – Grounding and Bonding

Bonding Jumper, Supply-Side. A conductor installed on the supply side of a service or within a service equipment enclosure(s), for a separately derived system, that ensures the required electrical conductivity between metal parts required to be electrically connected.

Article 310 – Conductors for General Wiring

Electrical Ducts. Electrical conduits, or other raceways round in cross-sections, that are suitable for use underground or embedded in concrete.

Thermal Resistivity. As used in this *Code*, the heat transfer capability through a substance by conduction. It is the reciprocal of thermal conductivity and is designated Rho and expressed in the units °C-cm/W.

Article 390 – Underfloor Raceways

Underfloor Raceway. A raceway and associated components designed and intended for installation beneath or flush with the surface of a floor for the installation of cables and electrical conductors.

Article 406 – Receptacles, Cord Connectors, and Attachment Plugs (Caps)

Child Care Facility. A building or structure, or portion thereof, for educational, supervisory, or personal care services for more than four children ages 7 years old or less.

Article 422 - Appliances

Vending Machine. Any self-service device that dispenses products or merchandise without the necessity of replenishing the device between each vending operation and is designed to require insertion of coin, paper currency, token, card, key, or receipt of payment by other means.

Article 480 – Storage Batteries

Battery System. Interconnected battery subsystems consisting of one or more storage batteries and battery chargers; can include inverters, converters, and associated electrical equipment.

Nominal Battery Voltage. The voltage of a battery based on the number and type of cells in the battery.

Informational Note: The most common nominal cell voltages are 2 volts per cell for the lead-acid systems, 1.2 volts per cell for alkali systems, and 4 volts per cell for Li-ion systems. Nominal voltages might vary with different chemistries.

Sealed Cell or Battery. A cell or battery that has no provision for the routine addition of water or electrolyte or for external measurement of electrolyte specific gravity and might contain pressure relief venting.

Article 500 – Hazardous (Classified) Locations, Classes I, II, and III, Divisions 1 and 2

Combustible Dust. Any finely divided solid material that is 420 microns (0.017 in.) or smaller in diameter (material passing a U.S. No. 40 Standard Sieve) and presents a fire or explosion hazard when dispersed and ignited in air.

Article 517 – Health Care Facilities

Battery-Powered Lighting Units. Individual unit equipment for backup illumination consisting of the following:

(1) Rechargeable battery
(2) Battery-charging means
(3) Provisions for one or more lamps mounted on the equipment, or with terminals for remote lamps, or both
(4) Relaying device arranged to energize the lamps automatically upon failure of the supply to the unit equipment

Article 555 – Marinas and Boatyards

Marine Power Outlet. An enclosed assembly that can include equipment such as receptacles, circuit breakers, fused switches, fuses, watt-hour meter(s), distribution panelboards, and monitoring means approved for marine use.

Ground-Fault Protection. The main overcurrent protective device that feeds the marina shall have ground fault protection not exceeding 100 mA. Ground-fault protection of each individual branch or feeder circuit shall be permitted as a suitable alternative.

Article 600 – Electric Signs and Outline Lighting

LED Sign Illumination System. A complete lighting system for use in signs and outline lighting consisting of light-emitting diode (LED) light sources, power supplies, wire, and connectors to complete the installation.

Neon Tubing. Electric-discharge luminous tubing that is manufactured into shapes to illuminate signs, form letters, parts of letters, skeleton tubing, outline lighting, other decorative elements, or art forms and filled with various inert gases.

Informational Note: Where used in illumination systems for signs, outline lighting, skeleton tubing, decorative elements, or art forms, cold cathode luminous tubes are neon tubing as defined by this article.

Article 610 – Cranes and Hoists

Festoon Cable. Single- and multiple-conductor cable intended for use and installation in accordance with Article 610 where flexibility is required.

Informational Note: Festoon cable consists of one or more insulated conductors cabled together with an overall jacket. It is rated 60°C (140°F), 75°C (167°F), 90°C (194°F), or 105°C (221°F) and 600 V.

Article 625 – Electric Vehicle Charging System

Electric Vehicle. An automotive-type vehicle for on-road use, such as passenger automobiles, buses, trucks, vans, neighborhood electric vehicles, electric motorcycles, and the like, primarily powered by an electric motor that draws current from a rechargeable storage battery, fuel cell, photovoltaic array, or other source of electric current. Plug-in hybrid electric vehicles (PHEV) are considered electric vehicles. For the purpose of this article, off-road, self-propelled vehicles, such as industrial trucks, hoists, lifts, transports, golf carts, airline ground support equipment, tractors, boats, and the like, are not included.

Plug-In Hybrid Electric Vehicle (PHEV). A type of electric vehicle intended for on-road use with the ability to store and use off-vehicle electrical energy in the rechargeable energy storage system, and having a second source of motive power.

Rechargeable Energy Storage System. Any power source that has the capability to be charge and discharged.

Informational Note: Batteries, capacitors, and electro-mechanical flywheels are examples of rechargeable energy storage systems.

Article 645 – Information Technology Equipment

Critical Operations Data System. An information technology equipment system that requires continuous operation for reasons of public safety, emergency management, national security, or business continuity.

Information Technology Equipment (ITE). Equipments and systems rated 600 V or less, normally found in offices or other business establishments and similar environments classified as ordinary locations, that are used for creation and manipulation of data, voice, video, and similar signals that are not communications equipments as defined in Part I of Article 100 and do not process communications circuits as defined in 800.2.

Informational Note: For information on listing requirements for both information technology equipment and communications equipment, see UL 60950-1, *Information Technology Equipment – Safety – Part I: General Requirements*.

Information Technology Equipment Rooms. A room within the information technology equipment areas that contains the information technology equipment.

Remote Disconnect Control. An electric device and circuit that controls a disconnecting means through a relay or equivalent device.

Zone. A physically identifiable area (such as barriers or separation by distance) within an information technology equipment room with dedicated power and cooling systems for the information technology equipment or systems.

Article 680 – Swimming Pools, Foutains, and Similar Installations

Low Voltage Contact Limit. A voltage not exceeding the following values:

(1) 15 volts (RMS) for sinusoidal ac
(2) 21.2 volts peak for nonsinusoidal ac
(3) 30 volts for continuous dc
(4) 12.4 volts peak for dc that is interrupted at a rate of 10 to 200 Hz

Article 690 – Solar Photovoltaic (PV) Systems

Monopole Subarray. A PV subarray that has two conductors in the output circuit, one positive (+) and one negative (-). Two monopole PV subarrays are used to form a bipolar PV array.

Subarray. An electrical subset of a PV array.

Article 694 – Small Wind Electric Systems.

Charge Controller. Equipment that controls dc voltage or dc current, or both, and that is used to charge a battery or other energy storage device.

Diversion Charge Controller. Equipment that regulates the charging process of a battery or other energy storage device by diverting power from energy storage to dc or ac loads, or to an interconnected utility service.

Diversion Load. A load connected to a diversion charge controller or diversion load controller, also known as a dump load.

Diversion Load Controller. Equipment that regulates the output of a wind generator by diverting power from the generator to dc or ac loads, or to an interconnected utility service.

Guy. A cable that mechanically supports a wind turbine tower.

Inverter Output Circuit. The conductors between an inverter and an ac panelboard for stand-alone systems, or the conductors between an inverter and service equipment or another electric power production source, such as a utility, for an electrical production and distribution network.

Maximum Output Power. The maximum 1-minute average power output a wind turbine produces in normal steady-state operation (instantaneous power output can be higher).

Maximum Voltage. The maximum voltage the wind turbine produces in operation, including open-circuit conditions.

Nacelle. An enclosure housing the alternator and other parts of a wind turbine.

Rated Power. The wind turbine's output power at a wind speed of 11 m/s (24.6 mph). If a turbine produces more power at lower wind speeds, the rated power is the wind turbine's output power at a wind speed less than 11 m/s that produces the greatest output power.

Informational Note: The method for measuring wind turbine power output is specified IEC 61400-12-1, *Power Performance Measurements of Electricity Producing Wind Turbines.*

Tower. A pole or other structure that supports a wind turbine.

Wind Turbine. A mechanical device that converts wind energy to electrical energy.

Wind Turbine Output Circuit. The circuit conductors between the internal components of a small wind turbine (which might include an alternator, integrated rectifier, controller, and/or inverter) and other equipment.

Wind Turbine System. A small wind electric generating system.

Informational Note: See also definitions for interconnected systems in Article 705.

Article 700 – Emergency Systems

Emergency Systems. Those systems legally required and classed as emergency by municipal, state, federal, or other codes, or by any governmental agency having jurisdiction. These systems are intended to automatically supply illumination, power, or both to designated areas and equipment in the event of failure of the normal supply or in the event of accident to elements of a system intended to supply, distribute, and control power and illumination essential for safety to human life.

Informational Note: Emergency systems are generally installed in places of assembly where artificial illumination is required for safe exiting and panic control in buildings subject to occupancy by large numbers of persons, such as hotels, theaters, sports arenas, health care facilities, and similar institutions. Emergency systems may also provide power for such functions as ventilation where essential to maintain life, fire detection and alarm systems, elevators, fire pumps, public safety communications systems, industrial process where current interruption would produce serious life safety or health hazards, and similar functions.

Relay, Automatic Load Control. A device use to energize switched or normally-off lighting equipment from an emergency supply in the event of loss of the normal supply, and to de-energize or return the equipment to normal status when the normal supply is restored.

Informational Note: For requirements covering automatic load control relays, see ANSI/UL 924, *Emergency Lighting and Power Equipment.*

Article 705 – Interconnected Electric Power Production Sources

Power Production Equipment. The generating source, and all distribution equipment associated with it, that generates electricity from a source other than a utility supplied service.

Informational Note: Examples of power production equipment include such items as generators, solar photovoltaic systems, and fuel cell systems.

Article 770 – Optical Fiber Cables and Raceways

Cable Routing Assembly. A single channel or connected multiple channels, as well as associated fittings, forming a structural system that is used to support, route, and protect high densities of wires and cable, typically communications wires and cables, optical fiber, and data (Class 2 and Class 3) cables associated with information technology and communications equipment.

Conductive Optical Fiber Cable. A factory assembly of one or more optical fibers having an overall covering and containing non-current-carrying conductive member(s) such as metallic strength member(s), metallic vapor barrier(s), metallic armor, or metallic sheath.

Nonconductive Optical Fiber Cable. A factory assembly or one or more optical fibers having an overall covering and containing no electrically conductive materials.

Optical Fiber Cable. A factory assembly of one of more optical fibers having an overall covering that transmits light for control, signaling, and communications.

Optical Fiber Raceway. An enclosed channel of nonmetallic materials designed for holding optical fiber cables in plenum, riser, and general-purpose applications.

Article 800 – Communications Circuits

Communications Raceway. An enclosed channel of nonmetallic materials designed for holding communications wires and cables in plenum, riser, and general-purpose applications.

Article 840 – Premises-Powered Broadband Communications Systems

Fiber-to-the-Premises (FTTP). Conductive or nonconductive optical cable that is either aerial, buried, or through a raceway and is terminated at an optical network terminal (ONT), establishing a communications network.

Optical Network Terminal (ONT). A device that converts an optical signal into component signals, including voice, audio, video, data, wireless, and interactive service electrical, and is considered to be network interface equipment.

Premises Communications Circuit. The circuit that extends voice, audio, video, data, interactive services, telegraph (except radio), and outside wiring for fire alarm and burglar alarm from the service provider's ONT to the customer's communications equipment up to and including terminal equipment, such as a telephone, a fax machine, or an answering machine.

Premises Community Antenna Television (CATV) Circuit. The circuit that extends community antenna television (CATV) systems for audio, video, data, and interaction services from the service's provider ONT to the appropriate customer equipment.

Annex C
Reference Standards

NFPA 70B, *Recommended Practice for Electrical Equipment Maintenance*

NFPA 70E, *Standard for Electrical Safety in the Workplace*

NFPA 79, *Electrical Standard for Industrial Machinery*

NFPA 496, *Standard for Purged and Pressurized Enclosures for Electrical Equipment*

NFPA 497, *Recommended Practice for the Classification of Flammable Liquids, Gases, or Vapors and of Hazardous (Classified) Locations for Electrical Installations in Chemical Process Areas*

NFPA 499, *Recommended Practice for the Classification of Combustible Dusts and of Hazardous (Classified) Locations for Electrical Installation in Chemical Process Areas*

OSHA – Occupational Safety and Health Administration

ROC – Report on Comments

ROP – Report on Proposals

UL – Underwriters Laboratories

Chapter 5
Special Occupancies

1. (B)	500.8(E)(1)
2. (A)	500.8(E)(1), Ex.
3. (D)	501.17(3)
4. (A)	502.6
5. (C)	503.6
6. (B)	504.30(B)
7. (C)	505.16(C)(1)(b)
8. (A)	506.9(E)(2)
9. (B)	517.13(B)(3)
10. (D)	517.20(A)(2)
11. (C)	520.44(C)(3)
12. (C)	525.23(A)(2)
13. (A)	553.4
14. (D)	590.6(A)(1)
15. (B)	590.6(A)(2)

Chapter 6
Special Equipment

1. (D)	600.6
2. (A)	600.33(A)
3. (C)	645.5(B)(1)
4. (B)	645.10(A)(2)
5. (B)	680.21(C)
6. (A)	680.26(B)(2)
7. (A)	680.73
8. (D)	682.14(B)
9. (C)	690.4(B)(3)
10. (D)	690.4(B)(4)
11. (C)	690.10(E)
12. (A)	690.16(B)
13. (B)	690.31(E)(1)
14. (D)	690.31(E)(4)
15. (B)	694.12(B)(2)
16. (A)	694.22(C)(4)
17. (C)	694.70(B)(1)
18. (B)	695.6(A)(2)(d)(1)
19. (D)	695.6(H)(1)
20. (B)	695.14(F)(1)

Chapter 7
Special Conditions

1. (A)	700.12(B)(6), Ex.
2. (B)	701.6(D)
3. (C)	705.60(B)
4. (B)	760.41(B)
5. (D)	770.106(A)(1)
6. (B)	770.106(A)(2)
7. (A)	770.113(B)(1)
8. (C)	770.113(C)(1) thru (C)(4)
9. (A)	770.113(D)(1) thru (D)(3)
10. (D)	770.113(E)(3)

Chapter 8
Communications Systems

1. (C)	800.100(A)(3)
2. (B)	800.100(A)(3)
3. (C)	800.113(B)(1)
4. (D)	800.113(C)(1) thru (C)(4)
5. (D)	800.113(E)(1) and (E)(2)
6. (A)	800.113(G)(2)
7. (B)	800.113(H)(1)
8. (C)	800.113(J)(2)
9. (C)	820.44(A)(4), Ex.
10. (A)	820.100, Ex.